Scalable Interactive Visualization

Special Issue Editors

Achim Ebert

Gunther H. Weber

MDPI • Basel • Beijing • Wuhan • Barcelona • Belgrade

MDPI

Special Issue Editors

Achim Ebert
University of Kaiserslautern
Germany

Gunther H. Weber
University of California
USA

Editorial Office
MDPI
St. Alban-Anlage 66
Basel, Switzerland

This edition is a reprint of the Special Issue published online in the open access journal *Informatics* (ISSN 2227-9709)) in 2017 (available at: http://www.mdpi.com/journal/informatics/special_issues/interactive_visualization).

For citation purposes, cite each article independently as indicated on the article page online and as indicated below:

Lastname, F.M.; Lastname, F.M. Article title. *Journal Name* **Year**, *Article number, page range.*

First Edition 2018

ISBN 978-3-03842-803-9 (Pbk)
ISBN 978-3-03842-804-6 (PDF)

Cover photo courtesy of Achim Ebert.

Table of Contents

About the Special Issue Editors v

Preface to "Scalable Interactive Visualization" vii

Daniel Haehn, John Hoffer, Brian Matejek, Adi Suissa-Peleg, Ali K. Al-Awami, Lee Kamentsky, Felix Gonda, Eagon Meng, William Zhang, Richard Schalek, Alyssa Wilson, Toufiq Parag, Johanna Beyer, Verena Kaynig, Thouis R. Jones, James Tompkin, Markus Hadwiger, Jeff W. Lichtman and Hanspeter Pfister
Scalable Interactive Visualization for Connectomics
doi: 10.3390/informatics4030029 1

Sathish Kottravel, Riccardo Volpi, Mathieu Linares, Timo Ropinski and Ingrid Hotz
Visual Analysis of Stochastic Trajectory Ensembles in Organic Solar Cell Design
doi: 10.3390/informatics4030025 32

Mark Taylor
TOPCAT: Desktop Exploration of Tabular Data for Astronomy and Beyond
doi: 10.3390/informatics4030018 52

Sizhe Wang, Wenwen Li and Feng Wang
Web-Scale Multidimensional Visualization of Big Spatial Data to Support Earth Sciences—A Case Study with Visualizing Climate Simulation Data
doi: 10.3390/informatics4030017 70

Boris Kovalerchuk and Dmytro Dovhalets
Constructing Interactive Visual Classification, Clustering and Dimension Reduction Models for n-D Data
doi: 10.3390/informatics4030023 87

Björn Zimmer, Magnus Sahlgren and Andreas Kerren
Visual Analysis of Relationships between Heterogeneous Networks and Texts: An Application on the IEEE VIS Publication Dataset
doi: 10.3390/informatics4020011 114

Steffen Frey
Sampling and Estimation of Pairwise Similarity in Spatio-Temporal Data Based on Neural Networks
doi: 10.3390/informatics4030027 134

Han Kruiger, Almoctar Hassoumi, Hans-Jrg Schulz, Alex Telea, Christophe Hurter
Multidimensional Data Exploration by Explicitly Controlled Animation
doi: 10.3390/informatics4030026 154

Jun Wang, Alla Zelenyuk, Dan Imre and Klaus Mueller
Big Data Management with Incremental K-Means Trees–GPU-Accelerated Construction and Visualization
doi: 10.3390/informatics4030024 175

Di Jin, Aristotelis Leventidis, Haoming Shen, Ruowang Zhang, Junyue Wu and Danai Koutra
PERSEUS-HUB: Interactive and Collective Exploration of Large-Scale Graphs
doi: 10.3390/informatics4030022 190

Joris Sansen, Gaëlle Richer, Timothé Jourde, Frédéric Lalanne, David Auber and Romain Bourqui
Visual Exploration of Large Multidimensional Data Using Parallel Coordinates on Big Data Infrastructure
doi: 10.3390/informatics4030021 . . . . **213**

About the Special Issue Editors

Achim Ebert, Prof. Dr., Achim Ebert holds a degree and a doctor in Computer Science. He is the co-head of the Computer Graphics and HCI lab at the University of Kaiserslautern. He is also a member of the lead personnel of DFG's International Research Training Group (IRTG) "Physical Modeling for Virtual Manufacturing Systems and Processes". His current research topics include information visualization, immersive scenarios, and human-computer interaction. He participated or led several national and international research projects in the area of visualization and HCI, both with an academic or industrial focus. He has founded and is co-heading the IFIP working group 13.7 on Human-Computer Interaction and Visualization. He has published more than 200 refereed publications. Achim Ebert acts as a member of many international program committees (e.g., ACM and IEEE) and as a reviewer for several journals and conferences.

Gunther H. Weber, Prof. Dr., Gunther H. Weber received a Ph.D. in computer science, with a focus on computer graphics and visualization, from the University of Kaiserslautern, Germany in 2003. He is currently a Staff Scientist in the Computational Research Division at the Lawrence Berkeley National Laboratory (LBNL), where he serves as Deputy Group Lead of the Data Analysis and Visualization Group in the Data Science and Technology Department. Gunther Weber is an Adjunct Associate Professor of Computer Science at the University of California, Davis. His research interests include computer graphics, scientific visualization, data analysis with using topological methods, parallel and distributed computing for visualization and data analysis applications, hierarchical data representation methods, and bioinformatics. He has extensive experience in working with researchers from diverse science and engineering fields, including applied numerical computing, combustion simulation, gene expression, medicine, civil engineering, cosmology, climate and particle accelerator modeling. Dr. Weber has authored or co-authored over 80 publications, five of which won best paper awards. He has served as principal investigator (PI) or Co-PI on several Department of Energy (DOE) and National Science Foundation (NSF) projects. He is a reviewer for major funding agencies (DOE, NSF), conference proceedings and journals. Dr. Weber served as co-organizer, co-chair and program committee member of more than 40 internationally recognized conferences.

Preface to "Scalable Interactive Visualization"

Data available in today's information society is ever growing in size and complexity-i.e., unstructured, multidimensional, uncertain, etc.-making it impossible to survey and understand this data. Traditionally, most of these datasets are stored and depicted as huge tables, hindering efficient retrieval of salient information-similarities, outliers, structures, origin, etc. Interactive visualization provides an interface to this data that can help gleaning valuable information from it, thus supporting better data understanding by significantly reducing cognitive load on the analyst. Two fundamental concepts, visualization and interaction, form the basis of the underlying scientific methods. Combining these concepts connects two key research areas in computer science: visualization and human-computer interaction (HCI) and brings together practitioners from many disciplines. The result is highly multi-disciplinary work with significant impact.

Interactive visualization has virtually unlimited applications, including analysis of complex data sets ("big data"), virtual reality environments, augmented reality, mobile environments, cooperative work, computer-supported surgery, large-scale simulations, experimental and observational data, and sensor networks. However, truly interactive visualizations are hard to design and implement. Researchers have to solve multiple problems, e.g., transforming complexity into simplicity, efficient algorithms and implementations, guaranteeing real-time performance, scaling to multiple platforms and user types, minimizing and managing data transfer, and efficient parallel implementations.

This Special Issue covers recent work in the field of Interactive Visualization as well as trends for future development. It contains several examples of applying interactive visualization to spatial and abstract data understanding problems. Furthermore, it describes new visualization techniques and metaphors addressing growing data size and complexity, and it introduces multiple systems developed for visualizing Big Data. The single contributions presented in this special issue can be categorized into one of the four subjects: "Applying Interactive Visualization to Spatial Data" (Haehn et al., Kottravel et al., Taylor, and Wang et al.), "Applying Interactive Visualization to Abstract Data" (Kovalerchuk et al. and Zimmer et al.), "New Visualization Technqiues and Metaphors" (Frey, Kruiger et al., and Wang et al.), and "Visualization Systems for Big Data" (Jin et al. and Sansen et al.).

While providing a snapshot of the current state of the art in Interactive Visualization, this special issue can only outline some of the most prominent problems and potential solutions in this area. Instead of providing a final answer to these issues, it offers a glimpse into an exciting area of growing importance.

Achim Ebert and Gunther H. Weber

Special Issue Editors

informatics

MDPI

Article

Scalable Interactive Visualization for Connectomics

Daniel Haehn [1,2,*], John Hoffer [1,2], Brian Matejek [1,2], Adi Suissa-Peleg [1,2], Ali K. Al-Awami [3], Lee Kamentsky [1,2], Felix Gonda [1,2], Eagon Meng [1,2], William Zhang [1,2], Richard Schalek [2], Alyssa Wilson [2], Toufiq Parag [1,2], Johanna Beyer [1,2], Verena Kaynig [1,2], Thouis R. Jones [1,2], James Tompkin [1,4], Markus Hadwiger [3], Jeff W. Lichtman [2] and Hanspeter Pfister [1,2]

[1] Harvard Paulson School of Engineering and Applied Sciences, Harvard University,
    Cambridge, MA 02138, USA; john@hoff.in (J.H.); bmatejek@seas.harvard.edu (B.M.);
    adisuis@seas.harvard.edu (A.S.-P.); lee_kamentsky@g.harvard.edu (L.K.); fgonda@g.harvard.edu (F.G.);
    emeng@college.harvard.edu (E.M.); williamzhang26@hotmail.com (W.Z.); paragt@g.harvard.edu (T.P.);
    jbeyer@g.harvard.edu (J.B.); vkaynig@gmail.com (V.K.); thouis@gmail.com (T.R.J.);
    james_tompkin@brown.edu (J.T.); pfister@seas.harvard.edu (H.P.)
[2] Harvard Brain Science Center, Harvard University, Cambridge, MA 02138, USA;
    rschalek@fas.harvard.edu (R.S.); wilson4@fas.harvard.edu (A.W.); jeff@mcb.harvard.edu (J.W.L.)
[3] Computer, Electrical and Mathematical Sciences and Engineering,
    King Abdullah University of Science and Technology, Thuwal 23955, Saudi Arabia;
    ali.awami@kaust.edu.sa (A.K.A.); markus.hadwiger@kaust.edu.sa (M.H.)
[4] Computer Science Department, Brown University, Providence, RI 02912, USA
*   Correspondence: haehn@seas.harvard.edu

Academic Editors: Achim Ebert and Gunther H. Weber
Received: 7 July 2017; Accepted: 24 August 2017; Published: 28 August 2017

Abstract: Connectomics has recently begun to image brain tissue at nanometer resolution, which produces petabytes of data. This data must be aligned, labeled, proofread, and formed into graphs, and each step of this process requires visualization for human verification. As such, we present the BUTTERFLY middleware, a scalable platform that can handle massive data for interactive visualization in connectomics. Our platform outputs image and geometry data suitable for hardware-accelerated rendering, and abstracts low-level data wrangling to enable faster development of new visualizations. We demonstrate scalability and extendability with a series of open source Web-based applications for every step of the typical connectomics workflow: data management and storage, informative queries, 2D and 3D visualizations, interactive editing, and graph-based analysis. We report design choices for all developed applications and describe typical scenarios of isolated and combined use in everyday connectomics research. In addition, we measure and optimize rendering throughput—from storage to display—in quantitative experiments. Finally, we share insights, experiences, and recommendations for creating an open source data management and interactive visualization platform for connectomics.

Keywords: scientific visualization; connectomics; electron microscopy; registration; segmentation; proofreading; graph analysis

1. Introduction

The grand challenge of connectomics is to completely reconstruct and analyze the neural "wiring diagram" of the mammalian brain, which contains billions of interconnected nerve cells [1–4]. Deciphering this vast network and studying its underlying properties will support certain aspects of understanding the effects of genetical, molecular, or pathological changes at the connectivity level. This may lead to a better understanding of mental illnesses, learning disorders, and neural pathologies, as well as provide advances in artificial intelligence [5]. As such, the field of connectomics is rapidly growing, with hundreds of neuroscience labs world-wide eager to obtain nanoscale level descriptions of neural circuits. However, there are many problems in analyzing the individual synaptic connections between nerve cells and signal transmission to other cells, not least of which is scale: connectomics occurs at both the *nano* and *peta* scales, as electron microscopy (EM) data at a resolution sufficient to identify synaptic connections produces petabytes of data.

The process of producing a wiring diagram suitable for interactive visualization and analysis at this scale has many steps, and each step brings its own challenges (Figure 1). Throughout this process, interactive visualization is key to helping scientists meet these challenges.

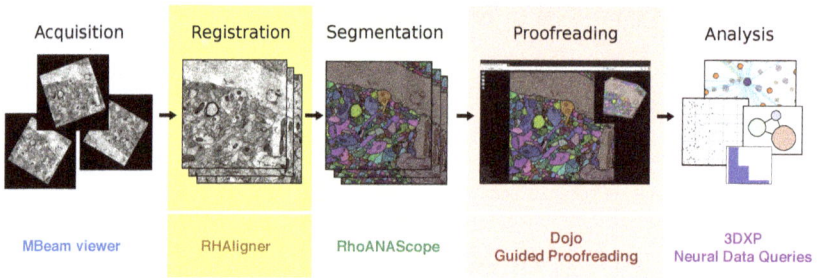

Figure 1. The typical connectomics workflow includes several steps: image tiles of brain tissue are *acquired* using an electron microscope, *registered* in 2D and 3D, and automatically *segmented* into neurons. Since the output of the automatic segmentation is not perfect, it is mandatory to *proofread* the result prior to any *analysis*. Each step of the workflow requires visual exploration, for which we have developed open source software tools (listed below each step).

Acquisition: The process begins with a brain sample, which is embedded into resin, cut into slices, and imaged with an electron microscope. The imaging process is fallible, and can cause severe noise and contrast artifacts. Fast and scalable 2D visualizations enable quick signal-to-noise ratio and contrast assessments across image tiles during acquisition. Rapid progress in automatic sample preparation and EM acquisition techniques make it possible to generate a 1 mm^3 volume of brain tissue in less than six months, with each voxel of size $4 \times 4 \times 30$ nm^3 resulting in 2 petabytes of image data [6,7].

Registration: Each brain slice is imaged in tiles independently in 2D, and so the resulting images must be aligned into a larger 2D section. Then, a stack of sections must be aligned into a 3D scan. Here, visualizing stitched tiles and sections allows quick human assessment of alignment quality in addition to computing quantitative measures [6].

Segmentation: Given the stack of sections, the cell membrane borders and synaptic connections between cells must be discovered, and this requires both manual and automatic labeling methods [8,9]. The resulting segmentations are stored as label volumes which are encoded as 64 bits per voxel to support the labeling of millions of nerve cells (neurons) and their connections (synapses). Visualization of sections of the aligned scan in 2D and renderings in 3D helps us assess segmentation and classification quality.

Proofreading: All available segmentation methods make mistakes [10], and so the results must be proofread by humans before any biological analysis occurs [11–13]. For proofreading, intelligent interactive visualization tools are key to minimizing the time committed.

Analysis: At this point, our connectomics data consists of cell membrane annotations for all neurons, and synaptic annotations where neurons connect. This information can be modeled as a graph. Nodes in the graph represent individual neurons and the edges between nodes resemble synaptic connections. The edges can be weighted by quantitative measures, e.g., number of connections or neuron type. This connectivity graph can be flattened and visualized in 2D or rendered in 3D. Typical analysis tools for connectomics range from abstract visualizations which focus on higher level aspects to lower level biologically-correct visualizations [14,15].

Each step from acquisition to analysis requires interactive visual exploration: both to check the quality of the results and to explore the data for insights. However, these tools must be able to scale to the large data at hand.

To this end, we present the scalable BUTTERFLY middleware for interactive visualization of massive connectomics datasets. This system integrates Web-based solutions for data management and storage, semantic queries, 2D and 3D visualization, interactive editing, and graph-based analysis.

Our systems were developed in collaboration with neuroscientists, working on the only-slightly-more-modest goal of imaging a whole rodent brain (*Mus musculus*, BALB/c strain, females of age P3, P5, P7, P60, with a fresh volume of 1 cm^3, brainstem upwards; as well as *Rattus norvegicus*). Like humans, mice and rats are vertebrate mammals which learn. Many of the brain structures in these rodents are found in humans, and many conditions in humans can be found in rodents at the genetic level [16]. Currently acquired EM image stacks represent only a fraction of a full mouse brain (1 μm^3, 0.01%) but are multiple terabytes in size, which provides us with many difficult scalable interactive visualization problems. Our solutions to these problems enable the analysis of the connectome at nano scale and can help lead to scientific discoveries, e.g., Kasthuri et al. [5] recently disproved that physical proximity is sufficient to predict synaptic connectivity.

The Butterfly middleware unifies connectomics tools from each stage of the pipeline. It integrates with the following visualization applications for multi-user environments:

- **MBeam viewer**, a tool to quickly assess image quality and contrast during acquisition.
- **RHAligner**, visualization scripts for debugging the registration process.
- **RhoANAScope**, a visualizer for image data and label overlays during segmentation.
- **Dojo**, an interactive proofreading tool with multi-user support [11].
- **Guided Proofreading**, a machine learning proofreading tool to correct errors quickly [12].
- **3DXP**, a 3D visualization of neuron geometries.
- **Neural Data Queries**, a system for semantic queries of neurons and their connections.

All visualization components are Web-based and are part of a multi-user environment. This avoids duplication of the massive connectomics datasets since the majority of the data stays on the server and only a small subset is transferred for each user interaction. To maintain code quality, we use continuous integration and automated testing of each code base change. All software tools are available as free and open source software. In addition, we provide a VirtualBox distribution including all reported visualization tools and the Butterfly middleware. The distribution includes various test data such that the visualization tools can be used in an existing network environment with minimal configuration.

This paper describes the motivation and design decisions for data management and visualization, as well as implementation details of the Butterfly integration with our front-end applications. We relate the stages of the connectomics pipeline, the end-user tools, and the Butterfly middleware. We report on performance and scalability of our visualization landscape, and describe a series of everyday use-cases with neuroscience experts. Finally, we provide insights and recommendations for creating an open source data management and interactive visualization platform for connectomics.

2. Related Work

An overview of existing visualization tools for connectomics is given by Pfister et al. [17]. The article describes visualizations at different scales: (a) macroscale connectivity, with data coming from functional magnetic resonance imaging, electroencephalography, magnetoencephalography, and diffusion tensor imaging [18]; (b) mesoscale connectivity, obtained from light and optical microscopy; and (c) microscale connectivity, which enables imaging at the nanometer resolution using electron microscopes. Recent advances in sample preparation have further enabled nano-imaging [19] and a description of many techniques in this emerging field is given by Shaefer [20]. While our work targets microscale connectomics, in this section we also relate to works at the other two scales.

Visualization tools for connectomics mainly focus on three different areas: (1) visualization to support the segmentation and proofreading of volumes; (2) visualization to explore high-resolution segmented volume data; and (3) visualization to analyze neuronal connectivity. Most visualization tools for connectomics are standalone applications, requiring high-performance workstations and modern GPUs [21–25]; thus, it is harder for these tools to achieve general scalability across both compute and users.

Visual proofreading of segmentations is supported by several tools [21–23,26], mostly targeting expert users and offering many parameters for tweaking the proofreading process. However, none of these tools run in a distributed setting and allow non-expert users to correct erroneous segmentations. Several applications allow distributed and collaborative segmentation of connectomics volumes. EyeWire [27] is an online tool where novice users participate in a game to earn points for segmenting neuronal structures using a semi-automatic algorithm. D2P [28] uses a micro-labor workforce approach via Amazon Mechanical Turk where boolean choice questions are presented to users and local decisions are combined to produce a consensus segmentation. Both tools are designed for non-expert users. For experts, Catmaid [29] and the Viking Viewer [30] are collaborative annotation frameworks which allow users to create skeleton segmentations for large data sets. More recently, Neuroblocks [13] proposed an online-system for tracking the progress and evolution of a large-scale segmentation project.

Visualization tools for exploring high-resolution segmented volume data typically run on powerful GPU systems and employ complex multi-resolution strategies. Hadwiger et al. [24] present a volume visualization system for large EM volumes, which was later extended to segmented EM volumes [25]. Several systems, such as Neuron Navigator [31] and Connectome Explorer [15] support interactive or visual queries to further explore these typically very large data sets.

Most visualization tools for connectomics focus on the analysis step and explore the connectivity between neurons that is extracted from the segmented data volumes. Several systems [32,33] are based on WebGL [34] and run directly in a Web browser. Ginsburg et al. [32] propose a rendering system which combines brain surfaces with tractography fibers to render a 3D network of connected brain regions. Similar visualizations can be created using the X toolkit [35], which offers WebGL rendering for neuroimaging data, and SliceDrop [36], which is a Web-based viewer for medical imaging data including volume rendering and axis-aligned slice views. Neuroglancer [33] provides different 2D and 3D visualizations for large datasets.

Notable efforts to allow exploration and reusability of published connectome findings exist. Paired with the scalable brain atlas visualization tool [37], the CocoMac database contains findings on connectivity of the macaque brain [38]. Bota et al. [39] introduce the Brain Architecture Management System, which stores and infers relationships about nervous system circuitry. Query results are here visualized as network diagrams and represented as tabular data. The neuroVIISAS system provides similar visualizations and pairs them with slice-based renderings [40].

Other stand-alone viewers exist with similar network visualization features. The Connectome Viewer Toolkit [41] targets the analysis of macroscopic neuronal structures and brain region connectivity, whereas the Viking Viewer [30] displays a connectivity graph on a cell level. More recently, neuroMap [42] uses circuit wiring diagrams to represent all possible connections of neurons. For nanoscale connectomics, Neurolines [14] allows neuroscientists to analyze the connectivity on the level of individual synapses.

Some research and development has been conducted towards producing a central data management platform for the full connectomics workflow with visualization capabilities. Notable efforts are DVID [43], a centralized data service offering version control and distributed access, and The Boss [44], a cloud-based storage service. Both systems manage data sources in their own specific format and support Neuroglancer for visualization.

Our proposed Butterfly middleware is different from this approach as it provides a platform for creating visualizations for connectomics without having to deal with data wrangling or other low-level issues. Using the middleware, it is possible to concentrate development effort towards the front end rather than the back end. We also propose a series of integrated applications which allow users to visualize connectomics data and debug problems for every stage of the connectomics pipeline.

3. Overview

We designed the Butterfly middleware to integrate and unify data management of visualization tools for every step of the connectomics pipeline (Figure 2). Different data formats and their access are abstracted to support a variety of data queries. This reduces heterogeneity among tools, interfaces, and data formats. We use several data management and visualization concepts to enable scalable and interactive applications.

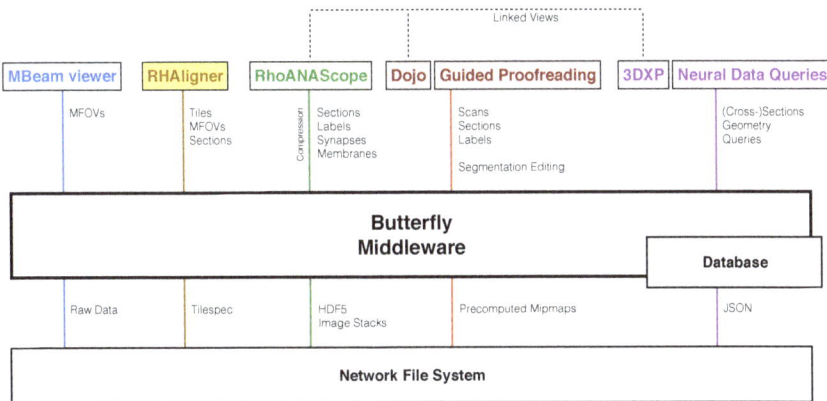

Figure 2. Our software system includes tools tailored to each stage of the connectomics pipeline, which requires exploring and interacting with data stored in different ways on a network file system (XFS storage): The MBeam viewer enables image quality assessment during image acquisition by rendering multi-beam-field-of-views (MFOVs) from raw microscope data. The RHAligner scripts allow for the monitoring and debugging of the alignment process by rendering the tilespec format which separates transformations and data. RhoANAScope visualizes sections and the corresponding neuron labels, synapse segmentations, and membrane probabilities during segmentation. Dojo and Guided Proofreading are two interactive applications to correct errors in the labeled sections and volumes. For analysis, 3DXP enables the visualization of neuron geometry by rendering meshes. The Neural Data Query application program interface (API) allows querying neurons for their synaptic connections by including JSON formatted data structures. The Butterfly middleware integrates and unifies these heterogeneous applications within a multi-user environment by abstracting data access and supporting a variety of data requests. Besides data integration, our middleware enables linked visualizations for several important use-cases between RhoANAScope, Dojo, and 3DXP (Section 11).

3.1. Volumetric EM Image Data

Volumetric EM image data is often organized hierarchically with different dimensions (Table 1). We demonstrate this with the Zeiss MultiSEM 505 electron microscope with which we capture our imagery. This microscope simultaneously captures 61 image *tiles*, each of roughly 3000×2700 pixels, to achieve acquisition rates of approximately 1 terapixel per hour [6,7]. Each simultaneous capture is called a *multi-beam field-of-view*, or *MFOV*. Multiple MFOVs are stitched into a *section*, which represents a two-dimensional slice of tissue. Thousands of sections are combined into a single 3D *scan* of tissue.

Table 1. Our electron microscopy data form a hierarchy of different types and dimensions. Multiple image tiles are acquired at nanometer resolution as an MFOV, and then stitched in sections. Thousands of sections are stacked into a scan. There is a large difference between image data and segmentation data due to the 64 bit encodings required to represent potentially billions of neurons. For segmentation data, we report average compression rates of $700\times$ using the Compresso scheme [45]. For images, we assume a resolution of $4 \times 4 \times 30$ nm^3 per voxel and 8 bit encoding. The Butterfly middleware and its applications have been tested on Scan A (100 μm^3), acquired using the Zeiss MultiSEM 505 electron microscope. Scan B is yet to be acquired using the same hardware.

Type	Digital Dimensions	Physical Dimensions	Image Size	Segmentation Size (Compressed)
Tile	3.1 k × 2.7 k pixels	0.78×0.68 μm^2	8.5 MB	68.2 MB (97.0 KB)
MFOV	30 k × 26 k pixels	115×100 μm^2	520 MB	4.2 GB (5.9 MB)
Section	200 k × 170 k pixels	810×700 μm^2	3.6 GB	29.1 GB (41.6 MB)
Scan A (100 μm^3)	26 k × 26 k × 3394 voxels	$100 \times 100 \times 100$ μm^3	2.2 TB	18.3 TB (26.2 GB)
Scan B (1 mm^3)	260 k × 260 k× 33,940 voxels	$1 \times 1 \times 1$ mm^3	2.2 PB	18.3 PB (26.2 TB)

3.2. Data Management Concepts

Connectomics data includes several different data types and data structures. This is driven by different use cases and the goal of performant random access.

Image formats: We acquire image volumes of the mammalian brain at nanometer resolution. Our data is typically anisotropic across sections with in-plane section resolution roughly 7.5 times higher than between sections ($4 \times 4 \times 30$ nm per voxel). Each section of a scan is a gray-scale image typically with 8 bits per pixel. We can store sections individually in general formats such as JPEG, PNG, or TIFF. We also store a collection of sections as volumetric data as HDF5 or multi-page TIFF containers. HDF5 allows random access without loading the full volume into memory.

Tiled storage: A single section of a volume can be many gigabytes in size. For scalable and efficient processing and storage, a section is typically stored split into multiple files using a row/column scheme. Individual tiles are usually of fixed power-of-2 dimensions to enable GPU texture mapping without conversion (e.g., 1024 × 1024 pixels), but can also be arbitrarily sized. Thus, metadata is required to understand a tiled image format. MFOVs are an example of tiled storage.

Label formats: Our labels are segmentation volumes including cell membranes as well as pre- and post-synaptic connections. Neuron segmentations can contain billions of values to uniquely identify each cell. This requires encoding with 32 or 64 bits per pixel, for which HDF5 is the preferred data format. For visualization, we color each identifier using a look-up table with neighboring cells colored distinctly.

Mipmap structures: Mipmaps (i.e., image pyramids, or image quad-trees) are hierarchical sequences of an image at different resolutions [46]. These are created by iteratively downsampling each image by a factor of two in each dimension until the entire image is reduced to a single pixel. For images, we use bilinear downsampling, while for labels we use nearest-neighbor downsampling to not alter the identifiers. Mipmaps are usually pre-computed to allow fast data access, but require a storage overhead of 33 percent. We store each mipmap level as tiled images, generating no representations smaller than a single tile. This reduces some storage overhead and allows partial loading for scalability.

Database: We store relations between label structures with unique identifiers, as well as dimensions and statistics, in a object-oriented database for fast access. This way, we can use queries to explore relationships of neurons and synapses or to request metadata.

Distributed storage and computation: Connectomics requires distributed processing and storage mechanisms to deal with large data. Typically, data is stored on a network file system and interactively explored or edited from client workstations. This is important to avoid unnecessary copies of datasets; however, parallel access of data has to be handled through a transaction mechanism to avoid conflicts between users [47].

Compression: The massive connectomics image and label volumes require compression for storage and for transfer from server to client. We compress EM images using JPEG encoding with average compression ratios of 2–4×. To compress label volumes, we must consider two important components: the per-pixel labels and the per-segment shapes. We use Compresso [45], which is specifically tailored to compress label volumes as it decouples these two components and compresses each separately over congruent 3-D windows. Often, Compresso is paired with an additional general-purpose compression scheme to further reduce the data size; with LZMA, we can compress the label volumes by a ratio of 600–1000× on average.

3.3. Scalability Through Demand-Driven and Display-Aware Web Applications

Efficient connetomics visualizations must provide easy access to large data for many people working across the pipeline. We tackle these problems by building web applications backed by scalable server software. A web-based visualization removes the need for any client-side installation, and multiple users can access the applications at the same time across all kinds of devices such as phones, tablets, laptops, and workstations. Data management is handled by the server, with connectome data being much larger than the total available CPU or GPU memory on a single workstation. The challenge is to minimize data loading and transfer while allowing full-fledged exploration of any size of data. To enable interactive visualizations, we use display-aware and demand-driven rendering. These techniques only transfer and render data which is actually displayed to the user [48,49]. This enables scalability since the resolution of the viewport limits the required data access to a small subset of the full-resolution data. The Butterfly middleware implements this as follows:

Tiled image transfer and rendering: Image data is usually transferred from a server to a client for visualization. Similar to the concept of tiled storage, this transfer usually also involves sending requested data in chunks. These chunks are significantly smaller than the client display and are typically rendered once they arrive to reduce the waiting time for the user.

On-the-fly mipmapping: To avoid processing and reduce storage overhead, it is possible to create a mipmap structure of image and segmentation data on-the-fly. This means full resolution data (typically stored as image tiles) can be downsampled on demand to provide lower-resolution levels of the mipmap hierarchy. This includes a trade-off between how many image tiles need to be loaded from disk for the requested mip level, and how large the image tiles are for transfer and rendering. The MBeam viewer uses this concept. We measure throughput based on different tile sizes for storage and for transfer and rendering (Sections 4 and 9).

Cut-out service: Demand-driven rendering can be realized using a cut-out service. Such service receives a query for a part of a larger nano-scale image or segmentation volume. The first step is to calculate which image tiles and which mipmap levels are required to deliver the requested part. The cut-out service loads the relevant image tiles at a pre-computed mipmap or, if applicable, generates the mip-level on the fly. The final part of the cut-out logic involves conversion to a requested format including compression, and then responding with the data. We named the Butterfly middleware after a *balisong* knife, as the cut-out service is one of the core components.

Scalable Editing: Beyond simply providing data, the proofreading stage of the connectomics workflow also requires interactive data editing [11]. Dynamic data structures such as lookup-tables enable non-destructive editing by storing an edit decision list separately from the actual image data, which then informs the rendering process. This allows simple undo, and reduces the data processing and transfer required for edits, especially across users. Once changes are ready to be committed, the modifications target only a small sub-volume of the larger dataset via a cut-out service. This first edits at full resolution, then recomputes the mipmap structure.

The Butterfly middleware makes the use of these data management and scalability concepts transparent as it provides a central access point to connectomics data for our visualization tools. From a user perspective, data processing is no longer limited to (smaller) copies of datasets on single workstations, reducing data duplication and communication overhead. From a developer's perspective, this reduces maintenance costs and code duplication, and makes it easy to support new data sources and file formats.

4. Visualization during Acquisition

4.1. Motivation

The Zeiss MultiSEM 505 electron microscope can generate roughly 1 terapixel per hour split across many 61-tile MFOVs and sections [6]. Neuroscientists must look at each tile to assess whether acquisition was of sufficient quality, otherwise the relevant area of the physical sample needs to be located and tagged for re-scanning. Previously, this task was performed in our lab with the preview function of Windows Explorer on a local machine, which does not allow the tiles to be viewed in their natural MFOV or section layouts. This assessment procedure is inefficient, slows down the acquisition process, and limits the microscope's effective throughput.

4.2. Data

The microscope provides estimates of the 2D locations of image tiles and multiple MFOVs in a section (Figure 3). This alignment method uses a very low-resolution image of the physical sample to identify the borders of each MFOV in a section, and the positions of the acquisition beams are used to align tiles within an MFOV. Imprecise alignment arises from this low-resolution data, from reduced sample points, and from vibrations (which leads to mandatory re-calibrations of beams). The microscope moves the stage to acquire another MFOV. However, this method is fast, and a more sophisticated alignment process would slow down the throughput of the microscope. The alignment coordinates are written to text files for each MFOV and for each section. Acquired data is stored on a network filesystem, typically as JPEG files. In addition to full resolution tiles, the microsope stores thumbnail versions of all image tiles (840×744 pixels or less). For contrast normalization, the Zeiss MultiSEM 505 generates a pixel-wise lookup table across all image tiles of one MFOV. The lookup table is stored in text files as base64 encoded arrays.

4.3. The MBeam Viewer

To improve the tile review workflow, we developed the *MBeam Viewer* (The MBeam viewer is freely available as open source software at https://github.com/rhoana/mb). This is Web-based to allow both local and remote neuroscientists to inspect tiles. The viewer visualizes tiles within their contextual MFOV and section. As the microscope generates data from across the sample and places new tiles onto the filesystem, the Butterfly server sends them immediately to the MBeam viewer to update the rendering. The user is able to zoom, pan, and scroll through different sections of the acquired data. Seeing the broader context allows our neuroscientists to identify erroneous regions both within and across tiles. The most frequent errors are unfocused images due to specimen height variations, dirt on the specimen, and tape edge errors. Another error is non-uniform contrast and brightness within an MFOV: each of the 61 beams is an independent detector, and so histogram equalization is required between the output images. Due to image content differences between the beams, this equalization can fail. Further, if beam alignment fails, then gaps form between individual image tiles. All of these errors can be detected quickly with our system.

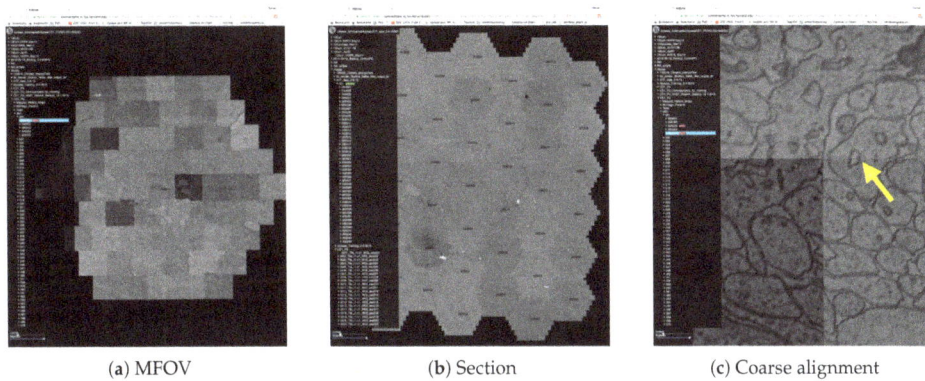

| (a) MFOV | (b) Section | (c) Coarse alignment |

Figure 3. The MBeam viewer visualizes data during image acquisition using the Zeiss MultiSEM 505 electron micoscope. (**a**) An individual multi-beam field-of-view (MFOV) is visualized, consisting of 61 images which are coarsely aligned using coordinates from the microscope. Each individual image is 3128 × 2724 pixels in size, resulting in roughly 30 k × 26 k pixels for each MFOV. Multiple MFOVs are stitched together as one section within an image volume. The MBeam viewer visualizes such sections with possibly hundreds of MFOVs. (**b**) 44 MFOVs stitched to roughly 200 k × 170 k pixels. Demand-driven rendering and on-the-fly mip-mapping enable user interaction in realtime. (**c**) The zoomed-in view shows the full resolution of the data and stitching artifacts due to coarse alignment (yellow arrow).

Observations of the traditional process of quality assessment during acquisition have led to the following design choices:

D1. Remote and collaborative visualization: Our data collection occurs four floors underground to protect the electron microscope from vibrations. It is important that each of the microscope technicians, pipeline developers, and neuroscientists are able to quickly assess scan quality. Hence, the MBeam viewer frontend is remotely accessible via a Web browser and supports region of interest sharing for collaborative viewing.

D2. Scalable visualization with on-the-fly mipmapping: Data from the microscope is stored on a high-performance network filesystem to optimize throughput and match the acquisition speed of the microscope. When viewing this data, the resolution of a single image tile (3 k × 2.7 k pixels) roughly matches the resolution of a typical off-the-shelf computer monitor. Assessing the quality of one tile at a time is simply not practical and contextual views of an MFOV or a section increase the assessment speed. Therefore, the Butterfly server creates mipmap representations of each tile to provide zoomed-out overviews. These representations are created online to avoid additional file input and output operations and to reduce storage overhead. Then, the viewer combines a caching mechanism with demand-driven rendering. When a tile is requested for display, we directly send it to the client and render it, resulting in a streaming effect of sequentially appearing tiles. Tile requests always start in the center of the current view and fan out, and are responsive to user pans, zooms, or scrolls. There is a trade-off between visualization response and mipmap generation time and we quantitatively evaluate this design choice in Section 9.

D3. Discovery of MFOVs, sections, and scans: The electron microscope writes data into a directory tree on a network filesystem. A directory holding a typical scan contains thousands of sub-directories for each section, with each of those holding hundreds of sub-directories for each MFOV. Listing such a hierarchical directory structure on a network filesystem is slow, and so data change detection is slow. This also adds many disk operations to the storage system. However, the MBeam viewer needs to detect when new data arrives and whether this data is of type MFOV, section, or scan. Instead of walking through all directories, we find new data by probing the coordinate text files written by the microscope. These files are stored in fixed relative locations and follow a hierarchical pattern which allows us to distinguish data types without directory traversal. This way, data can be written sequentially during acquisition, and we can detect when a researcher moves data around manually. We present the data structure in the frontend and indicate the data type which can be selected for visualization.

D4. Interactive overlays and image enhancement: Acquisition quality can differ between individual image tiles. The MBeam viewer overlays additional information such as tile, MFOV, and section identifiers on top of the image data. This way, the user is able to match poor quality images with the actual data on the network filesystem and in the physical sample to instantiate re-capturing. It is also possible to perform client-side contrast and brightness adjustments for the current view to aid in the visual identification of low-quality tiles.

Our collaborating neuroscientists use the MBeam viewer everyday, and the streaming overview provided is a significant improvement over simply using a file browser. Zeiss, the manufacturer of the MultiSEM 505 electron microscope, installed the MBeam viewer for internal use, liked the simplicity and the open source nature, and extended their microscope control software with similar functionality.

5. Visualization of Registration

5.1. Motivation

The acquired images need to be precisely aligned in 2D and 3D to enable further automatic and semi-automatic processing. For this, we apply a compute-intensive stitching mechanism [50]. This process needs to be monitored and debugged, and parameters need to be fine-tuned. During this process, we store each modification of the input data as a copy so that we can revert back to the original input.

5.2. Data

To compensate for non-linear distortions in our 2D and 3D images, we refine a rigid registration with an elastic process. First, we calculate a scale-invariant feature transform (SIFT) [51] per 2D image tile to detect local keypoints. Then, we minimize a per-tile rigid transformation distance between matched features in each pair of adjacent overlapping tiles.

To align the image stack in 3D, we use the method of Saalfeld et al. [50] to find a per-section elastic (non-affine) transformation. This process begins by finding an approximate affine transformation for each MFOV from a section to an adjacent section. This is followed by overlaying an hexagonal grid onto each section, and matching each vertex to a pixel in an adjacent section using a block matching algorithm: a small image patch around each vertex of the grid is compared against a constrained area in the adjacent section. These blocks are different than the tiles of an MFOV to incorporate oxverlap between tiles. Finally, a spring-based optimization process minimizes the deformation between grid points and outputs a transformation for each image tile. We expand this process to non-neighboring sections to overcome acquisition artifacts. Further, we create a parallel implementation and optimize the elasticity parameters for our datasets.

5.3. The RHAligner Plugin

We inspect MFOV alignment quality by clustering transformations based on the angles of one of the internal triangles of the hexagonal MFOV shape. Changes in these angles represent squash and shear effects. We visually inspect MFOVs whose clustered angles are far from a cluster center. For this, visualizations are key to monitor, debug, and fine-tune the alignment process. This process can fail if the data is noisy or when the optimization process does not find a solution. We have developed the RHAligner plugin (The RHAligner plugin is freely available as open source software at https://github.com/rhoana/rh_aligner) for Butterfly to support the visualization of the alignment steps. Several requirements for inspecting the registration lead to the following design choices:

D1. Store transformations and data seperately: Computed transformations during registration are stored for each image tile in a specification format called tilespec [52]. Meta information such as the dimensions of a tile, the file path, and the list of different transformations are stored as part of this JSON data model. This format resulted from a collaboration with neuroscientists at the Janelia Research Campus. Storing transformations separately allows flexibility during the alignment process since the image data itself is not modified. This flexibility is needed to tweak alignment of tiles within MFOVs, sections, and scans.

D2. Visualization of SIFT features: The first step of the alignment process is the computation of one scale-invariant feature transform [51] per image tile. If these features do not describe the image tile properly, all successive steps fail. We visualize the SIFT features as overlays on top of the original image tile as shown in Figure 4a.

D3. Illustration of overlap between tiles: We use SIFT features of overlapping tiles to replace the coarse alignment of image tiles within an MFOV. To debug and monitor this process, we visualize SIFT feature matches between tiles (Figure 4b).

D4. Online rendering of transformations: We added a tilespec reader to the Butterfly middleware. This lets us use the MBeam viewer to visualize the aligned data in 2D and scroll through the stack (Figure 4c). We inspect the alignment of a scan using a zoomed-out view of sections and concentrate on larger neurons while scrolling through the scan. The movement of large structures is relatively easy to track in grayscale images and interruptions due to false alignment are clearly visible. This allows us to find regions which require further refinement.

D5. Abstracted visualization of section alignment: We stitch MFOVs together in 2D sections and perform block matching with adjacent sections in 3D. An MFOV itself is roughly 520 megabytes in size and rendering a section containing hundreds of MFOVs is slow even with display-aware and demand-driven visualization. We add an abstracted visualization of stitched sections to the RHAligner (Figure 4d–f). It is possible to quickly assess the section alignment of adjacent sections using this technique.

D6. Visualize displacement of blocks within an MFOVs: Once we identify failing alignment in a section, we need to know which MFOVs are misaligned—specifically, which blocks of the MFOVs. For this, we visualize the displacement of blocks within each MFOV using vector fields (Figure 4g–i). These renderings are lower resolution to be scalable. However, this makes it difficult to see the arrows of the vector field for large datasets. Therefore, we color code the direction of each displacement according to its angle. If a displacement is larger than a user-defined threshold, we highlight the vectors in white.

Debugging the alignment is manual since the uniqueness of each dataset requires fine tuning of the alignment computation. The developed visualization scripts are used by experts to assess alignment quality and quickly find difficult areas which need parameter optimization. Once the data is fully aligned and the transformations are finalized, we harden the transformations and store the modified scans as volumetric data within HDF5 containers.

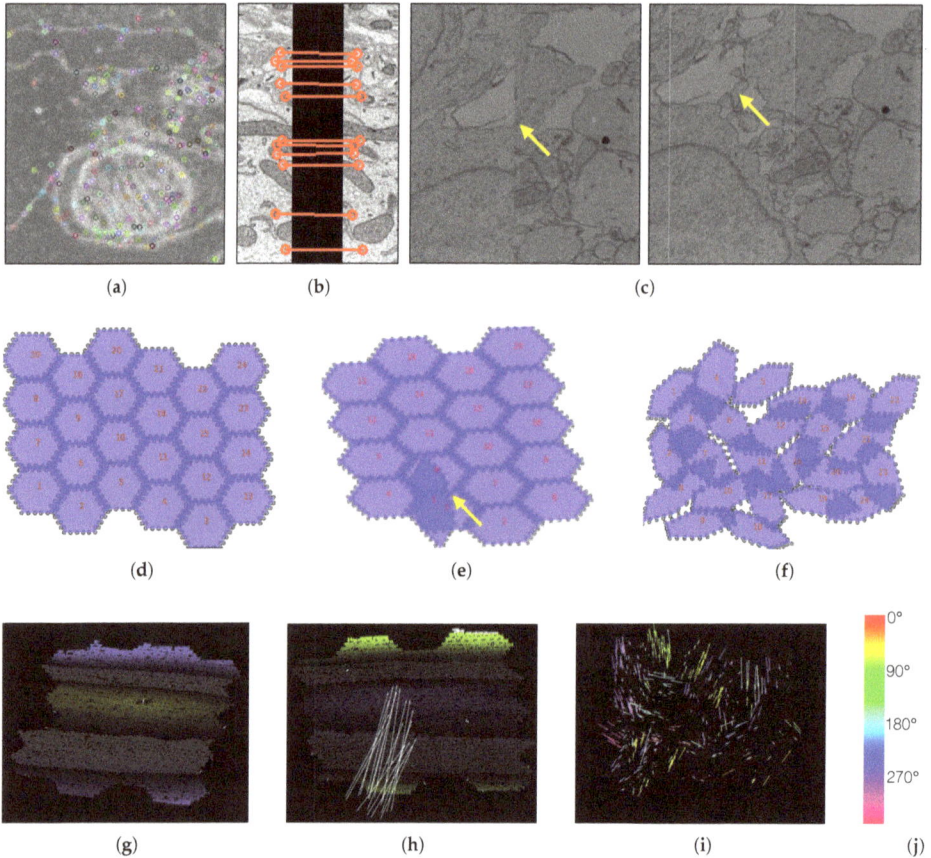

Figure 4. We create visualizations to monitor and debug the registration process. (**a**) We visualize scale-invariant feature transform (SIFT) features within an image tile. (**b**) We visualize these features when looking at overlapping tiles. (**c**) We render aligned MFOVs before and after the registration. (**d**) A correct transformation can be seen where the MFOVs appear to be slightly rotated clockwise. (**e**) Here we see an erroneous transformation for one MFOV. Further examination reveals that the mismatched MFOV (yellow arrow) is incorrectly matched with the features of the neighboring section. (**f**) This visualization shows a case where most MFOVs are incorrectly transformed. This is caused by a large rotation of the neighboring section, which has severe impact on the accuracy of SIFT features. The vector fields in the bottom row show the displacement of each single match between adjacent sections (after an affine alignment to align these two sections). We color code by angle. (**g**) A "normal" matching. The sections are more or less on top of each other, but after the affine alignment the top and bottom parts (purple) are stretched to the bottom left, and the middle (yellow) is stretched towards the top right. Some white arrows are outliers. (**h**) A case where one MFOV was not pre-aligned properly, and so all its arrows are white. (**i**) A case where the pre-alignment gave bad results. (**j**) The color map for visualizing the displacement angles.

6. Visualization of Segmentation

6.1. Motivation

Registered connectomics data is ready to be automatically segmented. We find cell membranes of neurons using an automatic segmentation pipeline [9]. This is a difficult and often error-prone task. These errors take two forms: *merge errors*, where two neurons are fused together, and *split errors*, where one neuron is split apart. These errors happen within a section and across sections. We report classification results on the SNEMI3D challenge [10] as variation of information scores of $V_{F-score}^{Info} > 0.9$ in Knowles-Barley et al. [9,53]. We also detect where neurons exchange information through synaptic connections. Errors can occur during synapse detection when synapses are not found or falsely labeled.

To find regions of error and to debug the segmentation pipeline, we need to look at feature maps such as membrane probabilities. We also need to visually assess segmentation quality and synapse detection since ground truth segmentations are sparse and time-consuming to generate. Similar to visualizations of the alignment process, visualizations aid the complicated process of parameter tuning.

6.2. Data

We use a state-of-the-art automatic labeling of neurons [8,9,53]. Cell membrane probabilities are generated using a convolutional neural network (CNN) based on the U-net architecture [54]. The probabilities are encoded as gray-scale images, then used to seed an implementation of the 3D watershed algorithm which generates an oversegmentation using superpixels. Then, we agglomerate superpixels with a parallel implementation [55] of Neuroproof [56,57]. The resulting segmentations contain labelings of millions of different neurons. For scalability, we run the segmentation steps in parallel as part of a distributed processing framework. We split a scan into blocks of fixed size which we segment in parallel. Once computed, the blocks are merged to create the segmentation of a full scan.

Synapse Detection. Concurrent with the automatic neuron segmentation, we also detect synaptic connections. For this, we use another U-net classifier to label the pre- and post-synaptic pixels near the synaptic connection. We find the synaptic polarity by combining this output with our cell membrane segmentation. Unfortunately, the combined segmentation and synaptic polarity detection method is not yet published and further details will appear in a future paper. Detection performance of $F1_{score} > 0.8$ can be expected [58].

For segmentations, HDF5 containers are our format of choice. This file format is widely supported and random access is possible without loading the entire volume into memory. Calculated features such as membrane probabilities and synaptic connections are also stored in HDF5 containers.

6.3. RhoANAScope

Visual inspection of segmentation output is required. For this, we developed RhoANAScope (RhoANAScope is freely available as open source software at https://github.com/rhoana/butterfly/bfly/static), a viewer for neuron and synapse segmentations. RhoANAScope visualizes sections and scans of gray-scale EM image data. It supports overlaying multiple layers to match image data to segmentation output or feature maps. Standard interactions such as zooming, panning, and scrolling through the stack enable the exploration of large datasets. To be scalable, we use Web-based demand-driven and display-aware rendering with GPU acceleration. We designed RhoANAScope as follows:

D1. Visualize grayscale images with multiple overlays: The inputs to our automatic segmentation pipeline are fully aligned scans. The pipeline uses these scans to compute membrane probabilities and to generate a cell membrane segmentation. The membrane probabilities and the membrane segmentation overlap spatially with the gray-scale EM scan. Our task is to overlay the data to debug and understand these outputs. RhoANAScope is designed to support multiple layers of image data blended using user-configured opacity values. This way, we can see the original image data while looking at probabilities and membrane output to understand the classifications. We show a section with cell membrane segmentation overlays in Figure 5a, and with synapse detection overlays in Figure 5b. We show membrane probabilities in Figure 5c.

D2. Colorize neighboring neurons: When rendering segmentation layers, we colorize the labelings to distinguish between neighboring neurons. The colorization involves a look-up procedure which maps the identifier to an RGB color. We use the following formulas to map labels to RGB color values:

$$R = ((107x) \bmod 700) \bmod 255 \tag{1}$$

$$G = ((509x) \bmod 900) \bmod 255 \tag{2}$$

$$B = ((200x) \bmod 777) \bmod 255 \tag{3}$$

For a given id x, we set each byte for red, green and blue to cycle approximately every seventh id ($\frac{107}{700} \approx \frac{1}{7}$) for R, every other id ($\frac{509}{900} \approx \frac{1}{2}$) for G, and every fourth id ($\frac{200}{777} \approx \frac{1}{4}$) for B, The exact values chosen preserve dramatic changes between neighboring values while adding subtle differences to allow 233,100 unique colors for sequential ID values. The resulting colorization is shown in Figure 5a.

D3. GPU accelerated rendering: Sections can be very large and any overlay doubles the amount of rendered data. We use demand-driven and display-aware rendering for scalability. However, we use GPU accelerated rendering to process the overlays. This way, we can add additional processing such as adding segment borders or applying a color map in cases where looping through the pixels on the CPU is too slow.

D4. Support multiple input formats: We mainly use HDF5 containers for storing segmentations to support random access without loading a full scan into memory. This is not always fast, and other connectomics tools aid scalability by storing sections separately as HDF5 or by providing a JSON type descriptor of the data (e.g., Neuroglancer [33]). We abstract the input format by using the butterfly middleware and support these different storage methods. Adding support for a new file format requires extending Butterfly by adding another derived input source.

D5. Index multiple data sources: During segmentation, different parameters result in different results. This means scans can have many alternate segmentations, and keeping track is difficult. RhoANAScope uses the Butterfly middleware to query directories for their datasets. This mechansism detects file types and meta information (such as dimensions, pixel encodings etc.) and creates a searchable listing of available datasets.

D6. Compression: Storing connectomics data can quickly add up to multiple terabytes or petabytes of data. We support the efficient and segmentation-map-specific Compresso algorithm for reading data and also for transferring data from butterfly to RhoANAScope [45].

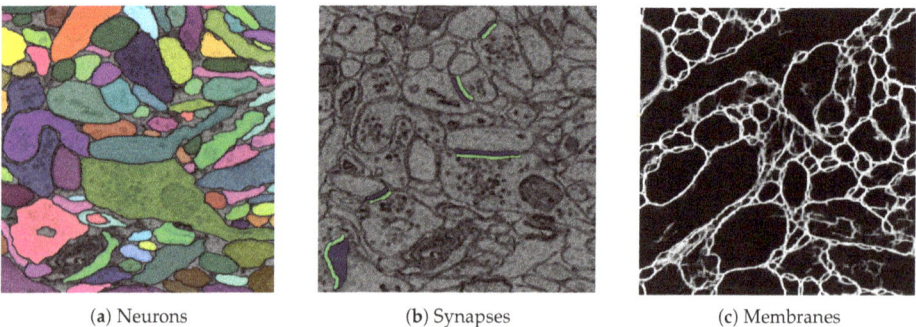

(a) Neurons   (b) Synapses   (c) Membranes

Figure 5. *Cont.*

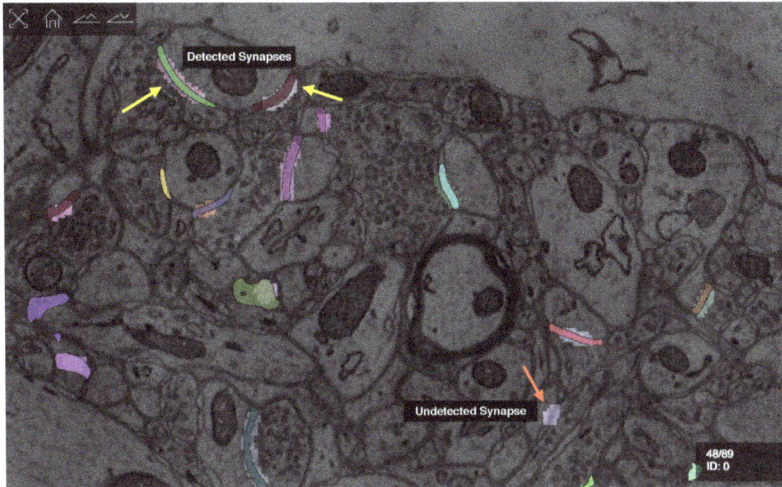

(**d**) Detected Synapses

Figure 5. We developed RhoANAScope: a Web-based visualizer for neuron and synapse segmentations at scale. RhoANAScope visualizes segmentations on top of image data using WebGL. (**a**) Individual neurons are colored according to their cell membranes. (**b**) Pre- and post-synaptic markers are colored as green and blue, and indicate information pathways between neurons. (**c**) We use membrane probabilities to debug the automatic segmentations. (**d**) RhoANAScope is scalable to large data by implementing display-aware and demand-driven rendering, and supports multiple overlays. Here, we match a groundtruth segmentation of synapses with the output of our automatic synapse detection to view undetected synapses (red arrow) versus detected synapses (yellow arrows). Please note that the synapse detection visualized here is not complete.

Automatic segmentation of 2D nanoscale images is a complex process. We use RhoANAScope to understand the output of our pipeline and to visually assess segmentation quality, but also as an every-day viewer for segmented connectomics data.

7. Interactive Visualization for Proofreading

7.1. Motivation

The output of our segmentation pipeline is not perfect. Labeled connectomics data, on average, requires hundreds of manual corrections per cubic micron of tissue [22]. As mentioned previously, the most common errors are *split errors*, where a single segment is labeled as two, and *merge errors*, where two segments are labeled as one. With user interaction, we can join split errors (Figure 6a), and we can define the missing boundary in a merge error (Figure 6b). This manual error correction is called proofreading.

select   click 1   click 2   click 3

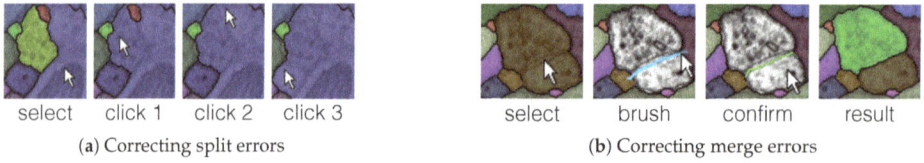

(**a**) Correcting split errors

select   brush   confirm   result

(**b**) Correcting merge errors

(**c**) Interactive proofreading using Dojo

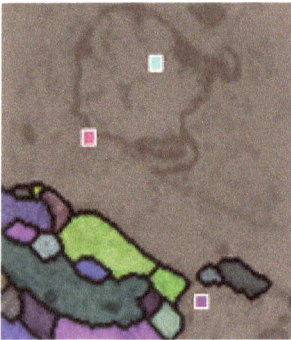

(**d**) Collaboration in 2D

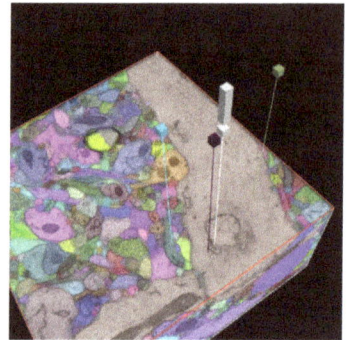

(**e**) 3D

Figure 6. The output of our automatic segmentation pipeline requires proofreading to correct split and merge errors. These errors can be fixed with simple user interaction: (**a**) Split errors can be corrected by joining segments; and (**b**) merge errors can be corrected by drawing the missing boundary. Including these interactions; (**c**) we build the interactive proofreading tool *Dojo* with a minimalistic user interface, display-aware and demand-driven visualization, and integrated 3D volume rendering. Proofreaders can collaborate, each with cursors in (**d**) 2D and (**e**) 3D. The exclamation mark in (**e**) is used to request help at a specific region within the data.

7.2. Data

After automatic segmentation, we compute mipmap representations of scans and their corresponding segmentation data. Each section of a scan is stored as a separate mipmap with each mip level stored as tiled images. This allows partial access when requesting zoomed out views of data. However, editing the segmentation then requires partial rebuilding of the mipmap if the image data is changed.

In practice, each mipmap representation includes meta information regarding the dimensions, tiling, data format, and encoding stored within an XML descriptor. The data is organized as a hierarchical directory structure, with zoom levels grouping sub-directories for each section. Then, each section is stored tiled as a row/column format.

7.3. Proofreading Applications

Proofreading is necessary to interactively correct segmentation errors before analysis. For this, we propose *Dojo* (Dojo is freely available as open source software at https://github.com/rhoana/dojo.), a Web-based proofreading application that supports multiple users. We evaluated Dojo and other proofreading software with novices trained by experts as part of a quantitative user study (between-subjects experiment) [11], and designed mechanisms for quality control [13].

7.3.1. Interactive Proofreading Using Dojo

Dojo enables proofreading in 2D with users able to zoom, pan, and scroll through the stack (Figure 6c). To visualize, we use GPU accelerated demand-driven and display-aware rendering in 2D, and we also incorporate 3D volume rendering. We made the following design choices:

D1. Minimalistic user interface: A proofreading tool has to be simple and easy to understand. This is especially true if we wish to support use by non-experts. The user interface of Dojo is designed to be parameter free and have limited options available through icons (Figure 6c). Textual information is presented small but still readable. This way, the proofreader can focus on the actual data and the task of finding errors.

D2. Interactive splitting and merging: Users of Dojo need to be able to correct split and merge errors efficiently. To correct split errors, the user clicks on one incomplete segment and then on the segments to be joined (Figure 6a). For merge errors, Dojo allows users to split a single segment into two or more by drawing a line across a segment (Figure 6b). Then, the best split line is calculated by using the user input as seed points for a watershed algorithm. Both interactions require minimal interaction and give immediate feedback.

D3. Three-dimensional volume rendering: In Dojo, proofreading happens when viewing a single 2D section. However, we also include 3D volume rendering to visualize segments in the context of the full scan. In a controlled experiment between novice users, we observed that this feature especially allowed non-experts to better understand the three-dimensional property of connectomics data [11].

D4. Dynamic merge table and partial mipmapping: Proofreading requires many corrections of segmentation data. Since our data results from an oversegmentation (Section 6), more split errors than merge errors exist. We use a segment remap table to allow merge operations without actually modifying the image data. Merging a segment with another segment is achieved by adding a look-up or redirection entry to the segment remap table. This dynamic data structure is stored separately from the data and is applied during rendering. For split errors, we calculate which parts of the mipmap need to be updated to properly adjust the segmentation data. This is scalable because each mip level is stored as individual image tiles.

D5. Collaborative editing: Proofreading of larger volumes can be sped up when multiple users correct the data at the same time. In Dojo, we synchronize the modifications of the segmentation data among all connected users via Websockets. This can result in many transfers depending on how many clients are connected. Therefore, we limit the transfer to coordinates and meta information and deliver updated segmentation tiles on request. If two or more users work on the same region of segmentation data, the other users' cursors are shown as small colored squares (Figure 6d). In 3D, the cursors are displayed as colored pins pointing to a position within the scan. In addition to cursor sharing, users can actively mark an area of the data to seek help from other users (Figure 6e).

Our experiments have shown that the majority of proofreading time is spent by users looking for errors. To reduce this time, we propose the Guided Proofreading system [12] which suggests errors and corrections to the proofreader.

7.3.2. Guided Proofreading

Using classifiers built upon a convolutional neural network (CNN), the Guided Proofreading system (Guided Proofreading is freely available as open source software at https://github.com/VCG/guidedproofreading) detects potentially erroneous regions in an automatic labeling. Then, we present the proofreader with a stream of such regions which include merge and split errors and their corrections. This way, proofreading can be hastened with a series of yes/no decision, which is faster than manual visual search using Dojo.

D1. Split error detection via CNN: We trained a split error classifier based on a CNN to check whether an edge with an automatic segmentation is valid. By choosing a CNN over other machine learning methods, we enable the classifier to learn features by itself rather than using hand-designed features. Our CNN operates only on boundaries between segments and, in particular, on a small patch around the center of such boundaries. We use the grayscale image, the membrane probabilities, a binary mask, and a dilated border mask as inputs. The architecture of our split error classifier uses dropout regularization to prevent overfitting (Figure 7a).

D2. Merge error detection using the split error classifier: We reuse the split error classifier to detect merge errors. We generate potential borders within a label segment using randomly-seeded watershed, and then test whether each border is a split error. If our CNN reports a valid split, we assume that this border should exist, and therefore we should split the label segment in two.

D3. Single-click corrections: We perform merge and split error detection as a pre-processing step and sort them by probability of error confidence. The Guided Proofreading system then presents the most likely errors one-by-one to the user and also shows a potential correction. Then the proofreader can decide whether to accept or reject a correction with a single click.

Our experiments show that Guided Proofreading reduces the average correction time of 30 s with Dojo to less than 5 s on average. Correcting segmentations of large connectomics datasets still takes a long time, but proofreading applications make this more feasible by supporting multiple users and simple operations. We are currently exploring new ways to proofread synaptic connections.

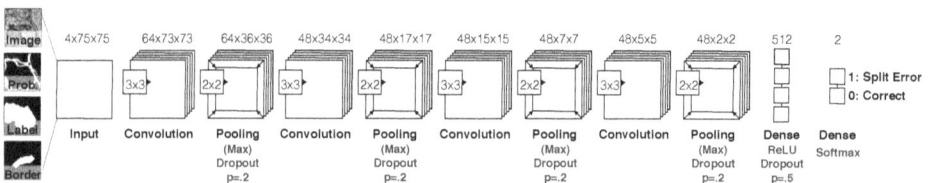

(**a**) A classifier for detecting split errors

Figure 7. *Cont.*

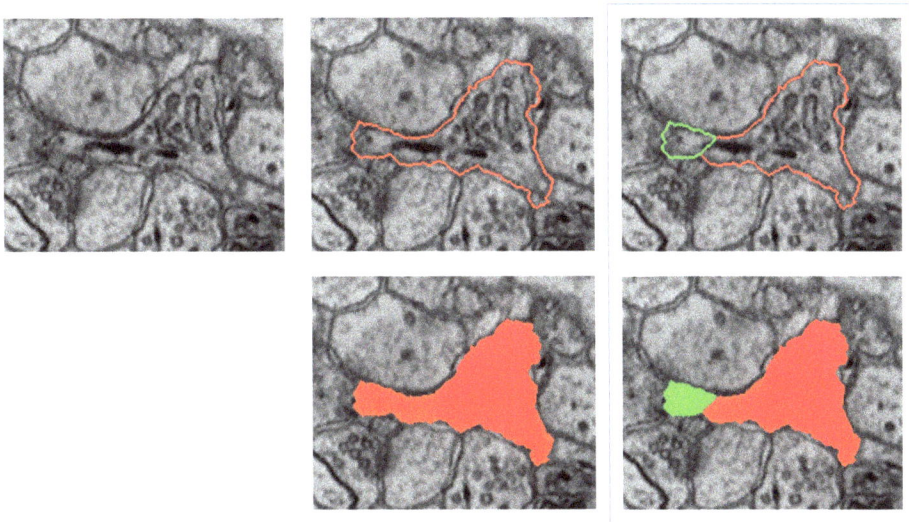

(**b**) The Guided Proofreading user interface

Figure 7. The Guided Proofreading system reduces the time spent finding potential errors by proposing candidate errors and corrections. (**a**) The system is informed by a convolutional neural network (CNN) with a traditional architecture. The CNN uses a patch of image data, membrane probability, binary cell mask, and cell border overlap to decide whether a split error between two neurons exists. (**b**) Then, we present the proofreader with a stream of regions and candidate corrections. Thus, proofreading can be hastened with a series of yes/no decisions, which is about 6× faster than using Dojo.

8. Network Analysis

8.1. Motivation

Analysis is the final step of the connectomics pipeline. Segmented and proofread data includes cell membrane annotations for all neurons and the synaptic connections between them. This information represents a (partial) wiring diagram of the mammalian brain, and this network can be modeled as a weighted graph (i.e., weighted by number of synaptic connections). Such a graph structure is three dimensional, dense, and can be hard to analyze due to the large size. To better understand the data, we need visualizations that render it in 3D and show the biological properties of neurons and their connections. However, typical analysis concentrates on a subset of our data. This means we also need sophisticated methods to query and filter our large wiring diagram.

8.2. Data

As previously mentioned, segmentation data and synaptic connections are typically stored as HDF5 files containing the full scan, or as tiled mipmap data structures. The generated neuron and synapse information is stored as a graph structure in JSON files. We parse these files and store the connection information in a database as part of the Butterfly middleware, so that we can perform efficient indexing and querying.

8.3. Tools for Network Analysis

8.3.1. 3DXP

To prepare for 3D analysis, we generate meshes representing neuron geometries by performing marching cubes [59] on our volumetric segmentation data. We have developed 3DXP (3DXP is freely available as open source software at https://github.com/rhoana/3dxp), a Web-based application for exploring volumetric image data and neuron geometries in 3D (Figure 8). This application is fully interactive and allows researchers to analyze individual neurons and their connections by zooming, panning, and scrolling through the scan. We designed 3DXP with the following choices in mind:

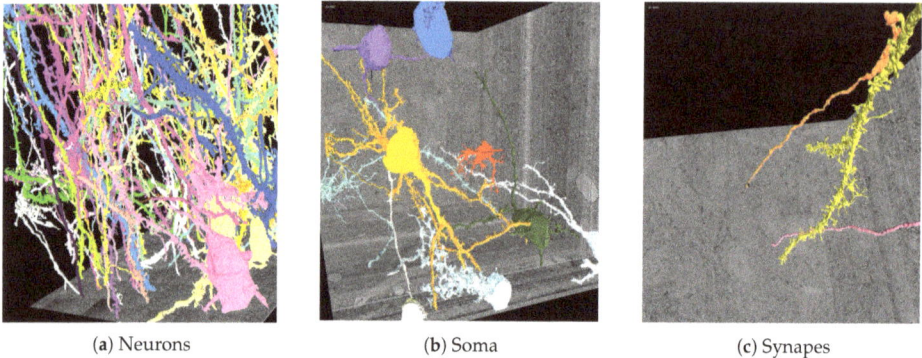

(a) Neurons   (b) Soma   (c) Synapes

Figure 8. Three-dimensional polygonal mesh reconstructions of automatically-labeled connectomics data using our Web-based 3DXP software (downsampled to 3 k $\times$ 3 k $\times$ 1.6 k voxels). Since the reconstructions are hundreds of megabytes in size, we stream and render the geometries progressively. The displayed scenes show: (**a**) twenty neurons stretching through a 100 μm^3 volume; (**b**) multiple cell bodies (soma) visualized; (**c**) a dendrite with two synaptic connections. All scenes are interactive with zoom, pan, and scroll interactions. It is also possible to mouse click on a mesh region to open other visualizers for further data exploration, e.g., in Dojo.

D1. Progressive rendering: For collaborative research, a shareable dynamic visualization allows more interactivity with the data than statically-generated images or videos. While the visuals must be simple enough to transfer over an internet connection, the most valuable information from detailed connectomics emerges when we can show highly detailed reconstructions. Progressive rendering provides such a solution through multiple meshes of varied levels of detail. 3DXP applies the existing format of POP Geometry [60] to direct bandwidth to the neural projections closest to the interactive camera.

D2. Parallel computation: Rendering meshes at several levels of detail takes too much time to compute on demand for each request. With 3DXP, we precompute meshes at multiple resolutions for all segmentation identifiers in the reconstructed volume. To reduce the number of days required to generate meshes for millions of neurons over trillions of voxels, we divide the full volume into a grid. The volumetric grid allows the parallelization of both mesh generation and conversion to the multiresolution format across any number of simultaneous connected machines with limited memory usage.

D3. Correspondence to EM imagery: The rendered meshes when viewed in isolation provide no indication of the raw data provided as input to the reconstruction. To show the position of a surface within a brain region, 3DXP displays the meshes alongside axis-aligned sections from the electron microscopy image volume. Our researchers need to ensure individual EM scans match at the corresponding depth of a given mesh, so we transfer highly downsampled images of each scan. The users freely move up and down through scans scaled to match the reconstructed meshes.

D4. Animation: Visualizing more neural projections in 3D corresponds to increasing visual complexity. An informative and visually appealing solution involves laying a single EM section to obscure all structures deeper than a given region of interest. The 3DXP viewer supports the creation of animations that interpolate between chosen camera positions, each linked to a single EM section. The user saves each given viewpoint as a keyframe linked to a single section. 3DXP can then step through each section at a constant rate in an animation, moving gradually through the saved viewpoints.

D5. Interoperability: While a 3D visualization is useful, it only partially supports a full understanding of a given volumetric reconstruction. Therefore, we designed the 3DXP viewer to work seamlessly with the 2D tiled image viewers in our system. A user can identify the coordinates of any point on a rendered mesh to immediately view it in 2D. In particular, the interoperability with the Dojo editor allows the discovery of segmentation errors in an exploratory 3D view to immediately facilitate manual corrections in a more focused and higher resolution editing environment.

8.3.2. Neural Data Queries

To query, filter, and parse the relationship of neurons and their connections, we developed the Neural Data Queries system (Figure 9). This system offers a well documented API to request the following information: all synaptic connections in a region of interest, the center of a specific synapse, pre- and post-synaptic neurons of a specific synapse, all neurons in a region of interest, the center of a neuron, and all synapses of a neuron. While the Neural Data Queries system returns numeric or text data, it is possible to use this information to render specific neurons and their connections using our visualizers. For this, we report the following design choices:

D1. Interoperability: To facilitate communication between 3D mesh and 2D image viewers, the Butterfly middleware supports neural data queries within a shared naming convention and single coordinate frame. For viewers where synapse data is not displayed, a query containing only the coordinates of a region of interest returns a list of the included synapses. Further requests return the information needed for a particular viewing task: either the specific coordinates, or the neuron segments joined by the synapse.

D2. Support for automation: The neural data queries consist of the number of elemental requests needed to express the spatially-embedded connectivity graph over many short and fast queries. Rather than attempt to enumerate all the complex queries of an embedded graph, the neural data queries often return a single property of a single entity. An automated client can request "the synapse location between neuron A and neuron B" in multiple neural data queries. If any ID value in "synapses of neuron A" occurs in the list of "synapses of neuron B", then a third request can be made for the location of the shared synapse.

D3. Informative feedback: The RhoANAScope viewer makes use of many small neural data queries to asynchronously index all tiled images available for viewing. A single request lists all experiments, which trigger requests for all samples per experiment, then all datasets per sample, and ultimately all tiled image channels per dataset. When manually querying the API, the Butterfly system facilitates manual indexing through helpful error messages for humans. Without any parameters, Butterfly suggests a list of experiments. In response to requests, including a misspelled channel, Butterfly suggests valid channels. The response is the same for any missing or invalid parameter such that humans can interact with the API without prior knowledge.

Query

```
/api/entity_feature?feature=synapse_parent&experiment=e&sample=s&dataset=d&channel=c&id=10
```

Response

```
{ "synapse_parent_pre": 888, "synapse_id": 10, "synapse_parent_post": 999 }
```

Query

```
/api/entity_feature?feature=synapse_keypoint&experiment=e&sample=s&dataset=d&channel=c&id=10
```

Response

```
{ "y": 12800, "x": 12800, "z": 1700 }
```

Figure 9. Two neural data queries and responses show separate properties of a given synapse. Both `synapse_parent` (**top**) and `synapse_keypoint` (**bottom**) queries request a field of data for synapse label 10 in dataset *d* of sample *s* in experiment *e*. The channel *c* refers to the tiled volume containing a reconstruction of synapse segments. The top request returns both neuron segment labels representing the presynaptic input and postsynaptic output of the provided synapse. The bottom request returns the coordinates of the center of the requested synapse.

We specifically designed the Neural Data Queries system together with neuroscientists to support targeted data exploration. Targeted data queries provide endpoints specific to neural reconstructions. API endpoints provide the center coordinates of any neuron ID with a cell body or any synapse ID value. Most importantly, we can return all synapses in a region, all synapses of a neuron, and both neurons for any given synapse. A configuration file listing file paths matches tiled images with the corresponding files for center coordinates and synapse connectivity. An efficient database integrated in Butterfly ensures that each query response is generated in less than 5 s.

9. Performance and Scalability Experiments

All 2D client visualization applications in our system request a view onto a scan volume from the Butterfly server. When a client requests a viewport that is smaller in pixel size than the files in storage, the server must load and transfer subsections from each file, e.g., for a zoomed-in view. When a client requests a viewport that is larger in pixel size than the files in storage, the server must load and combine many image files before transfer, e.g., for a zoomed-out view. For far-zoomed-out views, this combination requires loading many hundreds of files, which can be slow. To provide a faster response, the client sequentially requests only tiles within the total-requested viewport. These are independent of the tiles on network file storage. The server then sends these tiles as parts of the viewport of the client where they are displayed.

Each 2D client application opens with a zoomed-out view completely containing a full tiled image at the lowest available resolution. For an overview of the full volume at a low resolution, users can scroll through all full tiled images in the stack. The speed of this interactive overview depends on the time to load and transfer a full tiled image at a mipmap level within the viewport dimensions. By considering the most zoomed-out view, we compare transfer times without calculating the variable number of tiles needed for many possible zoomed-in views. We measure this tile transfer from the storage system as a second experiment, which applies to pre-computed and on-demand mipmap scenarios.

To optimize client tile transfers of arbitrary sizes from storage systems with an arbitrary numbers of files, we present results from two sets of experiments designed to separately test performance of client tile transfer and file storage. Given the optimal client tile size, the optimal bit rate to read files from a network file system reduces the overhead for each transfer. These experiments measure throughput from data on the network file system to a Web-based client viewer application, and are indicative of the general performance of all reported visualization tools. In this sense, they measure the scalability of our display-aware and demand-driven rendering (Figure 10). We measure both client tile size and file size by the pixel dimensions of the square tiles (denoted either as $N \times N$ or N^2 pixels.)

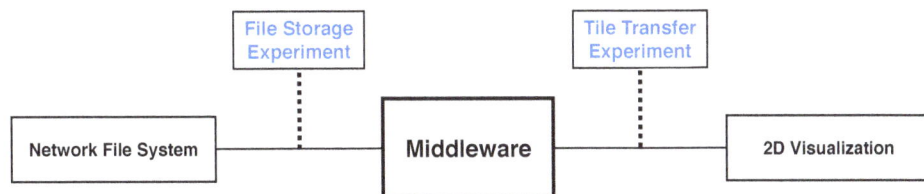

Figure 10. We record throughput of a generalized scalable visualization pipeline, to inform best-practice Butterfly parameter settings. This pipeline replicates the general setup of all reported 2D visualization tools. First, we load data from a network file system. Our middleware then processes these files and transfers tiled images to a 2D visualizer. Our experiments include the *file storage experiment* and the *tile transfer experiment*.

Client Tile Transfer Experiment: We measure both the time to transfer a single tile to an image viewer and the time to transfer data that is split into multiple tiles. This helps to answer the question: *Which tile size best enables streaming when transferring data from server to client viewer?* The size of the client-requested image tiles affects the transfer time from the server, where the optimal client tile size for each request should minimize both the delay between an individual tile appearing in the client and the duration until all tiles fill the client display.

File Storage Experiment: We measure loading speed for a full section of data using different-sized files in storage to answer the question: *Which file size is most efficient to read connectomics imagery in a network environment?* The size of the image files in storage affects the file read rate from network file storage, where the optimal tiled storage system should minimize the time to load tiles from the file system. For this experiment, we assume we want to read a fixed number of pixels from disk.

9.1. Client Tile Transfer Experiment

We measure the time to serve an image of 4096×4096 pixels from server memory. We send the full image to the viewer as one or more tiles, each fulfilling a separate image request. Because the total data transferred is constant, the time to serve the full image reflects the bit rate for data transfer to the viewer. For one tile of 4096^2 pixels, four tiles of 2048^2 pixels, 16 tiles of 1024^2 pixels, and 64 tiles of 512^2 pixels, we measure the total time from the start of the first request until the last response completes the full image. The full image transfer time divided by the total number of tiles directly gives the mean time to transfer a single tile to the client. The single tile transfer time also measures the delay between updates to the rendered image. With an increased mean time to transfer a single tile, the user sees tiles render at a slower frequency.

9.1.1. Experimental Setup

The client tile transfer experiment starts a Tornado [61] server from a Python 2.7.11 interpreter on a single CentOS Linux machine on the Harvard Odyssey research computing cluster (The Harvard Odyssey cluster is supported by the FAS Division of Science, Research Computing Group at Harvard University). The viewer contacts the server through an SSH tunnel on the Harvard network from Google Chrome v54 on an Ubuntu Xenial Linux distribution. Before sending data, the server divides a tile of 4096×4096 pixels from an existing EM image into all 64 tiles of 512×512 pixels needed for the first condition. In Chrome on the client, the first request opens HTML and JavaScript for an OpenSeadragon viewer. The viewer starts a timer when ready to make asynchronous requests. After the last image tile arrives, the viewer sends the full duration to the server.

9.1.2. Experimental Results

For the client tile transfer experiment, one-way ANOVA tests showed significance across all 500 repetitions. When measuring time for the full image ($F(3, 1996) = 21,119, p < 0.001$) or time per tile ($F(3, 1996) = 140,190, p < 0.001$), the four conditions of tile size show a significant effect at the $p < 0.05$ level for tiles of 512, 1024, 2048, or 4096 pixel edges.

Relative to a bulk transfer of 4096 × 4096 pixels, Figure 11b shows a longer time to transfer of 64 tiles of 512 pixels, and shorter transfer times for 16 tiles of 1024 pixels or 4 tiles of 2048 pixels. Post hoc comparisons (after Bonferroni correction) indicate that a full image transferred in tiles measuring 1024 pixels in significantly less time than in those measuring 2048 pixels ($t = -17.6, p < 0.001$), 4096 pixels ($t = -123, p < 0.001$) or 512 pixels ($t = -234, p < 0.001$). For all sizes measured, a full image of 4096 × 4096 pixels transfers the fastest in tiles of 1024 × 1024 pixels. The second fastest tiles of 2048 pixels transfer a full image in 8.0% more time on average than tiles of 1024 pixels.

Figure 11a shows that transfer time increases for larger tiles of 2048 pixel sides relative to smaller tiles of 1024 pixel sides. The mean delay between individual tiles greatly differs between tiles measuring 1024 and 2048 pixels. Post hoc comparisons after correction indicate that a single 1024 × 1024 pixel tile loads in much less time than a single 2048 × 2048 pixel tile ($t = -247, p < 0.001$). While any given image contains four times as many tiles of 1024 pixels as tiles of 2048 pixels, each tile of 2048 pixels takes on average 123 milliseconds longer to transfer than each tile of 1024 pixels.

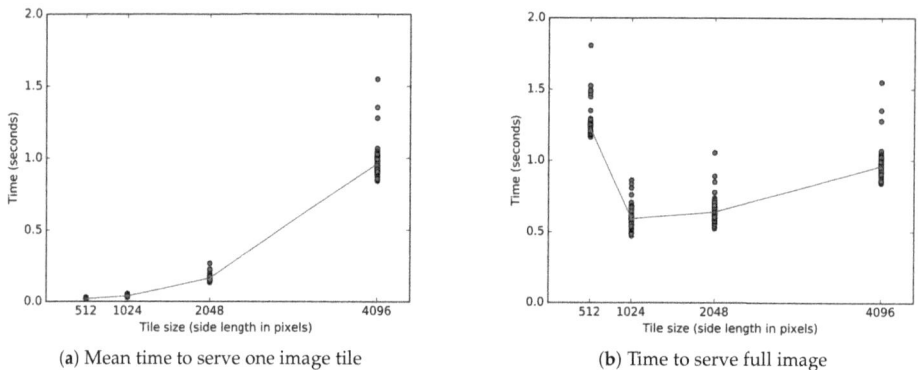

(**a**) Mean time to serve one image tile

(**b**) Time to serve full image

Figure 11. One experiment uses two metrics measured in seconds to evaluate tiled image transfer to an OpenSeadragon client. (**a**) The time to send a single client tile depends on the size of the tile and the bit rate. The sides of a single square tile range from 512 to 4096 pixels. Tiles of the two smaller sizes each arrive in less than 40 ms. Tiles measuring 2048 and 4096 pixels per side each arrive in approximately 160 ms and 960 ms, respectively. (**b**) Only the bit rate affects the time to send a full tiled image of 4096 pixel sides. Sending the full section in one tile of 4096 × 4096 pixels takes 960 ms, but a division into 64 square tiles of 512 × 512 pixels increases that time to 1200 ms. Tiles of 1024 and 2048 pixel sides reduce the time to 590 ms and 640 ms, respectively.

9.2. File Storage Experiment

We measure the bit rate to load from file system divisions of a full 32,768 × 32,768 pixel image, which is at the scale of a single MFOV. We measure the total time to load one file of $32,768^2$ pixels, four files of $16,384^2$ pixels, 16 files of 8192^2 pixels, or 64 files of 4096^2 pixels as square tiles ranging in length from 512^2 to 4096^2 pixels. This simulates a server that delivers tiles of a given size from multiple files of equal or larger size. Therefore, the solid lines in Figure 12 begin by loading each file as a single tile. This experiment isolates the effects on performance of repeated access to multiple files.

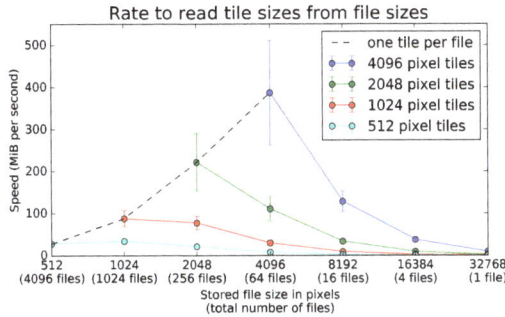

Figure 12. Rate to read various tile sizes from several file sizes. A server loads all tiles from an image stored on a network file system at a rate given by the number of separate TIFF files used to store the data. The server can use square tiles of variable area to load the full section of 32,768 × 32,768 pixels. The four different colored lines represent square tiles of length 512, 1024, 2048, and 4096. The negative slopes of all solid lines show loading speed decreases with larger files for any given tile dimensions. For any tile size above 512 pixels, the tiles load the fastest from files of the same size. The black dotted line shows that it is most efficient to read an entire file in a single section. Error bars represent one standard deviation above and below the mean over 75 trials.

9.2.1. Experimental Setup

The file storage experiment runs entirely from a Python 2.7.11 interpreter in parallel on fourteen similarly configured CentOS Linux machines on the Harvard Odyssey research computing cluster. For any given condition, the program first stores 32,768 × 32,768 pixels of random noise into a number of tiff files of sizes ranging from 512 × 512 pixels through 32,768 × 32,768 pixels. Each trial repeats all conditions in one uniquely labeled folder, and all file names contain an integer sequence unique within a given trial. After writing all files for a given condition to the network file system, the program separately measures the time to load each tile from part or all of the corresponding file. The sum of all loading times then gives the time to load a constant 32,768 × 32,768 pixels from the network file system regardless of the number of files or size of tiles.

9.2.2. Experimental Results

For the file storage experiment, one-way ANOVA tests showed significance at the $p < 0.05$ level for each line in Figure 12 across all 75 repetitions. When tile size equals file size, files from 512 through 4096 pixels significantly affect bit rate ($F(3, 296) = 367, p < 0.001$). For tiles of 512 pixels, files from 512 through 32,768 pixels significantly affect bit rate ($F(6, 518) = 947, p < 0.001$). For tiles of 1024 pixels, files from 1024 through 32,768 pixels significantly affect bit rate ($F(5, 444) = 1033, p < 0.001$). For tiles of 2048 pixels, files from 2048 through 32,768 pixels significantly affect bit rate ($F(4, 370) = 564, p < 0.001$). For tiles of 4096 pixels, files from 4096 through 32,768 pixels significantly affect bit rate ($F(3, 296) = 548, p < 0.001$).

The dotted black line in Figure 12 gives the bit rate when only one image tile loads from any given file. The rates increase along this line when loading from larger files. Relative to loading single tiles from 1024 pixel files, single tiles load more quickly from 2048 pixel files ($t = 16.2, p < 0.001$). Compared to 2048 pixel files, single tiles also load at faster rates from 4096 pixel files ($t = 10.0, p < 0.001$). As further analysis shows, the trend favoring large files only holds when the size of the tiles loaded at each instant can be arbitrarily increased.

Each solid line in Figure 12 gives the bit rate when one or more tiles of a constant size load from any given number of files. The rates decrease along these lines when loading tiles of fixed dimensions from smaller sections of larger files. Relative to loading 1024 × 1024 pixel tiles from files of 4096 pixels, tiles of the same size load more quickly from files of 2048 pixels ($t = 24.2, p < 0.001$). Compared to

25

files of 2048 pixels, tiles of 1024 × 1024 pixels also load at a slightly higher rate from tiles of 1024 pixels ($t = 3.61$, $p < 0.001$).

The tiled image transfer experiment shows tile dimensions of 1024 × 1024 pixels both optimally reduce delays between individual tiles and loading times for a viewport of 4096 × 4096 pixels. Given a constant tile size of 1024 pixels, the tiled image storage experiment suggests a division of the image volume into files also measuring 1024 pixels. However, when external constraints limit the division of the image volume into larger flies than 4096 × 4096 pixels, larger tile sizes on the scale of 2048 × 2048 pixels should be considered to prevent the limits of the network file system from hindering the transfer of image tiles to the viewers.

10. Implementations and Distribution

We choose the following implementations for the described applications.

The Butterfly middleware: We implement our middleware in Python and use the Tornado [61] Web framework to provide the server. We use OpenCV for image processing and mipmap generation.

MBeam viewer: This application is written in HTML5/Javascript and uses the OpenSeaDragon [62] rendering framework.

RHAligner: The alignment framework and visualization scripts are written in Python and use OpenCV.

RhoANAScope: This is an HTML5/Javascript Web frontend and uses the OpenSeaDragon [62] rendering framework combined with our developed viaWebGL (Rendering via WebGL in OpenSeaDragon is available as open-source software at http://github.com/rhoana/viawebgl/) plugin to use GPU accelerated rendering.

Dojo: The proofreading application Dojo uses a custom WebGL rendering engine and is written in HTML5/Javascript. We use Websockets to support collaborative editing and to synchronize any changes among all proofreaders. For volume rendering, we use the XTK WebGL library [35], which enables volume rendering of medical imaging data.

Guided Proofreading: This classifier is developed in Python using the Nolearn [63] machine learning library. The user interface is written in HTML5.

3DXP: This visualizer renders using the X3DOM WebGL library [64] and the user interface including keyframe recording is written in HTML5/Javascript.

Neural Data Queries: This API is written in Python and integrated into the Butterfly middleware. We use MongoDB [65] for the database.

10.1. Data Access API

The Butterfly middleware provides an application program interface to abstract data access by providing a cut-out service. This abstraction layer enables the requesting client application to not care whether data is stored as pre-computed mipmaps, or if different zoom levels need to be computed online. Further, the client is agnostic to file formats and data storage schemes. For example, segmentations can be stored with different bitrates depending on the number of encoded structures (Section 3.2). The data access API is the core feature of the Butterfly middleware and is documented online (The Butterfly data access API is documented at https://github.com/microns-ariadne/ariadne-nda/blob/master/specs/finished.md).

10.2. Distribution

All applications described in this paper are available as open source software and can be installed individually. However, we also provide a downloadable virtual machine image (Installation instructions for the Butterfly virtual machine image are given at https://github.com/Rhoana/bflyVM) based on Ubuntu linux, bundled with pre-configured installations of Butterfly and all visualization applications. This way, interested users can download the virtual machine, link to a network file system, and immediately access the bundled tools via a Web-browser from anywhere in the local network.

11. Use Case: Splitting Merged Somas

While we have designed each tool to function optimally for separate tasks in the analysis of connectomics imagery, it is also possible for information gained in one interface to inform the interaction in others. To demonstrate this, we present a method for solving the problem of splitting merged somas in nano-scale images, with linked views across several applications.

Gravitational centers of brain cells appear in the segmentation data as large uninterrupted round regions with a single identifier. These regions were segmented using membrane probabilities as described in Section 6. Given an identifier of a cell body within the scanned volume (soma, obtained through visual inspection), 3DXP opens the surface mesh of the identifier in an interactive 3D viewer. For a correct segmentation, we expect any mesh with a cell body to branch into thinner projections that continue past the edge of the volume or terminate in small synaptic connections. Any neuroscience researcher would immediately notice when 3DXP instead shows the projections of one soma grow seamlessly into another cell body.

A single surface mesh with two distinct globular masses indicates that the neuron segmentation contains two mistakenly-merged neurons. After noticing such a mesh, a researcher can visually search various views in 3DXP to inspect unlikely- and unevenly-shaped patterns in the thin projections. When clicking to identify a region of interest, the coordinates automatically open to the corresponding view in the Dojo proofreading tool. In Dojo, the user can follow the contours of the EM image to create a new segmentation label to separate the two merged neurons. After 3D agglomeration on the resulting segmentation, the researcher can separately analyze the two neuron segments.

Figure 13 shows one way a 3D view in 3DXP informs 2D proofreading in Dojo. In addition to improving the current segmentation through proofreading, the software presented here allows analysis of intermediate results to improve the algorithms behind future segmentation. The RhoANAScope viewer can maintain a central directory of all initial EM images, intermediate membrane potentials, and resulting segmentation volumes. Upon finding an unusual error while making corrections in Dojo, a researcher can open the corresponding membrane potentials in RhoANAScope to understand the source of the error in the automated process.

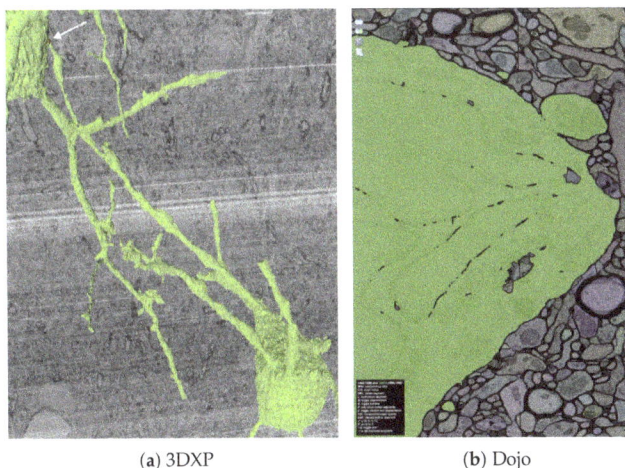

(a) 3DXP   (b) Dojo

Figure 13. *Cont.*

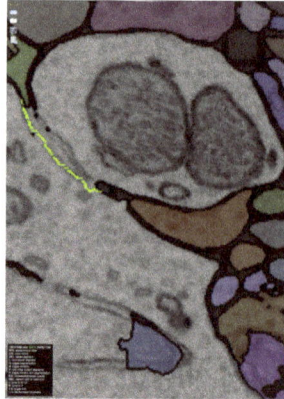

(**c**) Dojo—Split mode

Figure 13. Three steps in the use case of splitting connected cell bodies. In (**a**), a single green label includes two large cell bodies. Therefore, the one label contains the joined reconstruction of two neurons. A neural process runs from a broad base off the lower cell body. While the projection correctly passes narrowly behind one branch of the upper neuron, the tip of process mistakenly merges with the surface of the upper soma at a small point shown in red. In (**b**), the dojo editor displays the false merge between the large upper soma and the small neural process (upper right). Using the same segmentation available to the 3D view, both neurons display as parts of one green segment. In (**c**), a detailed view shows the same false merge with the segmentation label removed. The membrane between the large soma and the small projection displays as a darkened band in the EM image section. The green line, drawn by the user, allows the separation of the falsely merged neurons.

12. Conclusions

The Butterfly middleware makes working with connectomics datasets easier and more convenient. The simple application programming interface abstracts away the low-level problems that occur when working with massive datasets so that neuroscientists and computer scientists in this field can focus on connectomics as opposed to data management. We demonstrate the scalability and extendability of Butterfly with applications tailored towards every step of the connectomics workflow and provide all developments presented in this paper as open source software to the community. As the field of connectomics matures, more novel and sophisticated visualizations will be needed. We hope Butterfly will help us all to develop these future visualizations.

Supplementary Materials: The following are available online at https://github.com/Rhoana/butterfly/wiki/Supplemental-Material, UML Diagram of the Butterfly middleware and Video.

Acknowledgments: This research is supported in part by NSF grants IIS-1447344 and IIS-1607800, by the Intelligence Advanced Research Projects Activity (IARPA) via Department of Interior/Interior Business Center (DoI/IBC) contract number D16PC00002, and by the King Abdullah University of Science and Technology (KAUST) under Award No. OSR-2015-CCF-2533-01.

Author Contributions: Daniel Haehn is the principal architect of the landscape surrounding the Butterfly middleware and contributed to all presented software tools. John Hoffer performed major engineering of the presented applications. Brian Matejek worked on compression and provided theoretical insights. Adi Suissa-Peleg worked on alignment and performed software engineering. Eagon Meng developed the initial Butterfly image server. William Zhang worked on 3D visualization. Lee Kamentsky developed the segmentation framework. Richard Schalek and Alyssa Wilson acquired connectomics imagery. Ali K. Al-Awami, Felix Gonda, Toufiq Parag, Johanna Beyer, Verena Kaynig, Thouis R. Jones, and James Tompkin provided expert knowledge as well as guidance. Markus Hadwiger, Jeff W. Lichtman and Hanspeter Pfister supervised the project. All authors contributed to the paper.

Conflicts of Interest: The authors declare no conflict of interest.

References

1. Lichtman, J.W. The Big and the Small: Challenges of Imaging the Brain's Circuits. *Science* **2011**, *334*, 618–623.
2. Seung, S. *Connectome: How the Brain's Wiring Makes Us Who We Are*; Houghton Mifflin Harcourt: Boston, MA, USA, 2012.
3. Hagmann, P. From Diffusion MRI to Brain Connectomics. Ph.D. Thesis, Université de Lausanne de Nationalité Suisse et Originaire de Däniken, Lausanne, Switzerland, 2005.
4. Sporns, O.; Tononi, G.; Kötter, R. The Human Connectome: A Structural Description of the Human Brain. *PLoS Comput. Biol.* **2005**, *1*, doi10.1371/journal.pcbi.0010042.
5. Kasthuri, N.; Hayworth, K.J.; Berger, D.R.; Schalek, R.L.; Conchello, J.A.; Knowles-Barley, S.; Lee, D.; Vázquez-Reina, A.; Kaynig, V.; Jones, T.R.; et al. Saturated reconstruction of a volume of neocortex. *Cell* **2015**, *162*, 648–661.
6. Suissa-Peleg, A.; Haehn, D.; Knowles-Barley, S.; Kaynig, V.; Jones, T.R.; Wilson, A.; Schalek, R.; Lichtman, J.W.; Pfister, H. Automatic Neural Reconstruction from Petavoxel of Electron Microscopy Data. *Microsc. Microanal.* **2016**, *22*, 536–537.
7. Schalek, R.; Lee, D.; Kasthuri, N.; Peleg, A.; Jones, T.; Kaynig, V.; Haehn, D.; Pfister, H.; Cox, D.; Lichtman, J. Imaging a 1 mm^3 Volume of Rat Cortex Using a MultiBeam SEM. *Microsc. Microanal.* **2016**, *22*, 582–583.
8. Kaynig, V.; Vazquez-Reina, A.; Knowles-Barley, S.; Roberts, M.; Jones, T.R.; Kasthuri, N.; Miller, E.; Lichtman, J.; Pfister, H. Large-scale automatic reconstruction of neuronal processes from electron microscopy images. *Med. Image Anal.* **2015**, *22*, 77–88.
9. Knowles-Barley, S.; Kaynig, V.; Jones, T.R.; Wilson, A.; Morgan, J.; Lee, D.; Berger, D.; Kasthuri, N.; Lichtman, J.W.; Pfister, H. RhoanaNet Pipeline: Dense Automatic Neural Annotation. *arXiv* **2016**, arXiv:1611.06973.
10. IEEE ISBI Challenge: SNEMI3D—3D Segmentation of Neurites in EM Images. 2013. Available online: http://brainiac2.mit.edu/SNEMI3D (accessed on 21 August 2017).
11. Haehn, D.; Knowles-Barley, S.; Roberts, M.; Beyer, J.; Kasthuri, N.; Lichtman, J.; Pfister, H. Design and Evaluation of Interactive Proofreading Tools for Connectomics. *IEEE Trans. Vis. Comput. Graph.* **2014**, *20*, 2466–2475.
12. Haehn, D.; Kaynig, V.; Tompkin, J.; Lichtman, J.W.; Pfister, H. Guided Proofreading of Automatic Segmentations for Connectomics. *arXiv* **2017**, arXiv:1704.00848.
13. Al-Awami, A.K.; Beyer, J.; Haehn, D.; Kasthuri, N.; Lichtman, J.W.; Pfister, H.; Hadwiger, M. NeuroBlocks—Visual Tracking of Segmentation and Proofreading for Large Connectomics Projects. *IEEE Trans. Vis. Comput. Graph.* **2016**, *22*, 738–746.
14. Al-Awami, A.; Beyer, J.; Strobelt, H.; Kasthuri, N.; Lichtman, J.; Pfister, H.; Hadwiger, M. NeuroLines: A Subway Map Metaphor for Visualizing Nanoscale Neuronal Connectivity. *IEEE Trans. Vis. Comput. Graph.* **2014**, *20*, 2369–2378.
15. Beyer, J.; Al-Awami, A.; Kasthuri, N.; Lichtman, J.W.; Pfister, H.; Hadwiger, M. ConnectomeExplorer: Query-Guided Visual Analysis of Large Volumetric Neuroscience Data. *IEEE Trans. Vis. Comput. Graph.* **2013**, *19*, 2868–2877.
16. Lichtman, J.W.; Pfister, H.; Shavit, N. The big data challenges of connectomics. *Nat. Neurosci.* **2014**, *17*, 1448–1454.
17. Pfister, H.; Kaynig, V.; Botha, C.P.; Bruckner, S.; Dercksen, V.J.; Hege, H.C.; Roerdink, J.B. Visualization in Connectomics. *arXiv* **2012**, arXiv:1206.1428v2.
18. Margulies, D.S.; Böttger, J.; Watanabe, A.; Gorgolewski, K.J. Visualizing the human connectome. *NeuroImage* **2013**, *80*, 445–461.
19. Hayworth, K.J.; Morgan, J.L.; Schalek, R.; Berger, D.R.; Hildebrand, D.G.C.; Lichtman, J.W. Imaging ATUM ultrathin section libraries with WaferMapper: A multi-scale approach to EM reconstruction of neural circuits. *Front. Neural Circuits* **2014**, *8*, doi:10.3389/fncir.2014.00068.
20. Schaefer, H.E. *Nanoscience: The Science of the Small in Physics, Engineering, Chemistry, Biology and Medicine*; Springer: Berlin/Heidelberg, Germany, 2010; Charpter 2.
21. Janelia Farm. Raveler. 2014. Available online: https://openwiki.janelia.org/wiki/display/flyem/Raveler (accessed on 27 August 2017).

22. Knowles-Barley, S.; Roberts, M.; Kasthuri, N.; Lee, D.; Pfister, H.; Lichtman, J.W. Mojo 2.0: Connectome Annotation Tool. *Front. Neuroinform.* **2013**, doi:10.3389/conf.fninf.2013.09.00060.

23. NeuTu: Software Package for Neuron Reconstruction and Visualization. 2013. Available online: https://github.com/janelia-flyem/NeuTu (accessed on 20 May 2017).

24. Hadwiger, M.; Beyer, J.; Jeong, W.K.; Pfister, H. Interactive Volume Exploration of Petascale Microscopy Data Streams Using a Visualization-Driven Virtual Memory Approach. *IEEE Trans. Vis. Comput. Graph.* **2012**, *18*, 2285–2294.

25. Beyer, J.; Hadwiger, M.; Al-Awami, A.; Jeong, W.K.; Kasthuri, N.; Lichtman, J.W.; Pfister, H. Exploring the Connectome: Petascale Volume Visualization of Microscopy Data Streams. *IEEE Comput. Graph. Appl.* **2013**, *33*, 50–61.

26. Sicat, R.; Hadwiger, M.; Mitra, N.J. Graph Abstraction for Simplified Proofreading of Slice-based Volume Segmentation. In Proceedings of the 34th Annual Conference of the European Association for Computer Graphics, Girona, Spain, 6–10 May 2013.

27. Kim, J.S.; Greene, M.J.; Zlateski, A.; Lee, K.; Richardson, M.; Turaga, S.C.; Purcaro, M.; Balkam, M.; Robinson, A.; Behabadi, B.F.; et al. Space-time wiring specificity supports direction selectivity in the retina. *Nature* **2014**, *509*, 331–336.

28. Giuly, R.J.; Kim, K.Y.; Ellisman, M.H. DP2: Distributed 3D image segmentation using micro-labor workforce. *Bioinformatics* **2013**, *29*, 1359–1360.

29. Saalfeld, S.; Cardona, A.; Hartenstein, V.; Tomančák, P. CATMAID: Collaborative annotation toolkit for massive amounts of image data. *Bioinformatics* **2009**, *25*, 1984–1986.

30. Anderson, J.; Mohammed, S.; Grimm, B.; Jones, B.; Koshevoy, P.; Tasdizen, T.; Whitaker, R.; Marc, R. The Viking Viewer for connectomics: Scalable multi-user annotation and summarization of large volume data sets. *J. Micros.* **2011**, *241*, 13–28.

31. Lin, C.Y.; Tsai, K.L.; Wang, S.C.; Hsieh, C.H.; Chang, H.M.; Chiang, A.S. The Neuron Navigator: Exploring the information pathway through the neural maze. In Proceedings of the 2011 IEEE Pacific Visualization Symposium, Hong Kong, China, 1–4 March 2011; pp. 35–42.

32. Ginsburg, D.; Gerhard, S.; Calle, J.E.C.; Pienaar, R. Realtime Visualization of the Connectome in the Browser using WebGL. *Front. Neuroinform.* **2011**, doi:10.3389/conf.fninf.2011.08.00095.

33. Neuroglancer: WebGL-Based Viewer for Volumetric Data. 2017. Available online: https://github.com/google/neuroglancer (accessed on 29 May 2017).

34. Khronos Group. WebGL Specification. 2014. Available online: http://www.khronos.org/registry/webgl/specs (accessed on 31 March 2014).

35. Haehn, D.; Rannou, N.; Ahtam, B.; Grant, E.; Pienaar, R. Neuroimaging in the Browser using the X Toolkit. *Front. Neuroinform.* **2012**, doi: 10.3389/conf.fninf.2014.08.00101.

36. Haehn, D. Slice:Drop: Collaborative medical imaging in the browser. In Proceedings of the ACM SIGGRAPH 2013 Computer Animation Festival, Anaheim, CA, USA, 21–25 July 2013; p. 1.

37. Bakker, R.; Tiesinga, P.; Kötter, R. The Scalable Brain Atlas: Instant Web-Based Access to Public Brain Atlases and Related Content. *Neuroinformatics* **2015**, *13*, 353–366.

38. Stephan, K.E.; Kamper, L.; Bozkurt, A.; Burns, G.A.P.C.; Young, M.P.; Kötter, R. Advanced database methodology for the Collation of Connectivity data on the Macaque brain (CoCoMac). *Philos. Trans. R. Soc. Lond. B Biol. Sci.* **2001**, *356*, 1159–1186.

39. Bota, M.; Dong, H.W.; Swanson, L.W. Brain architecture management system. *Neuroinformatics* **2005**, *3*, 15–47.

40. Schmitt, O.; Eipert, P. neuroVIISAS: Approaching Multiscale Simulation of the Rat Connectome. *Neuroinformatics* **2012**, *10*, 243–267.

41. Gerhard, S.; Daducci, A.; Lemkaddem, A.; Meuli, R.; Thiran, J.; Hagmann, P. The connectome viewer toolkit: An open source framework to manage, analyze, and visualize connectomes. *Front. Neuroinform.* **2011**, *5*, doi:10.3389/fninf.2011.00003.

42. Sorger, J.; Buhler, K.; Schulze, F.; Liu, T.; Dickson, B. neuroMap—Interactive graph-visualization of the fruit fly's neural circuit. In Proceedings of the 2013 IEEE Symposium on Biological Data Visualization (BioVis), Atlanta, GA, USA, 13–14 October 2013; pp. 73–80.

43. DVID. Distributed, Versioned, Image-Oriented Dataservice. 2016. Available online: https://github.com/janelia-flyem/dvid/wiki (accessed on 14 January 2016).

44. The Boss: A Cloud Based Storage Service Developed for the IARPA MICrONS Program. 2017. Available online: https://docs.theboss.io/ (accessed on 29 May 2017).
45. Matejek, B.; Haehn, D.; Lekschas, F.; Mitzenmacher, M.; Pfister, H. Compresso: Efficient Compression of Segmentation Data For Connectomics. In Proceedings of the International Conference on Medical Image Computing and Computer-Assisted Intervention, Quebec City, QC, Canada, 10–14 September 2017.
46. Williams, L. Pyramidal parametrics. In Proceedings of the 10th Annual Conference on Computer Graphics and Interactive Techniques, Detroit, MI, USA, 25–29 July 1983; ACM: New York, NY, USA, 1983; Volume 17, pp. 1–11.
47. Kaiser, G.E. Cooperative Transactions for Multiuser Environments. In *Modern Database Systems*; ACM Press/Addison-Wesley Publishing Co.: New York, NY, USA, 1995; pp. 409–433.
48. Jeong, W.K.; Johnson, M.K.; Yu, I.; Kautz, J.; Pfister, H.; Paris, S. Display-aware image editing. In Proceedings of the 2011 IEEE International Conference on Computational Photography (ICCP), Pittsburgh, PA, USA, 8–10 April 2011; pp. 1–8.
49. Beyer, J.; Hadwiger, M.; Jeong, W.K.; Pfister, H.; Lichtman, J. Demand-driven volume rendering of terascale EM data. In Proceedings of the International Conference on Computer Graphics and Interactive Techniques, SIGGRAPH 2011, Vancouver, BC, Canada, 7–11 August 2011; p. 57.
50. Saalfeld, S.; Fetter, R.; Cardona, A.; Tomancak, P. Elastic volume reconstruction from series of ultra-thin microscopy sections. *Nat. Methods* **2012**, *9*, 717–720.
51. Lowe, D.G. Object Recognition from Local Scale-Invariant Features. In Proceedings of the Seventh IEEE International Conference on Computer Vision, Kerkyra, Greece, 20–25 September 1999; IEEE Computer Society: Washington, DC, USA, 1999; p. 1150.
52. Janelia Farm. The Tilespec JSON Data Model. 2015. Available online: https://github.com/saalfeldlab/render/blob/master/docs/src/site/markdown/data-model.md (accessed on 27 August 2017).
53. Nunez-Iglesias, J.; Kennedy, R.; Parag, T.; Shi, J.; Chklovskii, D.B. Machine Learning of Hierarchical Clustering to Segment 2D and 3D Images. *PLoS ONE* **2013**, *8*, doi:10.1371/journal.pone.0071715.
54. Ronneberger, O.; Fischer, P.; Brox, T. U-Net: Convolutional Networks for Biomedical Image Segmentation. In Proceedings of the International Conference on Medical Image Computing and Computer-Assisted Intervention (MICCAI), Munich, Germany, 5–9 October 2015; Springer: Berlin, Germany, 2015; Volume 9351, pp. 234–241.
55. Nguyen, Q. *Parallel and Scalable Neural Image Segmentation for Connectome Graph Extraction*; Massachusetts Institute of Technology: Cambridge, MA, USA, 2015.
56. Nunez-Iglesias, J.; Kennedy, R.; Plaza, S.M.; Chakraborty, A.; Katz, W.T. Graph-based active learning of agglomeration (GALA): A Python library to segment 2D and 3D neuroimages. *Front. Neuroinform.* **2014**, *8*, doi:10.3389/fninf.2014.00034.
57. Parag, T.; Chakraborty, A.; Plaza, S.; Scheffer, L. A Context-Aware Delayed Agglomeration Framework for Electron Microscopy Segmentation. *PLoS ONE* **2015**, *10*, doi:10.1371/journal.pone.0125825.
58. Santurkar, S.; Budden, D.M.; Matveev, A.; Berlin, H.; Saribekyan, H.; Meirovitch, Y.; Shavit, N. Toward Streaming Synapse Detection with Compositional ConvNets. *arXiv* **2017**, arXiv:1702.07386.
59. Lorensen, W.E.; Cline, H.E. Marching Cubes: A High Resolution 3D Surface Construction Algorithm. In Proceedings of the 14th Annual Conference on Computer Graphics and Interactive Techniques, Anaheim, CA, USA, 27–31 July 1987.
60. Limper, M.; Jung, Y.; Behr, J.; Alexa, M. The POP Buffer: Rapid Progressive Clustering by Geometry Quantization. *Comput. Graph. Forum* **2013**, *32*, 197–206.
61. Dory, M.; Parrish, A.; Berg, B. *Introduction to Tornado*; O'Reilly Media, Inc.: Sebastopol, CA, USA, 2012.
62. OpenSeaDragon. 2016. Available online: http://openseadragon.github.io/ (accessed on 27 August 2017).
63. Nouri, D. Nolearn: Scikit-Learn Compatible Neural Network Library. 2016. Available online: https://github.com/dnouri/nolearn (accessed on 27 August 2017).
64. Behr, J.; Eschler, P.; Jung, Y.; Zöllner, M. X3DOM: A DOM-based HTML5/X3D Integration Model. In Proceedings of the 14th International Conference on 3D Web Technology, Darmstadt, Germany, 16–17 June 2009; ACM: New York, NY, USA, 2009; pp. 127–135.
65. Chodorow, K.; Dirolf, M. *MongoDB: The Definitive Guide*, 1st ed.; O'Reilly Media, Inc.: Sebastopol, CA, USA, 2010.

informatics

MDPI

Article

Visual Analysis of Stochastic Trajectory Ensembles in Organic Solar Cell Design

Sathish Kottravel [1], Riccardo Volpi [2], Mathieu Linares [2], Timo Ropinski [3] and Ingrid Hotz [1,*]

[1]   Department of Science and Technology, Linköping University, 60174 Norrköping, Sweden;
      sathish.kottravel@liu.se
[2]   Department of Physics, Chemistry and Biology, Linköping University, 58183 Linköping, Sweden;
      ricvo@ifm.liu.se (R.V.); mathieu@ifm.liu.se (M.L.)
[3]   Institute for Media Informatics, Ulm University, 89081 Ulm, Germany; timo.ropinski@uni-ulm.de
*   Correspondence: ingrid.hotz@liu.se

Academic Editor: Gunther H. Weber and Achim Ebert
Received: 31 May 2017 ; Accepted: 26 July 2017; Published: 1 August 2017

Abstract: We present a visualization system for analyzing stochastic particle trajectory ensembles, resulting from Kinetic Monte-Carlo simulations on charge transport in organic solar cells. The system supports the analysis of such trajectories in relation to complex material morphologies. It supports the inspection of individual trajectories or the entire ensemble on different levels of abstraction. Characteristic measures quantify the efficiency of the charge transport. Hence, our system led to better understanding of ensemble trajectories by: (i) Capturing individual trajectory behavior and providing an ensemble overview; (ii) Enabling exploration through linked interaction between 3D representations and plots of characteristics measures; (iii) Discovering potential traps in the material morphology; (iv) Studying preferential paths. The visualization system became a central part of the research process. As such, it continuously develops further along with the development of new hypothesis and questions from the application. Findings derived from the first visualizations, e.g., new efficiency measures, became new features of the system. Most of these features arose from discussions combining the data-perspective view from visualization with the physical background knowledge of the underlying processes. While our system has been built for a specific application, the concepts translate to data sets for other stochastic particle simulations.

Keywords: stochastic trajectory ensemble visualization; organic solar cell design; charge transport

1. Introduction

In the quest to tap renewable energies, the development of organic solar cells plays an important role as they can be manufactured in high throughput at low prices. Additionally, the flexibility of these cells offers many benefits compared to conventional solar cells. Unfortunately, despite organic solar cells are already used in a few commercial products, their comparably low efficiency currently forbids a wide-spread use.

The efficiency of an organic solar cell is directly related to its molecular structure, which is usually formed by two aggregations of molecules (donor and acceptor) that are sandwiched between two electrodes. When photon absorption occurs it leads to the formation of excitons (electron-hole pairs), which are transported to the electrodes, whereby the donor transports the holes and the acceptor the electrons. The time to reach the electrodes is determined by the molecular structure of the donor as well as the acceptor, and inversely proportional to the efficiency of the cell. Thus, to improve the efficiency of organic solar cells, it is mandatory that the underlying physical principles regarding charge transport are better understood, and that an optimal molecular structure can be predicted. Kinetic Monte-Carlo simulation is a tool frequently used in this context to better understand the behavior of charge transport, by establishing a relation between the material structure and a solar cell's properties [1–3]. By simulating a multitude of charges traversing the sandwiched region large charge-transport trajectory ensembles are obtained. Understanding of these charge-transport trajectory ensembles and their connection to the molecular structure is key to be able to design more efficient organic solar cells [4].

In this paper, we propose an analysis system composed of a set of linked spatial visualizations together with plots of structure-aware trajectory measures. The structure of the data is similar to trajectories resulting from tracking of movement data and thus the exploration concepts are similar. However, an efficient exploration system requires a configuration targeted specifically toward the needs of the application. Accordingly, novel concepts were also needed for the proposed system. A central requirement for the charge trajectory analysis is relating the stochastic microscopic data to macroscopic efficiency measures. To achieve this, the concept of charge-flow lines has been introduced. They mimic the macro-level behavior of charges resulting in typical flow descriptors as flow direction and velocity. The morphology of the solar cell under investigation serves as context.

Thus, within this paper, we make the following contributions:

- We propose a set of linked visualization techniques that enable the investigation of dense charge-transport trajectory ensembles by exploiting trajectory abstraction and relating trajectories to a solar cell's morphology.
- We propose novel geometric measures to analyze the efficiency of individual trajectories and trajectory ensembles based on the concept of charge flow lines.
- We discuss how these components are integrated into a single visualization framework, which supports domain experts when visually analyzing organic solar cell simulations.

The remainder of the paper is structured as follows. In the next section, we briefly describe the application background and describe the visual analysis tasks we have identified as being essential when exploring the data at hand. In Section 2, we summarize the most important recent work that inspired the development of our framework. Section 3 starts with an overview of the proposed visualization framework, and introduces the applied visualizations and the novel efficiency measures. In Section 4, we describe the technical details. To demonstrate the effectiveness of our framework we apply it to simulation results with different levels of complexity, with respect to the underlying physical model, and discuss the findings made in Section 5. Finally, the paper concludes in Section 6.

1.1. Application Task Characterization

In the following, we will describe how the visual analysis tasks have been developed since they play a central role for the configuration of the system. The overall goal has been to gain a deeper understanding of the process of charge transport based on the simulation results. However, as it is often the case when scientists look at their data for the first time, there were no clear questions to start from and the analysis has been driven by the question: 'Let's see what we will find.' More specific tasks have then been gradually identified within a close collaboration between visualization experts and theoretical physicists who perform the simulations. The visualization system has been developing continuously by new hypotheses that have been developed during the visual exploration, see Figure 1.

The first task, which we call the *Overall Efficiency (OE)*, aims to give an overview of the data in its most original form. This means displaying the trajectory ensemble as a whole and allow simple interaction to inspire new questions to guide the further development of the system. This matches the visual-information seeking mantra: *Overview first, zoom and filter, then details-on-demand*. During the configuration of the system the visualization tasks have been shifting more and more from a microscopic to a macroscopic view. This reflects a generalization of the questions starting from the modeling perspective on the quantum mechanical level to questions related to large-scale properties as the efficiency of the probe. The macroscopic view has to a large extend been new to the physicists and triggered many new ideas for the design of the simulation. Understanding of the interaction between the scales is what finally paves the way for the further development of the technology.

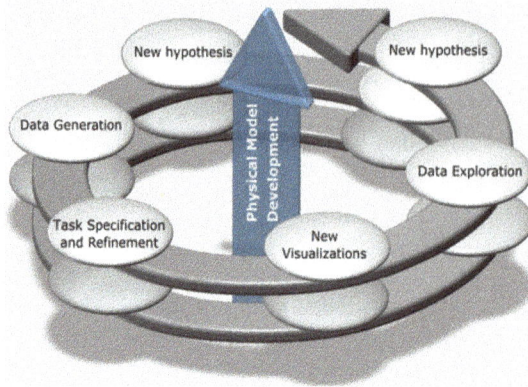

Figure 1. The visualization system has become an essential part of the scientific process in the applications and the specific tasks toward the system have been developing continuously. New hypotheses that are developed during the visual exploration trigger new tasks and new visualization methods as specific efficiency measures used for statistical plots and line abstraction.

The pertinent questions that arose during the development of the system can be summarized as follows. *Morphology Efficiency (ME)*: Understand the general distribution of charges and the impact of the morphology geometry on the distribution and the transport properties. Thereby the individual trajectories have not been considered as very interesting. *Charge Interaction (CI)*: A complementary question is the role of charge interactions for the charge transport. These questions involve the inspection of individual charge pair trajectories but also the morphology and especially the material interface as context. For these questions the fully detailed trajectories hide the trends of the transport and there is a demand for abstraction and macroscopic views and measures to quantify the efficiency. *Simulation Evaluation (SE)*: Orthogonal to the questions targeting toward understanding the underlying physics, is the evaluation of the performance of the simulation and its parameter settings. Therefore it is important to easily inspect the plausibility of the results and identify outliers. For this purpose almost all proposed visualizations are useful whereby simple geometric settings are of advantage.

The derived tasks suggest the employment of a two-dimensional *visualization parameter space*. One dimension pertains to the level of detail and abstraction ranging from a micro-level to a macro-level view. The second dimension relates to the number of trajectories that are investigated ranging from the entire ensemble to single trajectory analysis. We divide the parameter space into four quadrants as illustrated in Figure 2. The proposed methods and derived task are placed into this space to provide an overview.

1.2. Organic Solar Cell Design

To understand the benefits of the proposed visual analysis framework, some information regarding the application background needs to be provided first. The efficiency of an organic solar cell is determined by the efficiency of the different steps from photon absorption to charge collection. As these are directly related to the structure, different structures for such cells have been investigated. The simplest consisting of a layer of an organic semiconductor between two electrodes. However, the performance of a cell can be improved by having two layers of organic materials: the donor and the acceptor [5] (see Figure 3a,b). In a working solar cell, photons are absorbed generating excitons, which then diffuse toward the interface as illustrated in Figure 3a,b and form a charge transfer (CT) state. The CT states then split into free electrons and holes that can be collected at the electrodes. While Figure 3a shows this case for a single exciton, Figure 3b illustrates the existence of two excitons. After the charge carriers are freed, they may still move back to the interface and recombine. Here, two types of recombination can occur: geminate and nongeminate. Geminate recombination is when two charge carriers resulting from the absorption of the same photon recombine. Nongeminate recombination occurs when two free charge carriers originating from different photons recombine with each other at the interface.

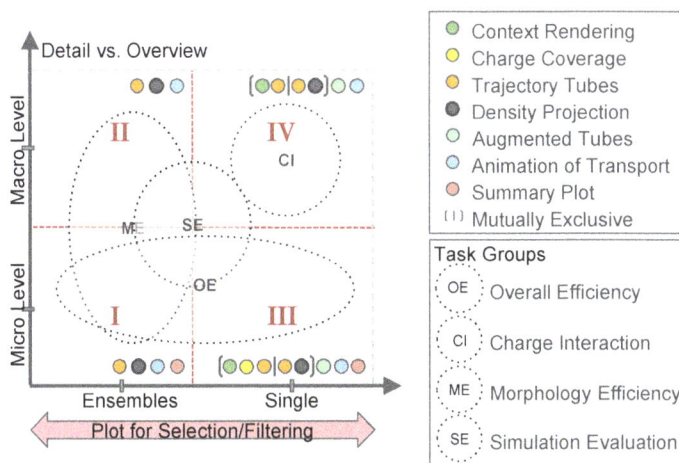

Figure 2. Our visualization parameter space can be roughly divided into four quadrants micro-level vs. macro-level of detail, ensemble vs. single trajectory). The parameter space can be investigated using several visualization techniques, which are associated with the four identified task groups. To move between the single and the ensemble level, brushing-and-linking is realized using plots.

The morphology of the donor-acceptor interface in an organic solar cell has a large impact on the efficiency of the solar cell. Excitons can only diffuse 10 nm before decaying, so the donor and acceptor should be sufficiently mixed, as otherwise the excitons could not reach the interface before decaying. An example of a more complex morphology is illustrated in Figure 3c,d. However, once separated, the charge carriers need pathways to their respective electrodes. If, for example, an electron is in an acceptor domain that is completely surrounded by the donor, there is no path for the electron to travel to the electrode. Consequently, it is important to establish a morphology-efficiency relationship and determine for instance how the domain size and tortuosity influence the different processes, such as transport of the exciton, dissociation of the CT, and free charge carriers transport.

Because of the amorphous nature of the material and the probabilistic nature of the competitive processes at play in a solar cell, stochastic methods such as the Monte-Carlo approach are applied. In our setup, a kinetic Monte-Carlo code is used where the hoping rates are calculated based on the Marcus Equation [6] using a multi-scale approach [1,2]. Based on the simulation parameters, these simulations result in a variety of data, whereby we focus on the analysis of the trajectory ensembles in combination with the morphology data. The trajectories are realizations of possible charge propagations based on a physically accurate transition probability from molecule to molecule. Each trajectory represents a sequence of discrete positions associated with one specific molecule and an associated dwell time. Along a trajectory, charges jump back and forth and may be trapped in some regions due to multiple physical fields interacting with the charges. To get a representative description of the charge movement an ensemble of trajectory-pairs representing one CT is computed, whereby each trajectory of the ensemble is represented as a discrete series of molecule identifiers and the dwell time at the respective molecule. All trajectories of one ensemble start at the same position. Thus, trajectories usually do not represent shortest paths within the constraints of the morphology. The morphology of the material consists of two materials, the acceptor and the donor material. It is represented as a volumetric data set generated by an ergodic process, whereby binary values (donor = 1, acceptor = 0) are used to mask the voxels.

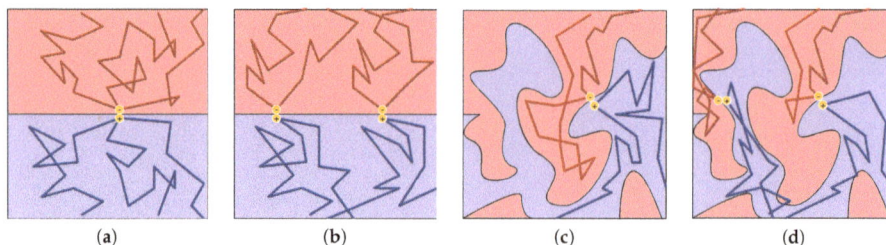

Figure 3. Illustration of the different organic solar cell setups. A simple organic solar cell consists of two layers while single (**a**) or multiple excitons can be considered (**b**). More complex morphologies reduce interface distances and can also be considered with single (**c**) or multiple excitons (**d**). We use a color scheme assigning red to donor material and trajectories and blue to acceptor material and trajectories for all visualizations.

1.3. Some Details about the Data

The data is a result of a kinetic Monte-Carlo simulation consisting of two parts: the geometry information of the material morphology and the charge trajectories. The morphology is represented as a volumetric data set, where each voxel encodes the material type, acceptor or donor, as a binary value. The interface between the acceptor and the donor is presented by an isosurface for the isovalue of 0.5. The morphology serves as a container for the donor and acceptor molecules. In the setting of our simulations the molecules are placed on a regular grid thus corresponding to the morphology data set. The morphology is the most important context information for the trajectories. For all visualizations we use a color schema assigning blue to donor material and electron trajectories and red to acceptor material and whole trajectories, if not stated differently.

The charges are always attached to one molecule. The transport is modeled as a probabilistic process for charges hopping from one molecule to the next according to a quantum-mechanical transition probability. Each charge trajectory thus consists of a series of molecule-IDs augmented with information such as dwell time and the type of the charge (electron or hole). Since the trajectories represent a stochastic process they are not smooth. Often, charges hop frequently back and forth between neighboring molecules. Each simulation run represents one possible path of a charge pair, a hole and an electron, which influence each other. The initial configuration for all trajectories of the entire ensemble are the same. This concerns the initial position of the charge pair and the morphology. The simulation assumes periodic boundary conditions, meaning a charge leaving the volume on one side will enter it again on the opposite side.

2. Related Work

In the following, we summarize previous work that is mostly related to our work. Thereby we focus on (i) previous visualization systems developed for similar applications in solar cell design; (ii) visualization and analysis of trajectory and movement data; (iii) rendering methods for lines; (iv) related ensemble visualization; and (v) efficiency measures for stochastic particle movements.

(i) Related applications. The work most closely related to our visualization system is the work by Aboulhassan et al. [7]. They are concerned with the same application, the design of efficient organic solar cells and the task of exploring the efficiency of the charge transport. However, from a data perspective of the system it differs a lot from our work. Their system has been designed to explore structural characteristics of the morphology [8] while we focus on the explicit charge trajectories resulting from a Monte-Carlo simulation. Therefore, they propose a topological approach for the

simplification of the morphology and distill a geometric backbone as simplification of the complex structure. Geometric bottlenecks for the charge transport are extracted from the backbone. Previously, the same authors developed a system for visual design of solar cell crystal structures [9]. To analyze these structures, the user can exploit semantic rules to define clusters of atoms with certain geometric properties. While the idea of knowledge-assisted exploration plays also an important role in our system, we focus on the exploration of the charge trajectories, which is a complementary task. Accordingly, we also do not discuss molecular visualization techniques, which would be required to explore the actual solar cell structure. Instead we refer to the recent state-of-the-art report by Kozlikova et al. [10], which covers most relevant techniques.

(ii) Analysis and visualization of trajectory and movement data. The analysis of trajectories is also a central task when dealing with motion tracking and movement data. Even though the applications are very different the data structure has some similarities. In both cases one deals with a large numbers of trajectories that are not smooth and allow crossings. Some challenges related to overplotting and clutter are similar. In an overview article about visual analytics of movement by Andrienko et al. [11,12] they classify the related work into four categories: Looking at trajectories, looking inside trajectories, bird's-eye view on movement, and investigating movement in context. These categories are also related to our parametrization of the visualization space. However, there are also some essential differences. The charge trajectories are three-dimensional and thus cannot easily be embedded in two-dimensional map representations. There are no interactions between trajectories for different ensemble members and the movements of the charges has a stochastic character. Therefore, filtering and efficiency measures are in general not transferable.

(iii) Trajectory visualization. Our application deals with a vast amount of trajectories, which need to be explored within the morphological context. Therefore, effective visualization of dense line sets is important. Several approaches to tackle similar problems have been developed for flow data or in medical context for fiber visualization. A typical approach is focus and context technique that enables an occlusion-free view into the trajectories, such that the trajectory under investigation becomes visible. An early work using this concept for flow data visualization has been presented by Doleisch et al. [13]. Flow features in focus are emphasized whereas the rest of the data are shown as context. Gasteiger et al. [14] applied the idea for the visualization of blood flow data. The focus and context technique employed by our system has been inspired by these approaches, whereby the morphology of the solar cell provides the context. Besides an occlusion-free view, an unambiguous perception of the visualized trajectories is important. There exist many approaches for rendering of large sets of lines. Much effort has been put on improving the spatial perception of occluding and overlapping lines. One way to approach this problem is to use illuminated lines [15,16]. Applying tubes or other geometries for the line rendering allows for more advanced methods. Techniques have been proposed reaching from the use of halos, ambient occlusion and the use of smart transparency. Such methods have been combined for enhanced molecular visualization by Tarini et al. [17]. Techniques exploiting halos have been frequently applied for the rendering of fibers in the medical field [18–21]. Schröder et al. [22] enhance illuminated lines with ambient occlusion in combination with transparency and halos to achieve a good depth perception and thus improve the visual quality of dense integral line rendering. A further trend to enhance the expressiveness of renderings in the use of illustrative visualizations. An overview of related methods for flow visualization is presented by Brambilla et al. [23]. To convey information about local flow properties, Everts et al. [24] proposed to augment flow lines with strips. We adopted this method for the visualization of properties of the charge flow lines derived from the charge trajectories. Another way to deal with large set of lines is to use filtering methods using line predicates as proposed by Salzbrunn et al. [25].

Most of the above described methods are however not appropriate for the rendering of the original charge trajectories, which are stochastic in their nature and non-smooth. Charges are hopping back and forth frequently between same spots, whereby the individual hops are not of particular interest in contrast to the dwell time in certain regions. A method that is well suited to highlight regions where the charges preferably stay is the method of trajectory density projection for vector field visualization by Kuhn et al. [26]. This is an efficient approach for large amount of trajectories reducing the clutter and occlusion due to the number of curves exploiting capabilities of modern graphics hardware. The rendering results in images giving a good impression of the distribution and density of trajectories. To combine the rendering of our charge flow lines, which are explicit geometry with the density distribution volume data, we intended to adopt an approach by Lindholm et al. [27]. They propose a hybrid data visualization method based on a depth complexity histogram analysis. But for sake of simplicity we used approach by Henning [28] since we assumed the geometry of charge flow lines to be opaque.

(iv) Ensemble visualization. An important aspect of our application is the interplay between the ensemble of trajectories and the individual lines. Ensembles receive more and more attention in the field of visualization, which is especially challenging for vector data. Typical visualizations are a combination of spaghetti plots of lines with appropriate statistical plots. Examples from the field of weather forecast can be found in Sanyal et al. [29] or Wilson et al. [30]. Ferstl et al. [31] use a clustering of flow lines, which are then visualized using variability plots representing the distribution of each cluster. These variability plots have some similarity with our charge coverage visualization.

(v) Efficiency measures. For the analysis and characterization of complex trajectories diverse measures have been used. Bos et al. [32] introduced angular statistics to reflect the multi-scale dynamics of pathlines in turbulent flows. Their measure reflects the multi-scale dynamics of high-Reynolds number turbulence. Savage et al. [33] also use an angular measure to characterize the diffusion process of charges in context with the analysis of perfluorosulfonic acid membranes. They investigate the caging effect of water and the hydrophobic moieties on the motion of the excess proton. In their method, they consider the relative angle between the vectors of motion for two successive time intervals as a probe of the directional changes in the diffusion process. Burov et al. [34] analyze random walks considering the distribution of relative angles of motion between successive time intervals, which provides information about the underlying stochastic processes. Some of these measures are related to our transport efficiency measures; however, none fits our setting of charges moving within a discrete regular grid within a constrained geometry. Instead of analyzing angles on multiple scales, we consider the effective distance and velocity of the charges on multiple scales.

3. Trajectory Exploration Framework

To explore the data on all levels, we have designed a framework that combines multiple spatial views on different levels of abstraction with statistical plots. It enables selection of trajectories and a detailed inspection of those. It allows to explore the data starting with overview representations and drilling down to more detailed visualizations in both dimensions of the visualization parameter space: moving from ensembles to individual trajectories and from macro-level to micro-level views. Thereby, we exploit typical visualization concepts like multiple linked views, focus and context visualization and brushing and linking. Figure 4 shows an example screen shot of the proposed system. In the following we first describe the various spatial views Section 3.1 then we discus the set of plots and efficiency measures that have been introduced Section 3.2.

3.1. Spatial Views

For each quadrant of the visualization parameter space a set of spatial visualizations are provided, which are described briefly in the following. For all visualization one can chose between the original simulation volume or an expended volume respecting the periodic boundary conditions unfolding the trajectories, see Figure 5. To encode the temporal aspect of the data we use color or animation, steered by a time slider.

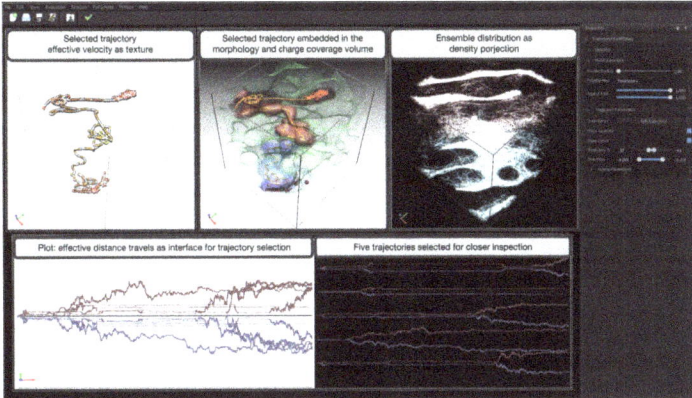

Figure 4. Screenshot of the system with annotations: It supports the exploration of the trajectory ensemble on different levels of detail. Single trajectories can be selected in qualitative plots (here effective distance a charge has traveled), which then can be inspected as isolated flow lines or with respect to the charge coverage volume within the morphology. An overview of the ensemble distribution provides contextual information.

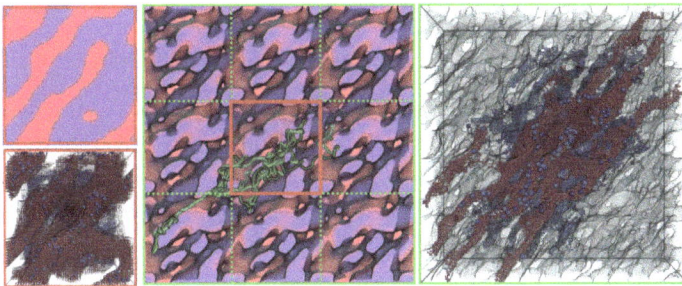

Figure 5. The simulation uses periodic boundary conditions, meaning that charges leaving the volume on one side will enter it again on the opposite side. Thus the trajectories are disrupted and the resulting density distribution misleading. The system supports a periodic continuation of the volume to get an untangled representation. The image on the left (bottom) represents raw data in the original simulation volume in contrast to the same data in the extended volume shown on the right. The image on the top left shows a slice through the morphology. The image in the center results from a periodic continuation of the morphology with one selected flow line. The original volume is highlighted by a red square.

3.1.1. Quadrant I: Ensemble Visualization, Microscopic View

The visualizations provide the most direct view on the data, Figure 6c. Thereby the morphology represents the context and the entire ensemble is in focus. Even though the individual trajectories are not of interest, the trajectories are still plotted in their original form as solid lines with all details. An example is shown in Figure 5. This visualization suffers heavily from over-plotting and is mostly useful for debugging purposes. However, it allows the domain scientists to quickly grasp the transport activity of charges in a material and has been used frequently to get a first impression of the data and its correctness.

3.1.2. Quadrant II: Ensemble Visualization, Macroscopic View

From an macroscopic ensemble perspective, Figure 6a, often the trajectory details are not of much interest. In that case, it makes sense to switch to the macroscopic view. It only displays the density distribution and the coverage of all trajectories highlighting regions in the morphology where the charges dwell for a longer time. We provide thereby two options for trajectory rendering. The *charge coverage volume* focuses on displaying coverage by generating a volume representing the frequency of charge-visits for each location in the morphology, see Figure 10a. The *trajectory density projection* [26] accumulates transparent slightly smoothed trajectories in one image, see Figure 10b for a simple morphology and Figure 13c for a complex morphology. These renderings allow conclusions about the transport efficiency with respect to the morphology and the detection of possible traps.

3.1.3. Quadrant III and IV: Inspection of Individual Trajectories

When exploring a few selected trajectories, Figure 6b,d–h, the context is not only given by the morphology but can also include the entire ensemble. Moving from microscopic to macroscopic view can be considered as a smooth transition from the stochastic data to smooth lines. For all these visualizations we consider pairs of trajectories consisting of an electron and a hole. They can influence each other and should always be inspected jointly. Thereby the lowest level of abstraction is the direct representation of the trajectory data. It renders every charge jump from one molecule to the next. The resulting trajectories are aligned to the grid structure defined by the molecular structure, Figure 6d. On the other end of the scale one can either look at the charge coverage volume for the selected trajectory or a gradual simplification of the trajectory. Due to the strong stochastic character of the trajectories, simple Gaussian smoothing does not give the desired results. Therefore, we introduced *flow lines* capturing the trend of the large-scale movement. Flow lines are motivated by the transition from the Brownian motion of water molecules to a continuous flow description. The detailed construction of the lines will be described in Section 4.1.

For any simplification level one can chose between tube and ribbon rendering. For a higher level of abstractions the lines can also be augmented with derived attributes emphasizing the large scale flow properties like velocity and flow direction, which are not well defined for the original data. For their visualization color, textures and arrows are used. The effective velocity is encoded using stripe patterns displaying equal time intervals.

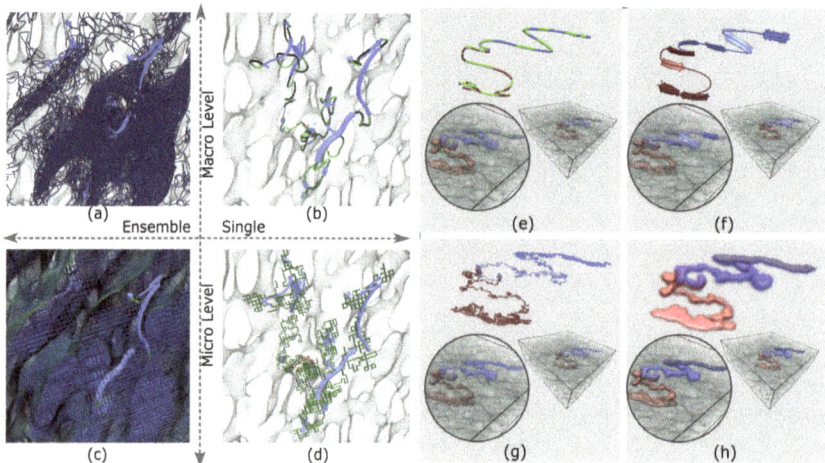

Figure 6. Trajectory visualization on different levels of abstraction. (**a,c**) visualization of multiple charge trajectories together with ribbon arrows giving a hint of the general trend of the charge movement. (**b,d**) show one selected trajectory with micro and macro abstraction level and rendering options. (**e–h**) show single trajectory abstractions. (**e**) The stripe pattern is a measure for the effective velocity of the charge. The time the charge needs for one stripe is constant. (**f**) Arrow representation added to simplified trajectory representation. (**g**) Direct rendering of raw trajectory represented using tube rendering. (**h**) Charge coverage volume visualization of single trajectory.

3.2. Statistical Plots and Efficiency Measures

The plots represent characteristic measures relevant for the assessment of the simulation data. They serve as a basis for the interaction and filtering of the data and are linked to the spatial renderings. Thereby trajectories of interest can be selected in the plots as well as in the spatial representations. As for the spatial plots we always consider a pair of electron and hole. To be able to distinguish the different charge types we assign positive values to electron-related measures and negative values to hole-related measures. *Trajectory plots* associate characteristic measures to the trajectories. The x-axis represents the straightened charge trajectories (hop-id or time), e.g., Figure 11a–d. *Parallel coordinates* relate different efficiency measures for the individual trajectories, Figure 11e. *Morphology Composition plots* allow to investigate the material composition in the neighborhood of a selected charge position, Figure 7.

Essential for the effectiveness of the statistical plots are the attributes that are displayed. Therefore, much emphasis has been put on the design of expressive measures for the efficiency of the charge transport. The derivation of the measures described below, has already been a result of the first visual exploration of the data in close collaboration with the physicists. The goal of the measures is to get a qualitative impression of the efficiency of the charge transport from creation to collection at the electrodes. The measures can be related to (i) individual trajectories, (ii) charge pairs, or (iii) the morphology. All measures can be explored on an ensemble or single trajectory basis.

(i) *Trajectory-based measures*

These measures have the purpose to equip the macro-level charge flow lines with measures that are commonly related to flow. A central measure is the effective velocity, which describes the macro-level charge velocity. The measures can be adapted to the chosen level of detail via a scale parameter r. The unit for r is intermolecular distance.

- *Escape time* $t_e(M_i, r)$. The escape time $t_e(M_i, r)$ of a charge from molecule M_i with respect to scale r is defined as the time the charge needs to leave the r-neighborhood of the molecule, see Figure 8c. It is high in regions where the charge is trapped for a longer time. The *dwell time* at a molecule corresponds to the escape time for $r = 1$.
- *Effective velocity* $v_e(M_i, r)$. The effective velocity $v_e(M_i, r) = r / t_e(M_i, r)$ is directly related to the escape time. Low velocity hints at low efficiency in the charge transport, this can be due to traps in the morphology or a strong inter charge interaction.
- *Effective distance traveled* $d_{eff}(t)$—The effective distance is the Euclidean distance of the current charge position to the start position as function of time. This measure is related to the escape time but allows a stronger focus on the geometry of the trajectory.
- *Tortuosity* $T(t)$—Tortuosity sets the actual path length $l(t)$ of the trajectory in relation to the effective distance traveled $T(t) = l(t) / d_{eff}(t)$.

(ii) *Charge pair related measures*

The morphology of the material is not the only critical aspect for the efficiency of the charge transport. There is also a strong interaction between individual charge pairs influencing their transport. If charge pairs come very close to each other, this comprises the risk of recombination, which means that the charge is lost for the entire process.

- *Pair distance* $d_p(t)$—This distance measure keeps track of the Euclidean distance of a hole and an electron created in one CT state. In the optimal case this would be a monotonously increasing function of time.
- *Minimal distance to charge of other kind* $d_{pmin}(t)$—In the case of multiple CT states a recombination is not only possible with the own 'partner' (geminate recombination) but with all charges of the complementary type (nongeminate recombination). In this way it is a generalization of d_p.

- *Minimal distance to charge of same kind d_{min}*—Charges of the same type interact with each other and can thus reduce the effective transport. This measure gives an overview over the distribution of the charges within the material. Charges of the same type interact with each other and can thus reduce the effective transport. This measure gives an overview over the distribution of the charges within the material.

(iii) *Morphology related measures*

The morphology is a critical parameter for the design of the solar cells. While a large interfacing surface is advantageous for the creation of CT states, a complex morphology can crate traps for the charge transport.

- *Distance to interface $d_{interface}(M_i)$*—This distance is the shortest distance of molecule M_i to the material interface. It is computed once for each morphology. As distance metric we use a Manhattan metric following the molecular grid structure. Thus the distance roughly corresponds to the minimal number of charge transitions necessary to reach the interface. Since recombination of charges only happens at the material interface it is favorable that the charges keep a certain distance to the interface.
- *Morphology Composition plots*—Through these plots the morphology composition (acceptor-donor ratio) can be investigated. It displays the acceptor-donor material ratio in the neighborhood surrounding a charge. For a single trajectory, the ratio is plotted for the entire transportation (Figure 7b). For ensembles, the morphology ratio at a specific time (Figure 7c) is plotted. The acceptor-donor ratio is computed for a spherical region. The spherical region is divided into eight octants illustrating the distribution in each octant. (Figure 7d).

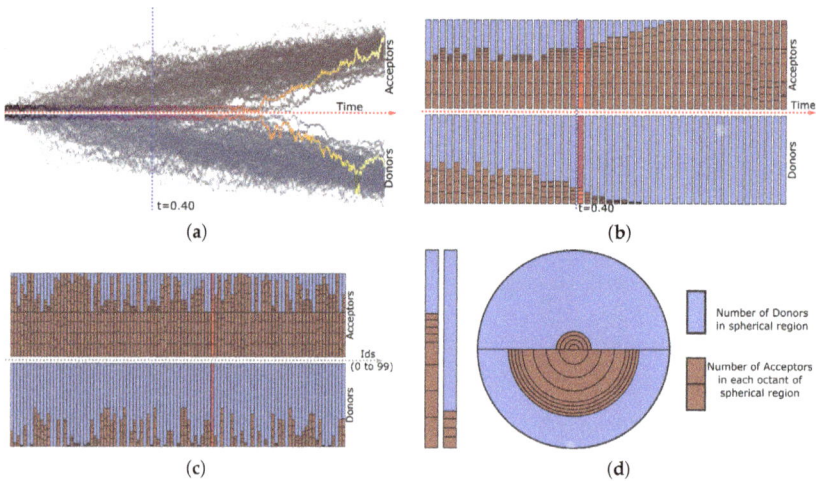

Figure 7. Interactive plots of morphology composition of single trajectory and ensemble. (**a**) selected distance measure at a certain time step, with one selected trajectory pair highlighted (**b**) *acceptor-donor ratio* along selected trajectory. (**c**) morphology composition of all trajectories at a specific time step $t = 0.4$. (**d**) stack and radial plot for octant *acceptor-donor ratio* of selected trajectory at specific time.

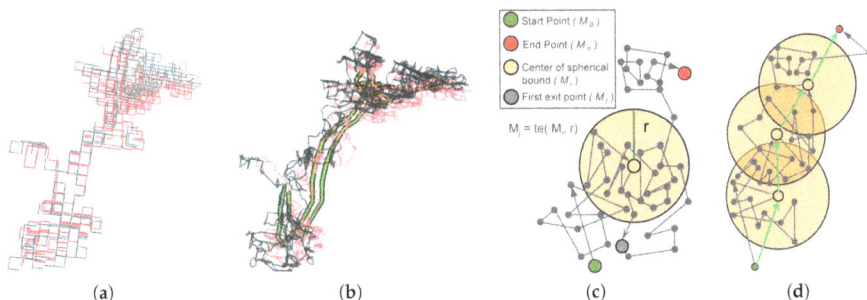

(a)   (b)   (c)   (d)

Figure 8. Different simplifications can be applied to a trajectory. The red lines are the acceptor trajectories and the grey-blue lines the donor trajectories. (**a**) One raw trajectory pair, (**b**) charge flow line with abstraction level $r = 4$ in comparison with Gaussian smoothing $n = 4$. The charge flow line is colored with respect to time (same color for donor and acceptor). (**c**) The escape time measures the time a charge needs to leave the r-neighborhood of a molecule for the first time. (**d**) charge flow line computation.

4. Technical Details

In this section we summarize few technical details that are necessary for the implementation and rendering of flow lines.

4.1. Charge Flow Lines

As the trajectories are the result from a Monte-Carlo simulation of charges jumping between discrete locations they are bound to the grid of the molecular structure and are very tortuous at the same time (Figure 8a). To get a better impression of the trend of the trajectories we propose two different approaches for simplification: smoothing and charge flow lines.

Trajectory smoothing corresponds to a simple Gaussian smoothing taking only every n-th sampling position into account. The parameter n can be interactively adjusted, which is especially useful for the density projection rendering where we typically used a value of $n = 4$. Gaussian smoothing mostly improves the rendering results; however, it is not suitable to highlight macroscopic flow properties as effective charge velocity. Since the complexity of the trajectories changes considerably along the trajectory no global smoothing factor would lead to satisfactory results. Therefore we introduced the notion of flow lines to convey the trend of the movement of the charge. It is defined on various abstraction levels, expressed by a scale factor r. The generation of flow lines can be interpreted in analogy to the transition of the stochastic Brownian movement of molecules in a flow and the macro-level description as a smooth line with direction and velocity. To generate a flow line the trajectory is sub-sampled ignoring all transitions within a sphere of radius r, compare Figure 8b. The time the charge needs to escape the sphere defines its effective velocity on the abstraction level r. The radius r is defined in units of the molecular grid cell size. As described above, the charge flow lines are rendered as solid lines with arrows encoding the direction of movement. An example of different abstraction levels for one charge trajectory is shown in Figure 8. The local trajectory complexity is reflected by its tortuosity and is also an important indicator for local transport efficiency.

4.2. Ribbon Computation

We use ribbons mostly for the representation of trends in the movement of charges based on the concept of flow lines introduced above. The ribbon computation uses a moving frame of reference for the flow lines, see Figure 9e,f. This frame is determined by its tangent, its normal, and its binormal. The normalized tangent $t_i = (P_i - P_{i-1})/|P_i - P_{i-1}|$ is approximated by the direction of the line segment. To obtain a stable normal computation, especially in region with low curvature, we introduce a weighted normal propagation

$$n_{i+1} = \frac{(1 - |t_i \times t_{i-1}|)n_i + (t_i \times t_{i-1})}{|(1 - |t_i \times t_{i-1}|)n_i + (t_i \times t_{i-1})|}.$$

The weight $|t_i \times t_{i-1}|$ is a measure for the stability of the normal. If it is close to 1 the previous normal has no influence, if it is close to zero (t_i is parallel to t_{i-1}) the previous normal is propagated. For placement of ribbon arrows we compute the curvature of the line and the high curvature segments are filtered out. Figure 9d,g,h, shows ribbon arrow segments that are placed using this approach.

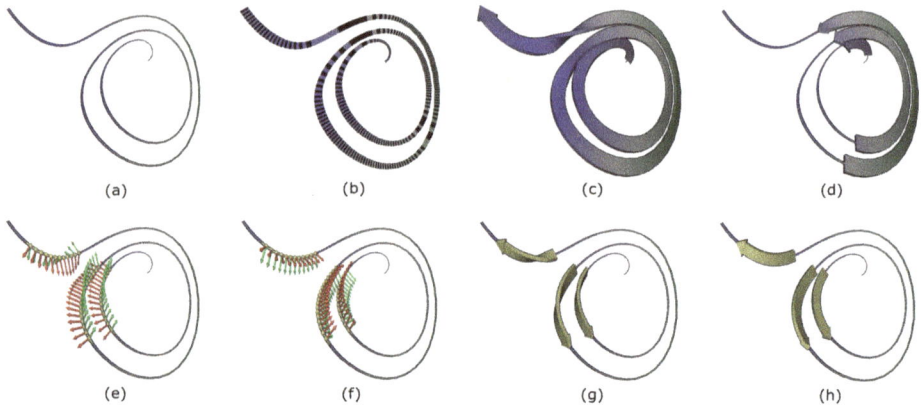

(a)   (b)   (c)   (d)

(e)   (f)   (g)   (h)

Figure 9. Top row illustrates various single trajectory representations on a synthetic data. Velocity is color mapped. **Bottom row** represents ribbon representation, which uses desired curvature range for placement of arrows and co-ordinate frame correction applied.

5. Use Cases

The visualization system became an integral part in the research project for organic solar cell design and the understanding of charge transport in complex materials. Typical for basic research projects is that questions and new tasks are developing at the same pace as getting new insights and answers. For that reason, there are only few tasks that are frequently performed in exactly the same way. In the following we describe two scenarios, in which the system has be used and insights that have been derived. We put the tasks into the context of the task classification introduced in Section 1.1. We start with a simple setting focusing on the simulation evaluation (Task SE). To illustrate scenarios where the exploration is driven by a domain specific task with the goal of gaining new insight (Tasks OE, CI and ME) we consider a more complex setting using a complex morphology. For all visualizations we use color-coding of red for acceptor and blue for donor material, the interface between the two regions is rendered as a green transparent surface. The observations discussed in the following sections summarize the reasoning of our partners when exploring the data with our system.

5.1. Scenario 1—Simple Planar Interface One CT

This scenario refers to a simple planar interface between the donor and acceptor material of the solar cell considering one CT, which corresponds to one charge pair. The ensemble consists of 100 different realizations of charge paths, whereby all paths start at the same location and diffuse toward the electrodes, which are placed on the top and the bottom of the volume. This configuration is of special interest for the evaluation of the simulation correctness and parameter setting (Task SE) since for this simple case there are clear expectations toward the results.

Derived insights: The distribution of the ensembles is expected to be approximately uniform, which is confirmed by the trajectory coverage visualization for the entire ensemble in Figure 10a. The slight variations that are visible in the charge coverage volume are due to the stochastic sampling of the space of possible charge trajectories using 100 realizations.

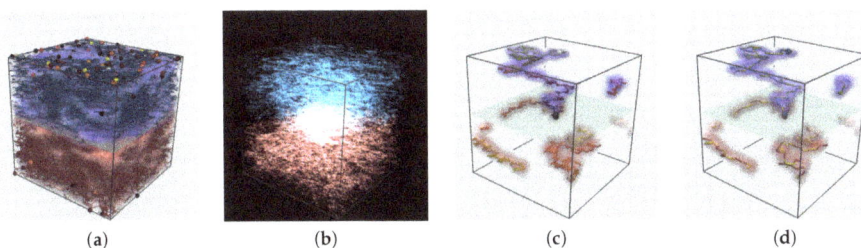

(a)	(b)	(c)	(d)

Figure 10. Scenario 1: Simulation evaluation (SE) of the flat morphology with one charge pair. The images show different rendering options for the entire ensemble (**a**) trajectories embedded in a charge coverage volume visualization, trajectories reaching the electrode are displayed as spheres colored by time they need to reach the electrode, (**b**) density projection of trajectories; for one selected trajectory embedded in the charge coverage volume (**c**) original trajectory colored by progression time (**d**) flow line displaying the effective velocity as stripe texture.

The spheres on the top and bottom of the volume highlight the locations where the charges reach the electrodes, their color displays the time-to-electrode for the respective trajectory. The ensemble density projection shown in Figure 10b allows a more detailed view into the volume highlighting preferred locations. As to expect for this setting, the region exhibiting the highest density is the area of the joint starting point in the center of the image. Figure 10c,d show one selected trajectory with a long time-to-electrode value. Even though the time the charge needed to get to the electrode is relatively high it is continuously moving in the expected direction and the path and its effective velocity appear plausible. The progression plots shown in Figure 11 confirm these findings. After a short time of strong interaction with the partner charge the charges continuously move toward the electrodes. All plots show that the two charges have a very symmetric behavior. Inspecting selected trajectory. plot (b) and (c) express the long interaction time of the charges until they finally separate and then quickly diffuse toward the electrodes. Plot (a) shows that during the interaction time the charges even visit their initial position again. A qualitatively similar behavior can also be observed for other charge pairs whereas the specific point in time where the charges start traveling independently strongly varies. This can also be seen in the parallel coordinates plot where it seems that there is no clear correlation between the time and the effective distance traveled. However, there is a strong correlation between the distance to the interface and the distance to the other charge. Plot (d) showing the escape time, expresses a general decrease of the escape time over time, which is confirmed in the parallel coordinate plot. The higher escape times at the beginning of the trajectory are a hint that the charge interaction traps the charges and slows their movement down due to attracting forces. All these measures that are represented in the parallel coordinates plot can be used to filter out and explore trajectories of interest. In summary, all the plots and spatial renderings support the reasoning and understanding of the most significant physical effects controlling the efficiency of the transport.

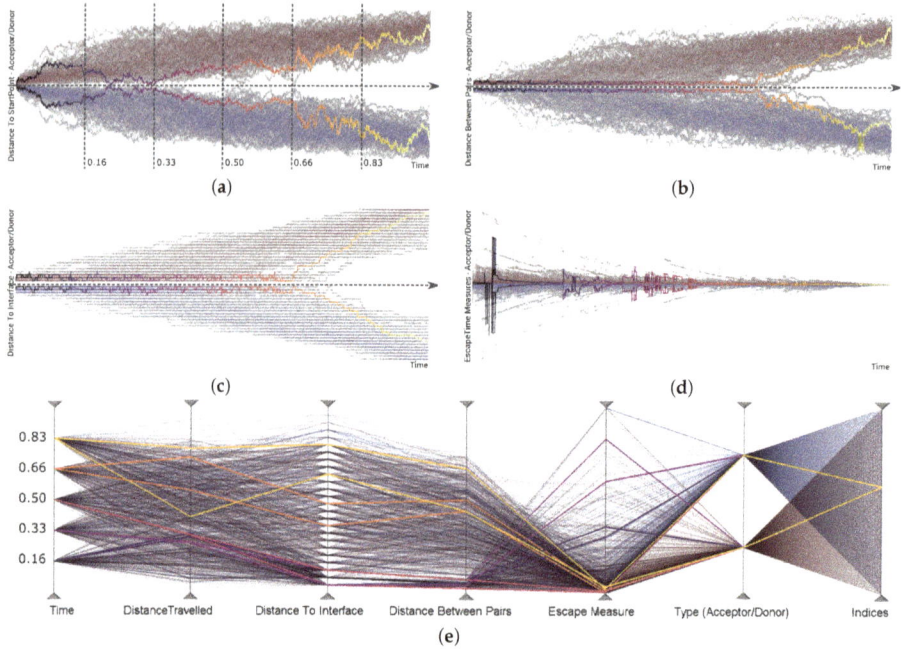

Figure 11. The plots of the derived measures for the ensemble highlighting one selected trajectory, the *x*-axis of all plots is time, which is also encoded in the color of the trajectories. The *y*-axis are (**a**) effective distance travelled from start point; (**b**) distance between the charge pairs; (**c**) shortest distance to the interface; (**d**) escape time for a radius of 10 units; (**e**) parallel coordinates of all the measures (**a**) through (**e**).

For some of the simulation runs a surprising observation has been made in the visualizations of the charge coverage volume. There, charges exhibited a tendency to stick close to the interface for a very long time never reaching the electrodes, Figure 12a. A closer inspection of the trajectories in temporal animations and in flow line representations made clear that this is due to the strong interaction between the charge pairs. The strength of this effect only became aware to the physicist through these visualizations. A similar observation was later also made for the complex morphology, Figure 12b. While it was possible to find an explanation of this effect, it was not expected in this clarity and it motivated us to make changes in the parameter setting for the external field that drives the diffusion process of the charges. This observation also lead to the introduction of a new efficiency measure the 'distance between charge pairs' shown in Figure 11b. These plots show the time it takes for charge pairs to separate and finally take off toward the electrodes. An inspection of the parallel coordinates plot of the efficiency measures also strengthened this explanation and makes it clear that the effect is not related to a long escape time.

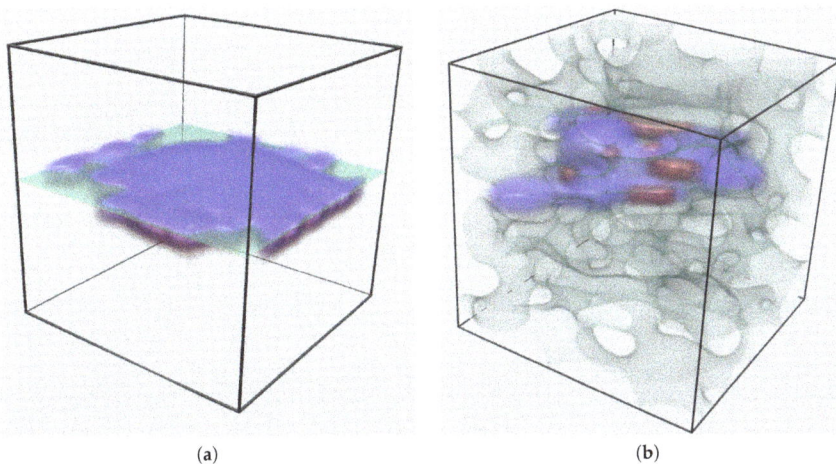

(a)   (b)

Figure 12. These images show examples of simulations ((**a**) Flat Interface, (**b**) Complex Interface) where the charge transport shows an unexpected behavior. The interaction between the two charges (donor in blue/acceptor in red) is so strong that they stick together for the entire simulation time. They never reach the electrode. The strength of this effect only became visible to the physicist through these visualizations. This observation led to a reconsideration of several simulation parameters and the introduction of the 'distance between charge pairs' as additional efficiency measure.

5.2. Scenario 2—Complex Interface Exploration

The second scenario is an exploration of the results of an ensemble simulation for a complex morphology considering one respective two CTs, which is one respective two charge pairs moving from the start point to the electrodes. The three tasks OE, CI and ME have been driving this exploration. We started the exploration with overview representations, Figure 13, for the entire ensemble and use one exemplary plot where some trajectories showed a behavior that draws interest. In the second part we explore these trajectories in more detail, Figure 14.

The collection of the entire ensemble within the morphology is displayed in Figure 13. The columns give an impression of the temporal evolution of the charge transport. The rows provide different ensemble visualizations for the four time steps. The top row (a) shows the charge coverage volume within the morphological context and the highlighted endpoints of the trajectories colored by time-to-electrode. The second row (b) shows only the trajectories without the morphology. The third row (c) uses the density projection plot highlighting preferred regions for the trajectories. The last row (d) shows the linked plots of the progression of the distance to the interface. They are composed of a summary plot overlying the evolution of the entire ensemble (left) individual trajectories that can be used for selection (right). A vertical line in these plots specifies the selected time step for the respective column.

Derived insights. The volume coverage visualization in row (a) shows that the charge paths of the ensemble are widely distributed within the volume. However, it also can be seen that they preferably move in the right direction, donor (blue) charges upward and acceptors (red) charges downward toward the respective electrode not being dragged in the wrong direction by the morphology, row (b). The locations where the charges reach the electrode do not have any preferred area on the electrode shown by the distribution of the spheres on the electrodes. Also the time-to-electrode value seems not to correlate to the location where the charge reaches the electrode. The density plots in row (c) highlight regions where the charges stay for a longer time. Besides the starting location and the electrodes there are some 'chambers' in the morphology, which show a higher density. The plots give insight into the charge interaction (Task CI) showing the trend that most trajectories initially stay close to the start position for some time until the charges escape the attraction of the partner charge and finally start moving toward the electrodes.

The patterns visible in the other plots are very different as compared to the simple morphology. While the dwell time in the simple morphology decays rapidly, the distribution of the dwell time for the complex topology stays the same for the entire time. The charges interact much longer with each other staying closely together and following similar paths. The escape time shows much higher peaks. A high escape time is a measure for a trapping of the charges. Symmetric behavior to the donor and acceptor hints at trapping due to CT effects and charge interaction. In the cases where we only see high values for one charge, the trapping must be related to other morphology effects.

Individual trajectory inspection—The second part of the exploration was focused on details of the data as required for task CI and ME. Three selected trajectories with peculiar behavior were detected in the effective-distance-travelled plot, Figure 14. The trajectories are analyzed using one of the trajectory plots. In Figure 14, time of charge transport is mapped to the x-axis and the distance from the start point of charge transport to the y-axis. Acceptor and donor can be distinguished in all images by the color red and blue respectively. The three rows refer to the three selected trajectories.

Derived insights. The plot gives hint about outlier trajectories and the CT can be examined within this context. Trajectory (1) in Figure 14 has expected charge transport, in contrast to trajectory (2) and (3). In trajectory (2), one of the charge pairs stopped propagation much earlier than the other. In trajectory (3), there was some transport in the beginning of the simulation and the propagation stopped before they started propagating again in the end. This indicates a long pause during the CT.

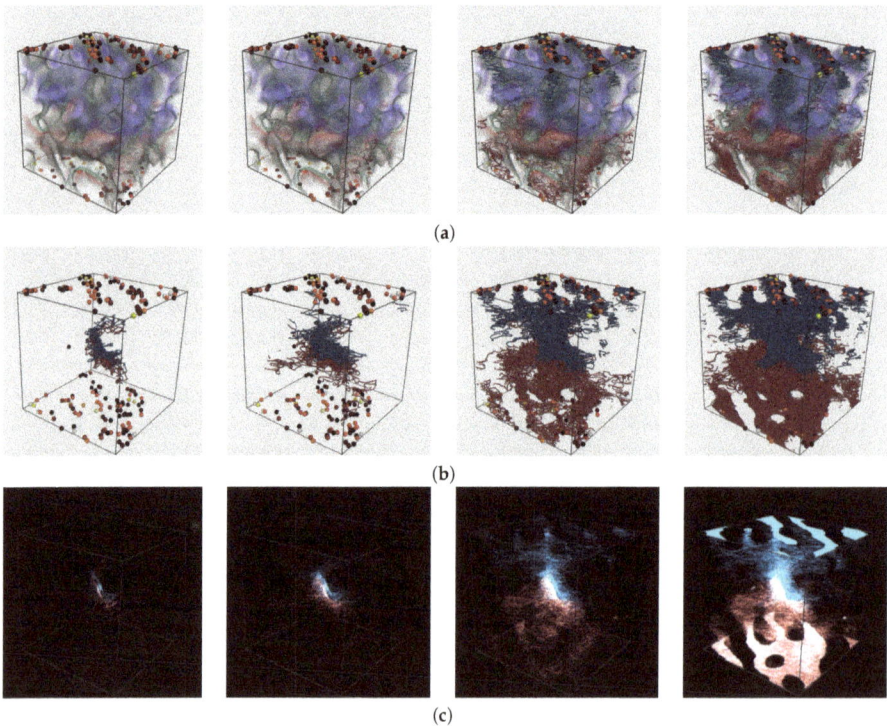

(a)

(b)

(c)

Figure 13. *Cont.*

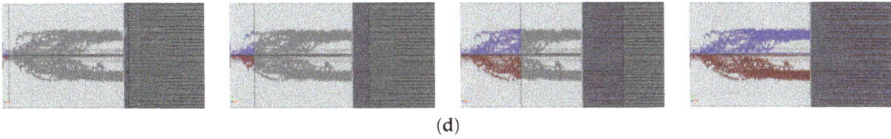

(d)

Figure 13. Overview visualizations for different time steps (columns). The selected time step is highlighted in the plots in the last row as vertical lines. The upper rows show different volumetric visualizations of the ensemble focusing on the different tasks. (**a**) charge coverage volume within context morphology (Task OE); (**b**) progression of the trajectory ensemble with time (Task OE, ME); (**c**) density projection of the trajectory ensemble (Task ME); (**d**) summary plot distance to start position (Task OE, CI).

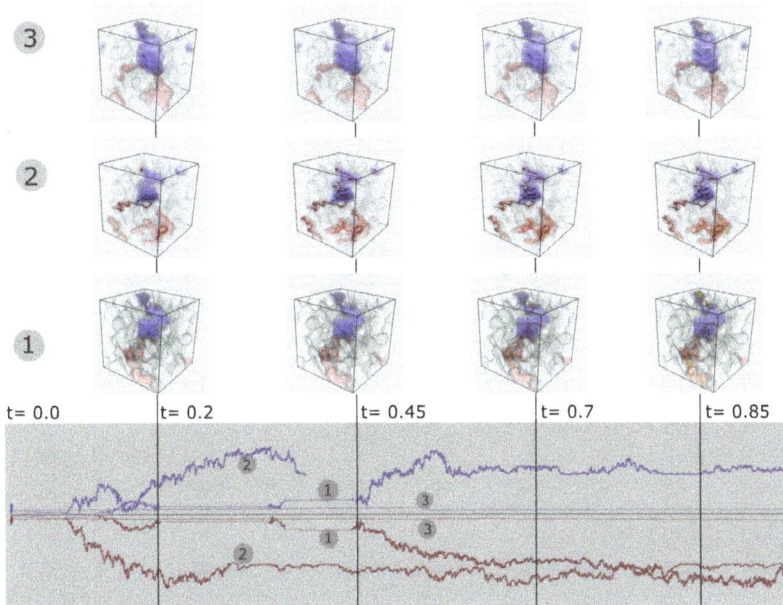

Figure 14. Three selected charge trajectories are analyzed as above using one of the summary plots. Plot uses time of charge transport along *x*-axis and *y*-axis as distance from the start point of charge transport (negative distance is used to differential acceptor and donor pairs in red and blue respectively). This measure allows the user to inspect if charge flow digresses. In this case Trajectory 1 wobbles in a small coverage region for some time before they split apart to reach the electrode. Trajectory 2, on the other hand has shorter transport time for the acceptor and longer transport time for the donor. Trajectory 3 jumps rapidly from start to end. Context rendering.

6. Conclusions

In summary, we have presented a framework for the exploration of ensembles of charge trajectories in the context of material morphology. Our partners have considered all individual visualizations and plots as useful; however, the special merit of the system lies in the combination of the plots and visualizations. This allows us to focus on individual trajectories as well as the ensemble as distribution. Thereby, the distribution gives insight into the overall efficiency of the solar cell design. The exploration of individual trajectories is useful to analyze the effect of the morphology on the charge transport. Many aspects of the data were accessible for the first time to our partners. Most importantly, all the plots and renderings have supported the reasoning about the characteristics of the charge transport in organic solar cells and the performance of the simulation itself. It inspired many vivid discussions, which help to understand the most significant physical effects controlling the efficiency of the transport. In several cases, the findings influenced the next simulation run and the adjustment of the simulation parameters. The discussions also inspired many new ideas for follow-up work. The efficiency measures that are now part of the system have been developed on the basis of the first versions of the visualization system and thus prove the usefulness of the visualization system. In the future will integrate other physical fields involved in the simulation into the exploration framework as context information. It is further planned to extend the system to less regular molecular structures and more diffuse interfaces, which is ongoing work for our collaboration partners. Many of the concepts derived for this application can also be of use for other applications that are concerned with ensembles of stochastic trajectories.

Acknowledgments: This work was supported through grants from the Swedish e-Science Research Centre (SeRC) and has been developed in the Inviwo framework (www.inviwo.org).

Author Contributions: All authors have been contributing essential in the development of the concepts of the visualization system and writing the paper. Sathish Kottravel is mainly responsible for the implementation of the visualisation system. Mathieu Linares and Riccardo Volpi are the Physicists that performed the simulation of the charge transport and are responsible for the data generation.

Conflicts of Interest: The authors declare no conflict of interest.

References

1. Jakobsson, M.; Linares, M.; Stafström, S. Monte Carlo simulations of charge transport in organic systems with true off-diagonal disorder. *J. Chem. Phys.* **2012**, *137*, 114901.
2. Volpi, R.; Stafström, S.; Linares, M. A consistent Monte Carlo simulation in disordered PPV. *J. Chem. Phys.* **2015**, *142*, 094503.
3. Volpi, R.; Kottravel, S.; Norby, M.S.; Stafström, S.; Linares, M. Effect of Polarization on the Mobility of C60: A Kinetic Monte Carlo Study. *J. Chem. Theory Comput.* **2016**, *12*, 812–814.
4. Volpi, R.; Linares, M. Organic Solar Cells. In *Specialist Periodic Reports—Chemical Modelling*; RSC: London, UK, 2016; Volume 13.
5. Tang, C.W. Two-layer organic photovoltaic cell. *Appl. Phys. Lett.* **1986**, *48*, 183–185.
6. Marcus, R.A. On the Theory of Oxidation, Reduction, Reactions Involving Electron Transfer. *J. Chem. Phys.* **1956**, *24*, 966.
7. Aboulhassan, A.; Baum, D.; Wodo, O.; Ganapathysubramanian, B.; Amassian, A.; Hadwiger, M. A Novel Framework for Visual Detection and Exploration of Performance Bottlenecks in Organic Photovoltaic Solar Cell Materials. *Comput. Graph. Forum* **2015**, *34*, 401–410.
8. Wodo, O.; Tirthapura, S.; Chaudhary, S.; Ganapathysubramanian, B. A graph-based formulation for computational characterization of bulk heterojunction morphology. *Org. Electron.* **2012**, *13*, 1105–1113.
9. Aboulhassan, A.; Li, R.; Knox, C.; Amassian, A.; Hadwiger, M. CrystalExplorer: An Interactive Knowledge-Assisted System for Visual Design of Solar Cell Crystal Structures. *EuroVisShort* **2012**, doi:10.2312/PE/EuroVisShort/EuroVisShort2012/031-035.
10. Kozlikova, B.; Krone, M.; Lindow, N.; Falk, M.; Baaden, M.; Baum, D.; Viola, I.; Parulek, J.; Hege, H.C. Visualization of Biomolecular Structures: State of the Art. *EuroVisSTAR2015* **2015**, doi:10.2312/eurovisstar.20151112.
11. Andrienko, N.; Andrienko, G. Visual analytics of movement: An overview of methods, tools and procedures. *Inf. Vis.* **2012**, *12*, 3–24.
12. Andrienko, G.; Andrienko, N.; Bak, P.; Keim, D.; Wrobel, S. *Visual Analytics of Movement*; Springer: Berlin, Germany, 2013.
13. Doleisch, H.; Hauser, H.; Gasser, M.; Kosara, R. Interactive Focus + Context Analysis of Large, Time-Dependent Flow Simulation Data. *Simulation* **2006**, *82*, 851–865.
14. Gasteiger, R.; Neugebauer, M.; Beuing, O.; Preim, B. The FLOWLENS: A focus-and-context visualization approach for exploration of blood flow in cerebral aneurysms. *IEEE Trans. Vis. Comput. Graph.* **2011**, *17*, 2183–2192.

15. Zöckler, M.; Stalling, D.; Hege, H.C. Interactive visualization of 3d-vector fields using illuminated stream lines. In Proceedings of the IEEE Conference on Visualization (Vis '96), San Francisco, CA, USA, 27 October–1 November 1996; pp. 107–114.

16. Schussman, G.; Ma, K.L. Anisotropic Volume Rendering for Extremely Dense, Thin Line Data. In Proceedings of the IEEE Conference on Visualization '04, Austin, TX, USA, 10–15 October 2004; pp. 107–114.

17. Tarini, M.; Cignoni, P.; Montani, C. Ambient Occlusion and Edge Cueing to Enhance Real Time Molecular Visualization. *IEEE Trans. Vis. Comput. Graph.* **2006**, *12*, doi:10.1109/TVCG.2006.115.

18. Everts, M.H.; Bekker, H.; Roerdink, J.; Isenberg, T. Depth-Dependent Halos: Illustrative Rendering of Dense Line Data. *IEEE Trans. Vis. Comput. Graph.* **2009**, *15*, 1299–1306.

19. Eichelbaum, S.; Hlawitschka, M.; Scheuermann, G. LineAO—Improved Three-Dimensional Line Rendering. *IEEE Trans. Vis. Comput. Graph.* **2013**, *19*, 433–445.

20. Isenberg, T. A Survey of Illustrative Visualization Techniques for Diffusion-Weighted MRI Tractography. In *Visualization and Processing of Higher Order Descriptors for Multi-Valued Data*; Springer International Publishing AG: Gewerbestrasse, Switzerland, 2015; pp. 235–256.

21. Diaz-Garcia, J.; Vazquez, P.P. Fast illustrative visualization of fiber tracts. In *Advances in Visual Computing*; Springer: Berlin/Heidelberg, Germany, 2012; pp. 698–707.

22. Schröder, S.; Obermaier, H.; Garth, C.; Joy, K.I. Feature-based Visualization of Dense Integral Line Data; OASIcs-OpenAccess Series in Informatics. In Proceedings of the IRTG 1131 Workshop 2011, Kaiserslautern, Germany, 10–11 June 2011; Volume 27.

23. Brambilla, A.; Carnecky, R.; Peikert, R.; Viola, I.; Hauser, H. Illustrative Flow Visualization: State of the Art, Trends and Challenges. *STARs* **2012**, 75–94, doi:10.2312/conf/EG2012/stars/075-094.

24. Everts, M.H.; Bekker, H.; Roerdink, J.B.; Isenberg, T. Interactive illustrative line styles and line style transfer functions for flow visualization. *arXiv* **2015**, arXiv:1503.05787.

25. Salzbrunn, T.; Garth, C.; Scheuermann, G.; Meyer, J. Pathline Predicates and Unsteady Flow Structures. *Vis. Comput.* **2008**, *24*, 1039–1051.

26. Kuhn, A.; Lindow, N.; Günther, T.; Wiebel, A.; Theisel, H.; Hege, H.C. Trajectory Density Projection for Vector Field Visualization, Eurovis Short Papers. In Proceedings of the EuroVis 2013, Leipzig, Germany, 17–21 June 2013.

27. Lindholm, S.; Falk, M.; Sunden, E.; Bock, A.; Ynnerman, A.; Ropinski, T. Hybrid Data Visualization Based On Depth Complexity Histogram Analysis. *Comput. Graph. forum* **2014**, *34*, 74–85.

28. Scharsach, H. Advanced GPU raycasting. In Proceedings of the CESCG 2005, Budmerice, Slovakia, 9–11 May 2005; pp. 69–76.

29. Sanyal, J.; Zhang, S.; Dyer, J.; Mercer, A.; Amburn, P.; Moorhead, R.J. Noodles: A Tool for Visualization of Numerical Weather Model Ensemble Uncertainty. *IEEE Trans. Vis. Comput. Graph.* **2010**, *16*, 1421–1430.

30. Wilson, A.T.; Potter, K.C. Toward visual analysis of ensemble data sets. In Proceedings of the 2009 Workshop on Ultrascale Visualization (UltraVis '09), Portland, OR, USA, 16 November 2009; ACM: New York, NY, USA, 2009; pp. 48–53.

31. Ferstl, F.; Bürger, K.; Westermann, R. Streamline Variability Plots for Characterizing the Uncertainty in Vector Field Ensembles. *IEEE Trans. Vis. Comput. Graph.* **2015**, *22*, 767–776.

32. Bos, W.J.T.; Kadoch, B.; Schneider, K. Angular Statistics of Lagrangian Trajectories in Turbulence. *Phys. Rev. Lett.* **2015**, *114*, 214502.

33. Savage, J.; Voth, G.A. Persistent Subdiffusive Proton Transport in Perfluorosulfonic Acid Membranes. *J. Phys. Chem. Lett.* **2014**, *5*, 3037–3042.

34. Burov, S.; Tabei, S.M.A.; Huynh, T.; Murrell, M.P.; Philipson, L.H.; Rice, S.A.; Gardel, M.L.; Scherer, N.F.; Dinner, A.R. Distribution of directional change as a signature of complex dynamics. *Proc. Natl. Acad. Sci. USA* **2013**, *110*, 19689–19694.

informatics

MDPI

Article

TOPCAT: Desktop Exploration of Tabular Data for Astronomy and Beyond

Mark Taylor

H. H. Wills Physics Laboratory, University of Bristol, Tyndall Avenue, Bristol BS8 1TL, UK;
m.b.taylor@bristol.ac.uk

Academic Editors: Achim Ebert and Gunther H. Weber
Received: 27 May 2017; Accepted: 24 June 2017; Published: 27 June 2017

Abstract: TOPCAT, the Tool for OPerations on Catalogues And Tables, is an interactive desktop application for retrieval, analysis and manipulation of tabular data, offering a powerful and flexible range of interactive visualization options amongst other features. Its visualization capabilities focus on enabling interactive exploration of large static local tables—millions of rows and hundreds of columns can easily be handled on a standard desktop or laptop machine, and various options are provided for meaningful graphical representation of such large datasets. TOPCAT has been developed in the context of astronomy, but many of its features are equally applicable to other domains. The software, which is free and open source, is written in Java, and the underlying high-performance visualisation library is suitable for re-use in other applications.

Keywords: interactive visualization; astronomy; tabular data; exploratory data analysis

1. Introduction

1.1. Source Catalogues

Astronomy is a discipline with a long history of collecting, storing and analysing data. This comes in various forms, including images, spectra and time series, but one of the most important is the *source catalogue*, a list of observed astronomical objects such as stars or galaxies, each with a fixed set of features. These features are mostly numeric and typically include quantities such as central sky coordinates, brightness in one or several wavebands, apparent size, ellipticity parameters and so on.

A catalogue is thus naturally represented as a table with a certain number of rows (one for each observed object) and columns (one for each feature). An early example is the Catalog of Nebulae and Star Clusters published by Charles Messier in 1781 [1], containing data in six columns for 103 celestial objects, each observed individually by eye. More recent examples are often, though not always, considerably larger; the features for modern catalogues are extracted automatically from image data obtained by highly sophisticated telescope instrumentation, and for the largest sky surveys can run to hundreds of columns such as the Sloan Digital Sky Survey [2], and/or the order of a billion rows such as the Gaia mission [3]. These numbers, of course, are expected to rise in the future, for instance in upcoming experiments such as ESA's Euclid satellite (http://sci.esa.int/euclid/) and the Large Synoptic Survey Telescope (https://lsst.org/). Many other catalogues however may contain only a few tens or thousands of objects identified as a particular astronomical type or of interest for a particular study. In some cases catalogues can be represented by a single table, in others as complex relational databases.

A number of technologies exist for accessing such data, including bulk download of full or partial tables in domain-specific (VOTable [4], FITS [5]) or generic (Comma-Separated Value) file formats, and remote access to relational databases using SQL-like query languages. A family of "Virtual Observatory" standards (see e.g., [6–8]), developed since 2002 by a group of interested parties known as the International Virtual Observatory Alliance (IVOA) (http://www.ivoa.net/), enables standardised access to many thousands of such catalogues, which are mostly available without usage restrictions, hosted by a large network of data servers around the world.

1.2. TOPCAT Application

TOPCAT (http://www.starlink.ac.uk/topcat/), the Tool for OPerations on Catalogues And Tables [9], is a desktop Java GUI application for retrieval, analysis, and manipulation of tables. It has been developed in the context of astronomy and the Virtual Observatory, with the aim of providing a toolkit for astronomers to perform all the mechanical operations they routinely require on catalogues, so they can focus on extracting scientific meaning from this hard-won data.

It has been under more or less continuous development since 2003, and is in 2017 a mature application with an active user base in the thousands, spread over six continents, including undergraduates, amateur astronomers, and research scientists. A number of factors have contributed to its popularity in the astronomy community, including astronomy-specific capabilities such as celestial coordinate system handling, table joins using sky positions, and Virtual Observatory data access, as well as more generic items such as its powerful expression language and support for large datasets, alongside a responsive development model, high-quality support, ease of installation and a relatively shallow initial learning curve.

This paper however describes just one aspect of TOPCAT's operation, its capabilities for exploratory visualisation, especially as applicable to generic (not necessarily astronomical) tabular data. Its distinctive combination of features compared to other visualisation applications includes:

- the ability to work with large datasets, without any special preparation of the data or prior assumptions about the visualisations required
- provision of many options to explore high-dimensional data, that can be adjusted interactively with rapid visual feedback
- meaningful representation of both high and low density regions of very large point clouds

The rest of the paper discusses these capabilities and outlines some of the underlying implementation. Section 2 describes TOPCAT's relatively unsophisticated approach to data access which can nevertheless, by using robust technologies such as file mapping, deliver high performance results, as well as providing a platform that is easy to deploy and install. Sections 3 and 4 explore the visualisation capabilities offered, principally representation of point clouds in one, two and three dimensions, including the optionally weighted "hybrid density map/scatter plot" which provides a unified view of high and low density regions in very crowded plots; the use of *linked views* for exploring high-dimensional data is also discussed. Sections 5 and 6 examine the difficult issue of providing a comprehensible user interface to control the highly configurable plots on offer, including mention of the command-line interface STILTS. Section 7 goes into some detail about the implementation of certain performance-critical parts of the code and Section 8 gives a few examples of its, currently minority, use in fields other than astronomy. Sections 9 and 10 conclude with information about availability of the software.

The visualisation framework described here corresponds to that introduced in version 4 of the application, released in 2013.

2. Application Overview

2.1. Data Access

TOPCAT uses a traditional model of data access for visualisation, in which the user identifies and retrieves to local storage one or more static tables, and then works with them. This makes it unsuitable for direct visualisation of extremely large datasets, but it turns out in most cases to be possible for users working with very large astronomical catalogues to preselect and download a subset of interest, by restricting for instance to a given sky region or class of astronomical object; TOPCAT provides a wide range of astronomy-specific options for selective acquisition of such data. In many other cases, users will be concerned with much smaller catalogues where data volume is not an issue.

Given this approach, it is important to be able to access large data files on local disk efficiently. TOPCAT's preferred input format is the FITS binary table [5]. This binary format lays out columns and rows in a predictable pattern on disk, so that files can be *mapped* into memory for sequential or random access, giving effectively instant load time and without encroaching on Java's limited heap space; for more explanation of this technique and its benefits in this context see [10]. However, other formats such as the more common Comma-Separated Values (CSV) are also supported. In this case direct file mapping is not useful, but for large CSV tables the data is copied on load into a temporary binary file which can itself be mapped, allowing similar access but with a significant load time. Another option is to use TOPCAT to convert from CSV to FITS format before use.

We note that more sophisticated data access models are in use by other visualisation applications, for instance running the computation on a remote data-hosting server and transmitting only the resulting images to be displayed in a browser or other desktop application (e.g., the Gaia archive visualisation service [11]), or retrieving and caching relevant data subsets on demand for local rendering as required by user navigation actions (e.g., Aladin-Lite [12]) or moving the user to the data using centralised high-performance visualisation facilities which may include large-scale display hardware alongside High Performance Computing capability (e.g., [13]). Such techniques can avoid wholesale transfer of an impractically large dataset without requiring the user to identify any particular subset of interest, and can deliver excellent interactive experiences. They also however suffer from some limitations. Visualisations which are intrinsically data intensive will have large resource requirements, consuming centralised resources while in progress which may be expensive and scale poorly to large numbers of users. In some cases, efficient server-side visualisation makes use of pre-computed data structures such as indexes or hierarchical multi-resolution maps, which may be expensive to compute but once in place can support rapid navigation and interaction. This can work well, but it generally requires prior information (or assumptions) about what visualisations are going to be required. In the case that there are many columns, and a user may want to plot not just any pair of columns against each other, but arbitrary functions based on available columns, it is typically not possible to ensure that the appropriate pre-calculated data structures are in place. As a general rule, coordinating client and server software adds a layer of complexity which can make software development slower and harder, and often impacts reliability. These techniques are also not suitable in the absence of network connectivity. TOPCAT's low-tech approach on the other hand has the benefits of reliability, network independence and above all flexibility in terms of the visualisation options available.

2.2. Usage Model

As well as its traditional approach to data access, TOPCAT supports a straightforward usage model: it is a standalone application running CPU-based code, and suitable for use on low-end desktop or laptop computers. The visualisation is multi-threaded to maintain GUI responsiveness, but does not currently distribute the bulk of its computation across multiple cores for efficiency (though it may do so in future). If GPUs are present they are not used except for normal graphical operations.

This generally low-tech approach can nevertheless deliver performant interactive visualisation for quite large datasets, and has the benefit that barriers to use are low.

2.3. Expression Language

One of the features of TOPCAT not directly related to visualisation is its provision of a powerful expression language which allows evaluation of simple or complex expressions involving column names. In general, wherever a coordinate is supplied for plotting, either a column name or an expression can be used, making it very easy to plot arbitrary functions or combinations of columns. The expression language can also be used to define row selections algebraically.

The implementation of this feature is based on JEL, the Java Expressions Language, available from https://www.gnu.org/software/jel/.

3. Visualising Point Clouds

A source catalogue may contain tens or hundreds of columns, and interesting relationships may be lurking between pairs or higher-order tuples of these features. Often, an interesting result is to identify a subset of rows occupying a particular region in some multidimensional parameter space whose axes may be table columns, or linear or non-linear combinations of columns. Physically, this corresponds to identifying a sub-population of observed astronomical objects sharing some physical characteristics, for instance a group of stars formed from the same primordial dust cloud. TOPCAT does not attempt to provide automated support for discovering such relationships, for instance by implementing data mining algorithms. Instead, it aims to provide the user with a flexible toolkit of options to display different aspects of the data, in order to pick out trends, associations or interesting outliers by eye.

Many, though not all, of these options are variations on the theme of a scatter plot in two or three dimensions of some point cloud in multi-dimensional space. A scatter plot can be an excellent tool for presenting a relationship between known variables, but it presents two main problems. First, if the dimensionality of an association is greater than that of the plot, the association may be masked. Second, if the number of points is large compared to the area on which they are plotted, data can be obscured. These issues can to some extent be addressed by providing a range of plotting and interaction options, and are discussed in the following subsections.

3.1. High-Dimensional Plots

To represent a relationship between two variables by plotting points on a two-dimensional plotting surface is straightforward. This can be extended to three dimensions by using various techniques for representing points in a 3-d space, though visual interpretation tends to be harder in this case. TOPCAT supports both options, though the 3-d representation is at present restricted to a 2-d projection whose 3-d nature only becomes apparent from user interaction with the mouse (rotation, zooming, navigation).

To visualise a higher-dimensional relationship however, spatial positioning is not enough, so TOPCAT provides various ways to modify the representation of each point according to additional features. Distinct sub-populations can be identified using markers of different colours, sizes or shapes, individual points can be labelled with per-object text labels, and additional numeric features can be encoded using:

- colour from a selected colour map
- marker size
- X/Y marker extent
- error bars aligned with the axes
- vector with magnitude and orientation
- ellipse primary/secondary radius and orientation

The user can combine these options freely; some examples are shown in in Figure 1.

In principle quite a large number of features can be encoded in this way, for instance one could represent seven dimensions on a 2-d scatter plot by marking coloured ellipses with text labels. In practice however, especially if the number of points is large, there are limits to what is visually comprehensible.

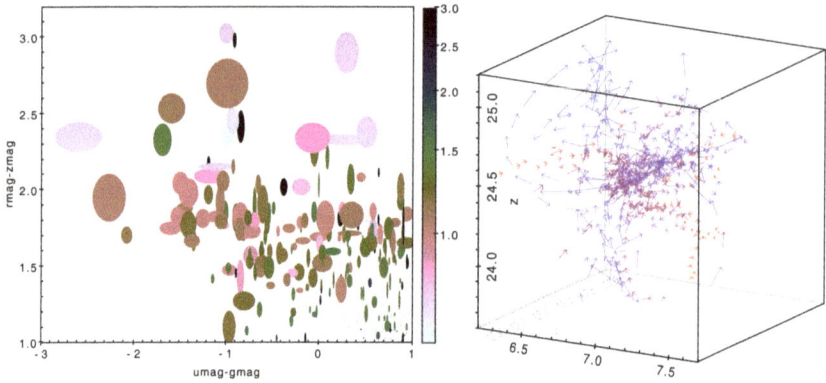

Figure 1. Options for high-dimensional visualisation. The left hand figure uses marker colour and shape to indicate three non-positional numeric features. The right hand figure uses arrows to represent points in six-dimensional phase space, the positions and velocities of simulated galaxies [14]; redshift is additionally shown by colour-coding.

3.2. Subset Selection and Linked Views

An alternative approach to understanding high-dimensional data is to extract sub-populations for further examination. A common workflow is to make a scatter plot in one parameter space, identify by eye a subset of points falling within a sub-region of that space and then replot the subset in a second parameter space to reveal some relationship evident in the subset but not the full dataset. This sequence may often be iterated to narrow down a sub-population of interest. This can be seen as a somewhat crude way to identify clusters that would be evident in a much higher dimensional space by combining multiple 2-dimensional views, but it is a powerful technique, and falls into the category of *linked views* [15,16].

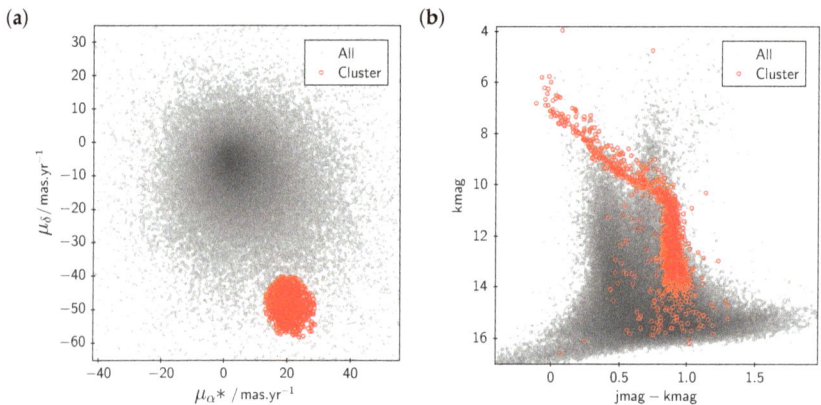

Figure 2. Linked views for row subsets. The user has identified the region in plot (**a**) by dragging the mouse, and the same subset of rows shows up in an interesting subregion of the different plot in (**b**). In this case each point represents a star in the region of the Pleiades open cluster [17]; (**a**) shows apparent velocity across the sky, while (**b**) characterises stellar classification. Those stars with similar motion, identified as the "Cluster" subset in (**a**), were formed in the same environment and therefore have similar physical characteristics, hence trace out a distinct path in (**b**).

TOPCAT allows the user to define subsets in various graphical and non-graphical ways, one of the most powerful being to drag out with the mouse an arbitrary shape over a region of a plot. Such subsets, once defined, can be plotted separately in any other plot of the same table, and also distinguished for other processing operations. An example is given in Figure 2.

3.3. Row Highlighting and Linked Views

Another aspect of linked plots in TOPCAT is that when the user highlights a plotted point by clicking on it, any point in other visible plots representing the same row is automatically highlighted. At the same time, if the underlying data is displayed in an table browser window, the corresponding row is highlighted so that the data values in all columns can easily be seen. The same operation works in reverse, so clicking a row in the table browser window will highlight any corresponding points in currently visible plots.

The application can also be configured so that some *Activation Action* takes place when a row is selected by user action in either of these ways. A typical activation action might be to display an image associated with the table row in question, for instance the original photograph that supplied the data, e.g., available from a URL column in the table. TOPCAT can perform basic image display internally, or communicate with external specialised display applications in order to achieve this kind of thing. This coordinated row-highlighting behaviour can be especially useful for investigating outliers.

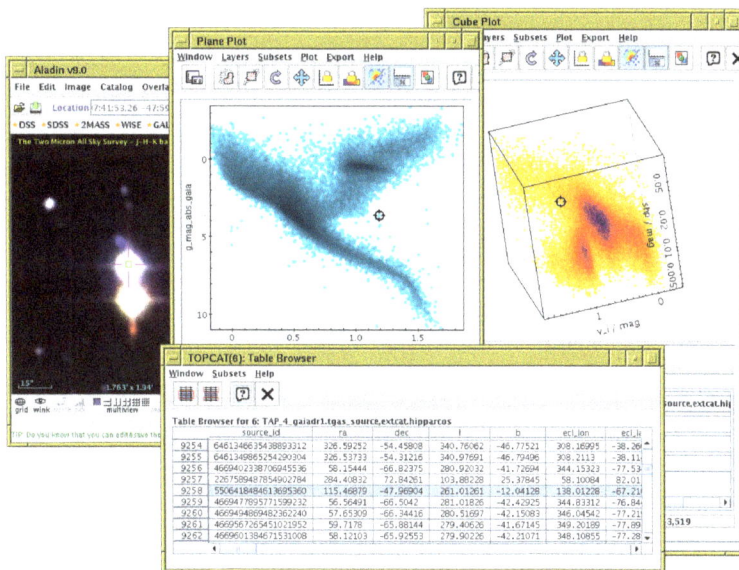

Figure 3. Linked highlighting of table rows. A user has clicked on an outlier in the 2-d plot (**center**), highlighting it with a "target" cursor. This automatically causes the same row to be highlighted in other ways: the target cursor marks the relevant point on the 3-d plot (**right**) and the row is flagged in the table data browser (**bottom**). In this case TOPCAT has also been configured to communicate with the external image display application Aladin [18] (**left**), which is caused to display a sky image corresponding to the highlighted row. This interaction makes it easy to see here that the relevant star is very close to another one; it is likely that this contamination of sources has led to a spurious brightness value, resulting in its anomalous position in the 2-d plot. Data show a Gaia-Hipparcos colour magnitude diagram and 2MASS colour image.

The communication with external display applications mentioned above, if required, can be done by invoking methods of Java classes supplied at runtime, or invoking system commands, or using the messaging protocol SAMP [19] implemented by a number of astronomy tools, including Aladin [18], SAOImage DS9 [20], Astropy [21] and others. This latter option in fact provides for inter-process exchange of both single rows and row subsets, allowing linked views *between* cooperating applications, as well as just within TOPCAT.

For an example of all this in action, see Figure 3.

3.4. High-Density Plots

TOPCAT aims to be able to explore tables containing many rows. The answer to the question, "how many?", is of course constrained by available resources of computation and user time, but in general the target is: as many as possible. The larger the dataset that can be explored interactively, the fewer decisions the user will need to take in pre-selecting data, and the more relationships are potentially available for discovery.

There are two main aspects to consider when attempting to satisfy this requirement. The more obvious concerns resource usage: will the computation require more memory than is available, and will it be fast enough to provide a fluid experience? These questions are discussed in Section 7. But there is also a question of what constitutes a visually faithful and comprehensible representation of a very large number of points. In particular, how can one represent a scatter plot when the number of points to plot exceeds the number of pixels available?

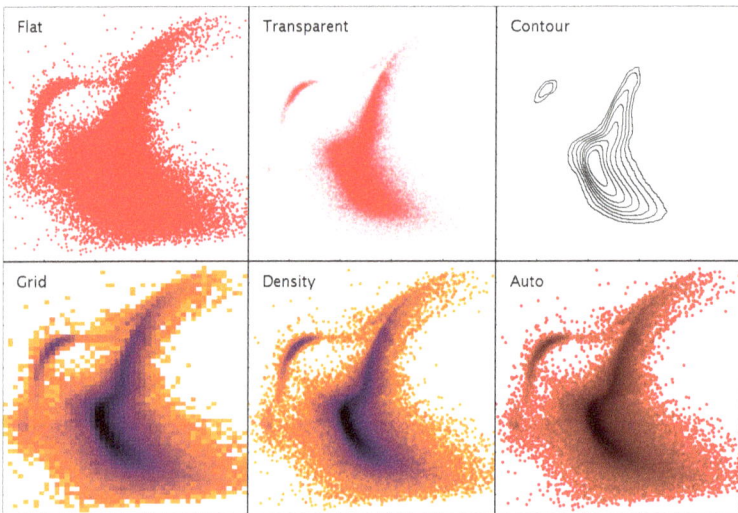

Figure 4. Various representations of a 2-dimensional point cloud available within TOPCAT. *Flat* simply plots markers at each point, obscuring the density structure. *Transparent* plots partially transparent markers, giving some indication of overdense regions, however regions over a certain density threshold are saturated. *Contour* is a somewhat smoothed contour plot. *Grid* is a 2-d histogram on a grid of fixed size rectangular bins; various options are available for combination within bins including mean, median, min, max etc. Density structure is clear, but resolution is lost and outliers are poorly represented. *Density* is a hybrid density map/scatter plot with a configurable colour map, showing both high-density structure and individual outliers. *Auto* is a standard profile of the hybrid map used by default, with a fixed colour map scaling that fades from a dataset's chosen colour to black; multiple overplotted datasets can be distinguished by using different base colours. Data represent a *V* vs. *B-V* colour-magnitude diagram of 139,000 stars from the globular cluster ω Centauri [22].

Simply plotting the points as opaque markers loses information where there are many points per pixel, since it is not possible to see how many points are overplotted. A number of options exist to address this, such as painting partially transparent markers, drawing contours, or binning data to generate colour-coded two-dimensional histograms (also known as density maps). Very often for source catalogues however, the outliers are just as important as the statistical trends, so both the low and high density regions of the plot must be represented faithfully. Contour plots and density maps do not work well for low-density regions, while transparent points are suitable for small variations in density but lose information at one or both ends of the density spectrum if there is a large density range. Use of a density map or contour plot also inhibits the row highlighting behaviour described in Section 3.3, since single points are not represented. To address this issue, a hybrid density map/scatter plot has been introduced, which is a convolution of a single-pixel density map with a shaped marker, and represents both high and low density regions of the plot well in the same display, resembling a smoothed density map at high density and a normal scatter plot at low density. This hybrid representation, which is the default plotting mode, also works particularly well when navigating a large point cloud: zooming in turns a high-density into a low-density region, so the plot transitions smoothly from a density map to a normal scatter plot. It is described in more detail in [23]. TOPCAT provides all these options and others for plotting large and small point clouds in two or three dimensions. Some examples are shown in Figure 4.

Colour-coding points to express extra dimensions as discussed in Section 3.1 presents additional issues in high-density regions, since multiple colours may be overplotted in the same pixel. To address this problem, TOPCAT offers a weighted generalisation of the hybrid density map/scatter plot, as illustrated in Figure 5.

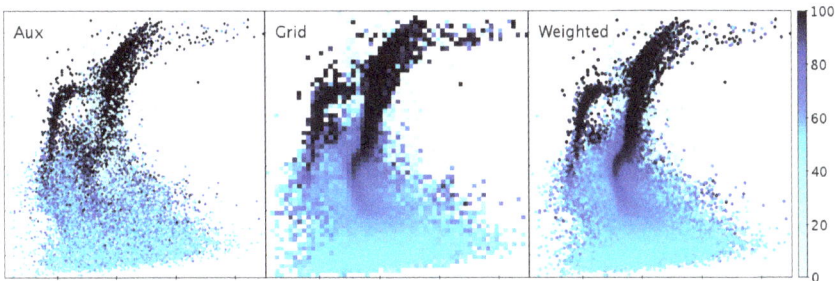

Figure 5. Various representations of a 2-dimensional point cloud with a third dimension indicated by colour. *Aux* paints opaque coloured markers, which works for low-density regions, but at high density the appearance is noisy and depends on row sequence, since points painted later obscure earlier ones. *Grid* is a 2-d histogram weighted by the third coordinate; points in the same bin are averaged, but some spatial resolution is lost. *Weighted* is a hybrid density map/scatter plot but with pixel bins weighted by the third coordinate, providing clarity in both low and high density regions. The weighting combination method in the latter two cases is configurable; in this case the median has been used. Data are as for Figure 4 with the third coordinate indicating probability of cluster membership.

3.5. Navigation

It is impossible to understand all the information in a point cloud consisting of millions of points from a static image, particularly if it is presented on a grid of, say, 100,000 pixels. But it is often possible to identify what might be a region of interest from a plot that represents the density structure and outliers appropriately, and zoom in on such a region for closer inspection, down to the features of individual objects. Interactive navigation of plots is therefore a crucial feature for exploratory visualisation.

The basic user interface for navigating 2-dimensional plots is fairly straightforward, namely that dragging a mouse around the screen will drag the plot with it, while rolling a mouse scroll wheel will zoom in or out around the current mouse position. Plots in TOPCAT in many cases have no natural aspect ratio, so various options are provided for anisotropic zooming: dragging the middle button will drag out a "window" rectangle with any required aspect ratio to become the new field of view, while dragging the right button stretches or shrinks the field of view in the X and Y directions independently according to the drag position. It is also possible to control the X or Y fields of view separately by performing the windowing or stretching gestures near the relevant axis. Keyboard modifiers are available for mice lacking all three buttons or a scroll wheel. The isotropic drag/zoom navigation gestures are generally intuitive for users of other GUI applications. The anisotropic zooming capabilities offer very flexible 2-d navigation, but they are less intuitive and advertising them well is difficult, so many users may not be aware of their proper use.

Allowing navigation through a 3-d scatter plot is a more difficult user interface problem. In this case, TOPCAT uses the default-button drag to rotate the visualised cube, and the mouse wheel to zoom in or out around the cube center. These gestures are intuitive. Translating the volume within the cube however is harder to do. Some mouse gestures are assigned to drag and stretch along the cube plane most nearly parallel to the screen projection plane, but it is difficult to use these to zoom in on a region of interest. More useful is the right-click; this takes the point under the current cursor position and translates it to the center of the cube, so that subsequent zoom actions will zoom in and out around it. However, since in 3-d the cursor position represents a line of sight rather than a unique point, this presents a problem: which of the positions under the cursor should be the new plot center? To break this degeneracy, the point chosen is the center of mass (mean depth) of all the points plotted along the line of sight. The effect is that when clicking on a single point, the point position is used, while when clicking on a dense region, the center of the region is used. This re-centering navigation works very well in practice for navigating to regions of interest in 3-d point clouds, though again the UI may not be obvious to all users.

Visual feedback is given as quickly as possible in all these cases. Typically the screen is updated at better than 10 frames per second up to a million or so rows for one of the standard scatter plots; for larger datasets or more complex plots the response may be more sluggish. To improve user experience, an adaptive "sketch" mode is in effect by default. If refreshing a frame takes more than a certain threshold time (0.25 s), fast intermediate plots based on a subsample of the data are drawn while a navigation action is in progress, the subsampling fraction being chosen depending on how long the full plot appears to take. When the user has stopped dragging/zooming and enough time has elapsed for the full plot to be drawn, the display is refreshed from the full dataset. In most cases this gives a good compromise between responsive and accurate behaviour, though for certain plot types the intermediate sketched frames can prove confusingly different from the final, correct, frame.

4. Other Plot Types

Although the main focus of the visualisation in TOPCAT is representing point clouds of various sorts, it has other visualisation capabilities too. One important category is depicting weighted or unweighted frequency data, which is a kind of point cloud in one dimension. The most common visualisation for this kind of data is a histogram, but a number of variations are available including smoothed representations with choices of fixed-width or adaptive smoothing kernels with various functional forms, a range of normalisation types, cumulative binning, control over bin width and phase, etc. Some of the options are illustrated in Figure 6.

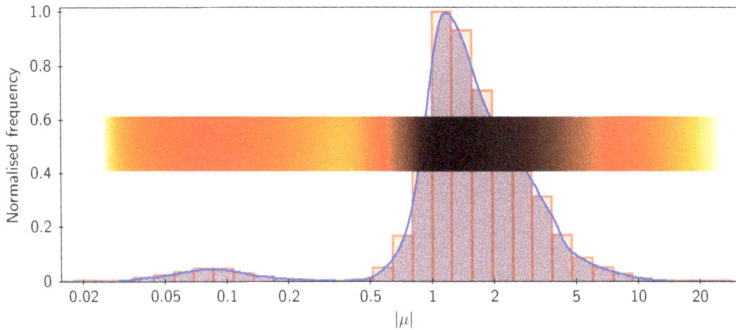

Figure 6. Some histogram-like plots. Shown here are three different representations of the same one-dimensional dataset: a traditional histogram, a Kernel Density Estimate which uses a smoothing kernel to avoid the quantisation implied by histogram binning, and a "densogram" that represents point density using a colour bar.

A number of other more specialised plot types are also offered, for instance analytic function plotting, line plots to trace samples against an independent variable, axes annotated with time coordinates, spectrograms, and some fairly basic data fitting algorithms.

There are also comprehensive facilities for plotting data with positions specified by latitude and longitude on the celestial sphere, including a range of sky projections, sky coordinate system grid annotations, and binning schemes suitable for spherical geometry. These sky plotting capabilities may also be used for data situated on other spheres, such as (an approximation to) the surface of the Earth.

5. Configuration User Interface

TOPCAT currently provides around thirty different plot layer types (fixed shape marker, variable size marker, error ellipse, histogram, ...) with seven different shading modes (flat, auto, transparent, weighted, ...), on half a dozen different plot geometries (2-d and 3-d Cartesian, spherical polar, celestial sphere, ...). Each of these options has typically 5–10 associated configuration variables. All of these options have been introduced to support anticipated, and in many cases actual, requirements for making sense of real user data.

This offers a great deal of flexibility for specifying a visualisation, but equally presents a serious problem of complexity: how does the user, especially the non-expert user, navigate all these options to look at some data? Packaging this flexibility in a comprehensible and usable GUI is perhaps the single most difficult problem in developing an application of this kind.

Addressing this from the user perspective, the application is designed on the principle that the user should always see some kind of reasonable plot with minimal effort.

In practice, this means that if the user hits one of the top-level *Plot* buttons in the application's main window, a reasonable plot is shown, if at all possible. If the user hits the *Plane Plot* button, a scatter plot is displayed of the two first numeric columns of the currently-selected table; an example is shown in Figure 7. The bounds of the plot region are automatically adjusted to display all the points in the selected dataset. The shading mode (*Auto* from Figure 4) is one which works well for both small and large datasets. This default plot is unlikely to be the one the user wants to see, but it is easier to take an existing GUI and modify its default settings, with instant visual feedback at every step, than to be presented with a lot of blank fields to fill in before any result is shown. The controls visible in the same plot window make it obvious how to choose a different table or different columns for the plot. It is somewhat less obvious how to modify the plot by changing marker characteristics, overplotting other datasets, adding error bars or contours, changing axis scaling or annotation etc, but by exploring the various tabs and list items the user can explore and adjust the various options in as much detail as they require. In this way each user can benefit from as much configuration effort as they are willing to expend, rather than being scared off by an initial need to understand the tool in detail. Comprehensive documentation is provided for each feature, accessible from a help button at the top of the window, though it is more of a pleasant surprise than a general expectation that confused users will seek and read this material.

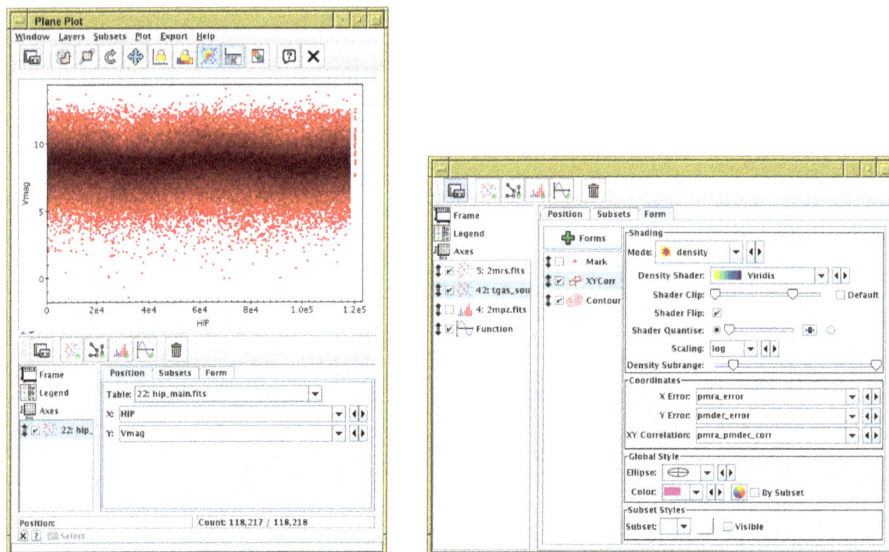

Figure 7. TOPCAT's Plane Plot window. The left hand panel is what appears as soon as the user opens the window; some default plot is displayed. It is easy to see how to change the data to be plotted. The right hand panel shows the control panel from the bottom of the plot window (here it has been expanded and floated out into its own window), as configured for controlling a much more complicated plot. Each of the controls towards the right can be adjusted interactively with instant effects on the displayed plot. The various tabs and list items provide more configuration options for other aspects of the plot.

This approach provides a GUI that is usable by novices to produce basic plots, but which can also be exploited by experts for very detailed control. However, the question of providing a usable interface for configuring complex plots is not a solved problem in TOPCAT; it also becomes more acute as additional plot types and options are added. While many users do manage to use the various combined features to perform sophisticated visualisations it is probably the case that only a minority understand the full range of available capabilities.

6. Alternative Interfaces

This paper mainly discusses the GUI application TOPCAT. However, it is possible to access the visualisation functionality in a number of ways from outside of that application, and many of the same remarks apply.

Alongside TOPCAT is a suite of command-line tools by the name of STILTS (STIL Tool Set) [24], which provides scriptable access to most of the functionality available from TOPCAT. Both are based on the table access library STIL (Starlink Tables Infrastructure Library), and all of these items are developed and maintained, with some other partially related software, in a single group of packages collectively known as Starjava. These packages are available at the from the URLs http://www.starlink.ac.uk/stilts/ and http://www.starlink.ac.uk/stil/.

STILTS provides commands that can generate all the visualisations available from TOPCAT's GUI, and in fact most of the figures in this paper were generated using STILTS, since its scriptable nature makes it more suitable for careful preparation of published figures than TOPCAT's point and click operation.

The output of STILTS, like that of TOPCAT, can be to various bitmapped or vector graphics file formats (including PNG, GIF, PDF and PostScript) or to an interactive window on screen that allows the same mouse-controlled navigation actions as TOPCAT.

The classes used for visualisation in both cases form a library known informally by the name *plot2*. These classes have not so far been formally packaged as a separate product, but are contained within the STILTS jar file and can be used independently of either the TOPCAT or STILTS applications, to provide high performance static or interactive visualisation within third party Java applications. Depending on what functionality is required, the code required for this is licensed under the LGPL or GPL. Some more background on this possibility is described in [25].

As explained in Section 7.1, plotting from STILTS is actually more scalable than that from TOPCAT. While there is no fixed limit on the size of tables loaded into TOPCAT, it is not really intended for use with tables more than a few tens of millions of rows; interactive use is generally sluggish on such data, and in some cases memory usage can be high. STILTS visualisation on the other hand is in most cases able to stream data within a fixed (and quite small) memory footprint, so it is possible to generate static plots of arbitrarily large data sets quite easily. Figure 8 shows an all-sky density plot generated from a 2 billion row table in about 30 min.

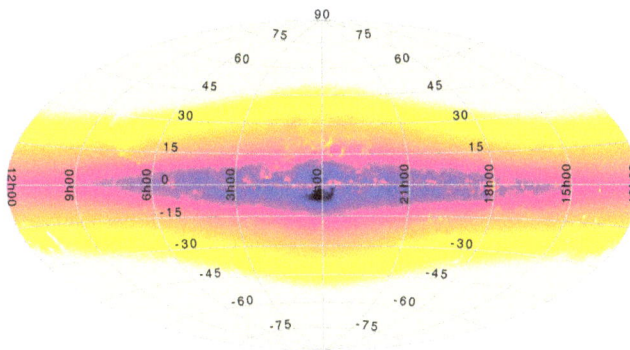

Figure 8. Plot of a large table: A map representing simulated density on the sky of stars in the Milky Way [26]. This figure (originally from [25]) was generated from a 2 billion row table in about 30 min on a normal desktop computer.

7. Implementation Notes

The user experience-driven requirements of usability with large datasets, fast navigation, flexible configurability and instant visual feedback place considerable constraints on the implementation. In this section we outline some of the strategies in place to deliver these features.

7.1. Scalability

The first requirement, built into the whole of the TOPCAT application and its underlying libraries, is to be able to process tables with large row counts N_{row} in fixed memory if at all possible. This means that generating a plot must not, unlike many off-the-shelf Java plotting libraries, allocate an object or other storage for each input row or plotted point. Instead, where possible, the plotting system uses data structures that scale with the number of pixels rather than the number of rows. This rule is violated in some cases, for instance if the number of rows can be determined to be small, or in a few cases where it is unavoidable such as z-stacking points for 3-d plots, but most of the common plot types obey it.

To support this model, the various plot layer implementations work with an abstraction of the data that simply iterates over each row, returning typed values (typically double precision scalars, though in some cases boolean, string or array values) for the required coordinates at each step. For each iteration they can then either paint directly to the graphics system or populate some limited-size data structure that will be used for graphics operations later in the rendering process.

The harnessing code can then decide how to deliver the iteration over the data values from the original table. The TOPCAT application reads the relevant values into in-memory primitive arrays once it is known what coordinates are required, ensuring maximal subsequent access speed, since actually extracting these values from the underlying loaded tables may be somewhat time-consuming. This does entail some N_{row}-scale memory usage, though usually at an acceptable level, e.g., only 16 bytes per row for a 2-d scatter plot. However the STILTS application writing to a static image file simply iterates over the rows of the underlying table without intermediate caching, thus requiring little additional storage. These different strategies have different benefits: TOPCAT, having prepared the data for a given visualisation at the expense of some memory usage can plot subsequent frames using the same data (e.g., the results of user navigation) as quickly as possible. STILTS (in its default configuration) may take longer for each frame of an animation sequence but can process arbitrarily large tables in a small memory footprint. In the common case where STILTS is painting to a single static frame, the benefits for subsequent replots would not be useful.

7.2. Responsive User Interface

As discussed in Section 5, the visualisation user interface contains many controls, each of which may change the appearance of some or all parts of the plot; the axes or one or more of the data layers contributing to a particular visualisation. A responsive user interface requires that whenever one of these controls is adjusted, the display is updated accordingly. However, regenerating the whole plot from scratch may be expensive for a large or complex plot, so this should be avoided where possible. In some cases (e.g., changing the plot colour of a currently hidden dataset) perhaps no replot is required at all. In other cases (e.g., axis annotations only are changed) some parts of the plot must be redrawn but the results of previous computations could be re-used. Or perhaps (e.g., coordinate data is replaced by a different table column) the whole thing needs to be redrawn. In general various parts of the plotting computations can be cached, and a great deal of effort goes into working out, whenever the controls are adjusted, which computations from the previous plot can be re-used.

The way this works is that every time a user control is adjusted, it triggers a *replot* action. This calculates a set of *label* objects for each of several plot characteristics such as the currently visible region of parameter space, the set of table data required for each plot layer, the per-layer configuration style options etc. Together, this set of labels completely characterises the plot. These label objects are small and cheap to produce, so multiple replot actions per second (for instance, as a user drags a slider) are in themselves easy to service, and this step can be done on Swing's Event Dispatch Thread (EDT) without impairing the responsiveness of the overall application GUI. If this is the first plot to be produced in a given window, these labels are stored for later reference, and then fed back to the plot components that generated them, providing instructions to draw the plot which is then calculated and displayed. However, on subsequent plots in the same window, the plotting system performs various comparisons of the labels for the new frame with those that specified the previous frame. Specifically, Java's `Object.equals` method is used for label comparison, so these label objects must be written with carefully implemented equality semantics. If the set is exactly equivalent to that for the previous frame, no replot needs to be done. In general, some recalculation or redrawing will be needed, but less than would be required for regenerating the whole plot from scratch. This computation prepares a new plot bitmap on a worker thread, which it passes on completion back to the EDT for display in the plot window. A queue of replot requests is maintained, and if a new one comes in while another is waiting to be performed, the older one is discarded. Whether requests in progress are aborted depends on some logic to decide whether the new request looks like a minor (navigation) or major (plot new data) configuration update.

The details of the selective caching that underlies this are quite complicated, but an example may be illustrative. A plot layer performs its plotting in two stages: in the *planning* stage it is given the opportunity to produce a *plan* object that may represent the results of expensive computations, and in the *painting* stage it is given back the same plan to use in order to perform actual graphical output. Management, including optional caching, of the plan objects is done by the plotting application and not the layer itself. If the layer can determine that a plan equivalent to the one it needs to produce for the currently requested plot is already available, because the management level has cached it from an earlier invocation, it can skip the planning stage and use the previously calculated plan for the painting stage instead. Hence: a density map layer might generate a plan containing a grid of bins populated by the expensive work of iterating over the table rows, and then in the painting stage simply transfer this grid to a bitmap using some configuration-determined colour map. If a subsequent invocation uses the same grid data but a different colour map, because the user has adjusted the colour map controls but not moved the grid, it can repaint the image (cheap) without requiring a rescan of the table data (expensive). The result is that the user can adjust colour map parameters with instant visual feedback even for a large dataset.

7.3. Configuration Option Management

As discussed above, many configuration options are available to control the data layers that combine to form a given visualisation, along with the details of the axis representation and annotation, legend display, plot dimensions, font selection etc.

In order to reduce the implementation complexity associated with these hundreds of options, each one is represented by a standard object known internally as a *ConfigKey*. Each of these keys can supply user-directed metadata (name, description), value type (which may just be a number or some more complex type like a colour map or marker shape), a sensible default value, a GUI component for specifying values, and methods for mapping between typed values and string representations. It is important that the default values of all keys taken together to specify a given plot will combine to give some reasonable default plot as discussed in Section 5.

Different components of the plotting system make use of these keys to build the user interface and gather configuration information without hard-coded knowledge of each plot and layer type. A harnessing application needs to establish which plot type and layers are in use, interrogate them for their ConfigKeys, and then acquire values for each key that can be fed back to the plot components to generate the plot. In TOPCAT's case it sets up plot window controls by stacking the relevant GUI components ready for adjustment by the user, while STILTS interrogates the list of name-value pairs supplied on the command line. Other front ends based on name-value pairs have also been implemented, including a cgi-bin interface for HTTP operation and a Jython front-end to STILTS; both are available and documented within the STILTS distribution itself, as the STILTS `server` task and the JyStilts application respectively.

The user documentation for each plot type can also be generated programmatically at documentation build time by interrogating each key for its user metadata; around 130 of the 400 pages in the PDF version of the STILTS user document are auto-generated from ConfigKey objects in this way.

The result of organising the configuration options in this uniform way is that new configuration options can be introduced easily by making only localised changes to plot type or layer type code; no corresponding updates to the UI code or hand-written documentation are required.

8. Use Beyond Astronomy

TOPCAT has been developed for astronomers, with the support of funding agencies whose responsibility is to the astronomy community. The large majority of its use to date has been within astronomy, mostly for use with source catalogues. It is also applied to other types of astronomical table such as time series and event lists, and within some related but distinct fields such as planetary and solar system science.

However, though it has much functionality that is specific to astronomy (understanding of sky coordinate systems, data access using astronomy file formats and Virtual Observatory protocols, table join techniques appropriate for the celestial sphere) many of its capabilities are suitable for any kind of tabular data, and some adventurous groups in other disciplines are also making enthusiastic use of it.

One example is the group of P. Pognonec from Université Nice, who use it for work investigating cellular events such as proliferation, mitosis and cell death. They report TOPCAT as their preferred option for visualising with dotplots and histograms the large tables (ten million cells with 50–100 parameters) of data produced by image analysis of high throughput microscopy acquisitions. An example is shown in Figure 9. Another is operational and laboratory work by H. Rydberg in the company Sustainable Waste and Water, City of Gothenburg in Sweden, where it is used especially for interactive analysis of long time series data concerning water quality; see Figure 10. In this case the capability to ingest large raw datasets without prior aggregation and navigate interactively makes it possible to clean data and identify trends over a wide range of timescales. The author has also had other informal reports of TOPCAT's sporadic use in bioinformatics, finance, urban transport planning and flight testing.

While the current funding arrangements do not prioritise support outside of astronomy, the author is very interested to hear of potential or actual uses in other domains, and willing to supply modest support, for either casual use or adapting the application or underlying libraries to other requirements. One missing feature that should be noted when considering applying the software more widely is its weak support for categorical data, which is not very common in TOPCAT's core use cases. However, enhancements in this area are possible in the future.

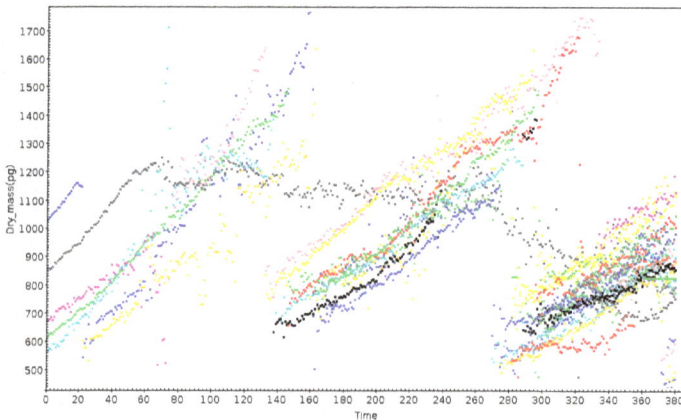

Figure 9. Growth and division of cells over time. Cell mass, represented by one colour for each distinct cell, increases until mitosis, when it is replaced by two daughter cells. In this plot the grey cell has become "blocked", losing mass slowly instead of dividing. *Credit:* P. Pognonec, Université Nice.

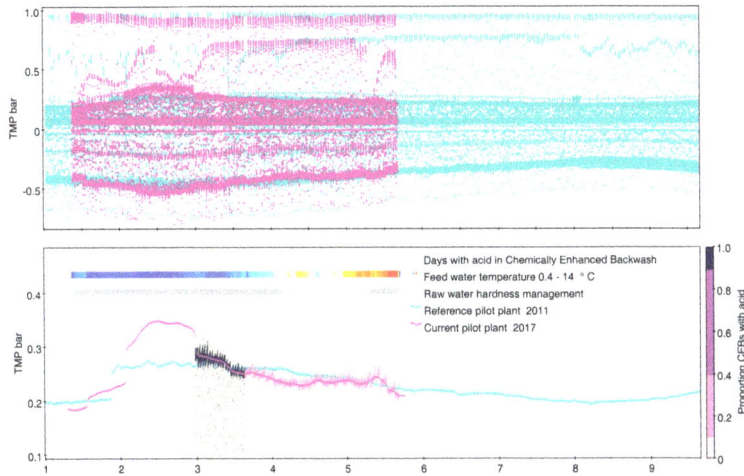

Figure 10. Time series (decimal month January–September) of Trans Membrane Pressure over two different ultrafiltration pilot plants in Gothenburg Sweden (7.7M rows). (**Top**) Two series of correct but noisy raw data, obviously not suitable for mean aggregations; (**Bottom**) Exactly the same data as top display, clarified by use of subsets, transparency, quantile smoothers, marking by auxiliary *Y* axis, histogram by time as event marking, and densograms for additional quantitative variable as well as operational information. *Credit:* H. Rydberg, Sustainable Waste and Water, City of Gothenburg.

9. Software Availability

TOPCAT is written in pure Java, and distributed as a single jar file depending only on the Java Standard Edition (Java SE), currently version 6 or later. The wide availability and excellent portability and backward compatibility characteristics of the Java platform mean that it can therefore be installed and run very easily on all widely used desktop and laptop computers. The jar file, as well as a MacOS DMG file, can be downloaded from the project web site http://www.starlink.ac.uk/topcat/. Other information including comprehensive tutorial and reference documentation, full version history, pointers to mailing lists etc can be found in the same place. The package has also recently been made available as part of the Debian Astro suite [27].

The software is available free of charge under the GNU Public Licence, and the source code is currently hosted on github (https://github.com/Starlink/starjava/).

10. Conclusions

TOPCAT is a GUI application for manipulating tables, that amongst other capabilities provides sophisticated visualisation capabilities for tabular data. It is a traditional desktop application, requiring neither exotic hardware nor server support. It has been developed within the context of astronomy and is widely used in that field, but is suitable, along with its command-line counterpart STILTS and underlying Java libraries, for visualising many other kinds of tabular data. The focus is on highly configurable interactive plots of both small and large (multi-million-row) tables, offering many variations on the representation of point clouds in one, two or three dimensions, with the aim of revealing expected and unexpected relationships at multiple scales in large and high-dimensional datasets.

Informatics **2017**, *4*, 18

Acknowledgments: Development of TOPCAT's current visualisation capabilities has been supported by a number of grants from the UK's Science and Technology Facilities Council. The features described here have benefitted greatly from advice, comments and feedback from its active user community. Special thanks to Henrik Rydberg and Philippe Pognonec for their input on use in non-astronomical contexts. The author also thanks the anonymous referees whose constructive comments have improved the paper.

Conflicts of Interest: The author declares no conflict of interest. The funding sponsors had no role in the design of the study; in the collection, analyses, or interpretation of data; in the writing of the manuscript, and in the decision to publish the results.

References

1. Messier, C. *Catalogue des Nébuleuses & des amas d'Étoiles (Catalog of Nebulae and Star Clusters)*; Technical Report; Memoirs of the Royal Academy of Sciences for 1771: Paris, France, 1781. (In French)
2. Stoughton, C.; Lupton, R.H.; Bernardi, M.; Blanton, M.R.; Burles, S.; Castander, F.J.; Connolly, A.J.; Eisenstein, D.J.; Frieman, J.A.; Hennessy, G.S.; et al. Sloan Digital Sky Survey: Early Data Release. *Astron. J.* **2002**, *123*, 485–548.
3. Brown, A.G.A.; Vallenari, A.; Prusti, T.; de Bruijne, J.H.J.; Mignard, F.; Drimmel, R.; Babusiaux, C.; Bailer-Jones, C.A.L.; Bastian, U.; Elteren, A.K.; et al. Gaia Data Release 1. Summary of the astrometric, photometric, and survey properties. *Astron. Astrophys.* **2016**, *595*, A2.
4. Ochsenbein, F.; Taylor, M.; Williams, R.; Davenhall, C.; Demleitner, M.; Durand, D.; Fernique, P.; Giaretta, D.; Hanisch, R.; McGlynn, T.; et al. VOTable Format Definition Version 1.3. IVOA Recommendation 20 September 2013. *arXiv* **2013**, arXiv:1110.0524.
5. Hanisch, R.J.; Farris, A.; Greisen, E.W.; Pence, W.D.; Schlesinger, B.M.; Teuben, P.J.; Thompson, R.W.; Warnock, A., III. Definition of the Flexible Image Transport System (FITS). *Astron. Astrophys.* **2001**, *376*, 359–380.
6. Arviset, C.; Gaudet, S.; IVOA Technical Coordination Group. IVOA Architecture Version 1.0. IVOA Note 23 November 2010. *arXiv* **2011** arXiv:1106.0291.
7. Dowler, P.; Rixon, G.; Tody, D. Table Access Protocol Version 1.0. IVOA Recommendation 27 March 2010. *arXiv* **2010**, arXiv:astro-ph.IM/1110.0497.
8. Plante, R.; Williams, R.; Hanisch, R.; Szalay, A. Simple Cone Search Version 1.03. IVOA Recommendation 22 February 2008. *arXiv* **2008**, arXiv:astro-ph.IM/1110.0498.
9. Taylor, M.B. TOPCAT & STIL: Starlink Table/VOTable Processing Software. In *Astronomical Society of the Pacific Conference Series, Proceedings of the Astronomical Data Analysis Software and Systems XIV, Pasadena, CA, USA, 24–27 October 2004*; Shopbell, P., Britton, M., Ebert, R., Eds.; Astronomical Society of the Pacific: San Francisco, CA, USA, 2005; Volume 347, p. 29.
10. Taylor, M.B.; Page, C.G. Column-Oriented Table Access Using STIL: Fast Analysis of Very Large Tables. In *Astronomical Society of the Pacific Conference Series, Proceedings of the Astronomical Data Analysis Software and Systems XVII, London, UK, 23–26 September 2007*; Argyle, R.W., Bunclark, P.S., Lewis, J.R., Eds.; Astronomical Society of the Pacific: San Francisco, CA, USA, 2008; Volume 394, p. 422.
11. Moitinho, A.; Krone-Martins, A.; Savietto, H.; Barros, M.; Barata, C.; Falcão, A.J.; Fernandes, T.; Alves, J.; Gomes, M.; Bakker, J.; et al. Gaia Data Release 1: The archive visualisation service. *Astron. Astrophys.* **2017**, in press.
12. Boch, T.; Fernique, P. Aladin Lite: Embed your Sky in the Browser. In *Astronomical Society of the Pacific Conference Series, Proceedings of the Astronomical Data Analysis Software and Systems XXIII, Waikoloa Beach Marriott, HI, USA, 29 September–3 October 2013*; Manset, N., Forshay, P., Eds.; Astronomical Society of the Pacific: San Francisco, CA, USA, 2014; Volume 485, p. 277.
13. Carbon, D.F.; Henze, C.; Nelson, B.C. Exploring the SDSS Data Set with Linked Scatter Plots. I. EMP, CEMP, and CV Stars. *Astrophys. J. Suppl.* **2017**, *228*, 19.
14. Springel, V.; White, S.D.M.; Jenkins, A.; Frenk, C.S.; Yoshida, N.; Gao, L.; Navarro, J.; Thacker, R.; Croton, D.; Helly, J.; et al. Simulations of the formation, evolution and clustering of galaxies and quasars. *Nature* **2005**, *435*, 629–636.
15. Tukey, J.W. *Exploratory Data Analysis*; Addison-Wesley: Boston, MA, USA, 1977.
16. Goodman, A.A. Principles of high-dimensional data visualization in astronomy. *Astron. Nachr.* **2012**, *333*, 505–514.

17. Altmann, M.; Roeser, S.; Demleitner, M.; Bastian, U.; Schilbach, E. Hot Stuff for One Year (HSOY). A 583 million star proper motion catalogue derived from Gaia DR1 and PPMXL. *Astron. Astrophys.* **2017**, *600*, L4.

18. Bonnarel, F.; Fernique, P.; Bienaymé, O.; Egret, D.; Genova, F.; Louys, M.; Ochsenbein, F.; Wenger, M.; Bartlett, J.G. The ALADIN interactive sky atlas. A reference tool for identification of astronomical sources. *Astron. Astrophys. Suppl.* **2000**, *143*, 33–40.

19. Taylor, M.B.; Boch, T.; Taylor, J. SAMP, the Simple Application Messaging Protocol: Letting applications talk to each other. *Astron. Comput.* **2015**, *11*, 81–90.

20. Joye, W.A.; Mandel, E. New Features of SAOImage DS9. In *Astronomical Society of the Pacific Conference Series, Proceedings of the Astronomical Data Analysis Software and Systems XII, Baltimore, MD, USA, 13–16 October 2002*; Payne, H.E., Jedrzejewski, R.I., Hook, R.N., Eds.; Astronomical Society of the Pacific: San Francisco, CA, USA, 2003; Volume 295, p. 489.

21. Robitaille, T.P.; Tollerud, E.J.; Greenfield, P.; Droettboom, M.; Bray, E.; Aldcroft, T.; Davis, M.; Ginsburg, A.; Price-Whelan, A.M.; Kerzendorf, W.E.; et al. Astropy: A community Python package for Astronomy. *Astron. Astrophys.* **2013**, *558*, A33.

22. Bellini, A.; Piotto, G.; Bedin, L.R.; Anderson, J.; Platais, I.; Momany, Y.; Moretti, A.; Milone, A.P.; Ortolani, S. Ground-based CCD astrometry with wide field imagers. III. WFI@2.2m proper-motion catalog of the globular cluster ω Centauri. *Astron. Astrophys.* **2009**, *493*, 959–978.

23. Taylor, M.B. Visualizing Large Datasets in TOPCAT v4. In *Astronomical Society of the Pacific Conference Series, Proceedings of the Astronomical Data Analysis Software and Systems XXIII, Waikoloa Beach Marriott, HI, USA, 29 September–3 October 2013*; Manset, N., Forshay, P., Eds.; Astronomical Society of the Pacific: San Francisco, CA, USA, 2014; Volume 485, p. 257.

24. Taylor, M.B. STILTS—A Package for Command-Line Processing of Tabular Data. In *Astronomical Society of the Pacific Conference Series, Proceedings of the Astronomical Data Analysis Software and Systems XV, San Lorenzo de El Escorial, Spain, 2–5 October 2005*; Gabriel, C., Arviset, C., Ponz, D., Enrique, S., Eds.; Astronomical Society of the Pacific: San Francisco, CA, USA, 2006; Volume 351, p. 666.

25. Taylor, M.B. External Use of TOPCAT's Plotting Library. In *Astronomical Society of the Pacific Conference Series, Proceedings of the Astronomical Data Analysis Software an Systems XXIV (ADASS XXIV), Calgary, AB, Canada, 5–9 October 2014*; Taylor, A.R., Rosolowsky, E., Eds.; Astronomical Society of the Pacific: San Francisco, CA, USA, 2015; Volume 495, p. 177.

26. Robin, A.C.; Luri, X.; Reylé, C.; Isasi, Y.; Grux, E.; Blanco-Cuaresma, S.; Arenou, F.; Babusiaux, C.; Belcheva, M.; Drimmel, R.; et al. Gaia Universe model snapshot. A statistical analysis of the expected contents of the Gaia catalogue. *Astron. Astrophys.* **2012**, *543*, A100.

27. Streicher, O. Debian Astro: An open computing platform for astronomy. *arXiv* **2016**, arXiv:astro-ph.IM/1611.07203.

Article

Web-Scale Multidimensional Visualization of Big Spatial Data to Support Earth Sciences—A Case Study with Visualizing Climate Simulation Data

Sizhe Wang [1,2], **Wenwen Li** [1,*] **and Feng Wang** [1]

[1]   School of Geographical Sciences and Urban Planning, Arizona State University, Tempe, AZ 85287-5302, USA;
       wsizhe@asu.edu (S.W.); fwang80@asu.edu (F.W.)
[2]   School of Computing, Informatics and Decision Systems Engineering, Arizona State University,
       Tempe, AZ 85281, USA
*   Correspondence: wenwen@asu.edu; Tel.: +1-480-727-5987

Received: 16 March 2017; Accepted: 24 June 2017; Published: 26 June 2017

Abstract: The world is undergoing rapid changes in its climate, environment, and ecosystems due to increasing population growth, urbanization, and industrialization. Numerical simulation is becoming an important vehicle to enhance the understanding of these changes and their impacts, with regional and global simulation models producing vast amounts of data. Comprehending these multidimensional data and fostering collaborative scientific discovery requires the development of new visualization techniques. In this paper, we present a cyberinfrastructure solution—PolarGlobe—that enables comprehensive analysis and collaboration. PolarGlobe is implemented upon an emerging web graphics library, WebGL, and an open source virtual globe system Cesium, which has the ability to map spatial data onto a virtual Earth. We have also integrated volume rendering techniques, value and spatial filters, and vertical profile visualization to improve rendered images and support a comprehensive exploration of multi-dimensional spatial data. In this study, the climate simulation dataset produced by the extended polar version of the well-known Weather Research and Forecasting Model (WRF) is used to test the proposed techniques. PolarGlobe is also easily extendable to enable data visualization for other Earth Science domains, such as oceanography, weather, or geology.

Keywords: virtual globe; octree; vertical profile; big data; scientific visualization

---

1. Introduction

The world is undergoing significant environmental and global climate change due to increasing population growth, urbanization, and industrialization [1–4]. These changes [5–7] are exemplified in the Earth's polar regions, as evidenced by melting sea ice [8] and glacier retreat [9], which significantly affect the living environment of wildlife and biodiversity in these areas. To better understand these climate phenomena and their driving mechanics, there exists an urgent need for new data, techniques, and tools to support scientific studies and the development of effective strategies to mitigate their negative influences [10].

Climate simulation has been considered a critically important means to address the aforementioned research challenges [11]. Global or regional climate models, such as WRF (Weather Research and Forecasting), are often used by the climate modeling community to unveil the historical climate trajectory and make projections for future changes. Through long-duration computations, these simulation models often generate very large climate data [12]. It is estimated that worldwide climate simulation data will reach hundreds of exabytes by 2020 [13]. Besides falling into the category of "big data" due to its size, climate data is multidimensional in nature. In other words, the time-series data not only spread across a geographic area on the Earth's surface (horizontal dimension), but they also occupy different altitudes with varying pressure levels (vertical dimensions).

Scientific visualization is considered an effective vehicle for studying such complex, big volume, and multiple dimension data [14]. By providing visual representations and analytics, visualization has the capability to validate hypothesis, uncover hidden patterns, and identify driving factors of various climate and atmospheric phenomena [15]. Nonetheless, the scientific visualization community still faces challenges in efficient handling of big data, the complex projection between the viewport coordinate system and the raw geospatial dataset, and finding innovative ways to present the voluminous data in order to reveal hidden knowledge. With the widespread adoption of Web technology, there is also an urgent demand for a Web-based visualization platform to allow web-scale access, visualization, and analysis of spatial dataset.

This paper introduces our PolarGlobe solution, a Web-based virtual globe platform that supports multi-faceted visualization and analysis of multi-dimensional scientific data. Built upon the popular Cesium 3D globe system, the PolarGlobe tool has the advantage of being seamlessly integrative with Web browsers, eliminating the need to install or configure any plug-ins before data viewing. In addition, an emerging graphics language (WebGL) is utilized to operate the GPU (Graphics Processing Unit) and develop functions for data rendering. The remainder of this paper is organized as follows: Section 2 reviews relevant works in the literature; Section 3 introduces the visual analytical techniques being applied to the PolarGlobe system; Section 4 demonstrates the PolarGlobe GUI (graphic user interface); Section 5 describes a number of experiments to test system performance; and Section 6 concludes this work and discusses future research directions.

2. Literature Review

In this section, we organize the review of previous works from two perspectives: (1) previously established visualization platforms and (2) key techniques employed to support such visualization.

2.1. Popular Visualization Platforms for Climate Research

Visualization has a long tradition in supporting climate research [16]. Standard 2D presentation techniques such as time/bar charts, 2D maps, and scatterplots are most frequently used in analyzing climate data [17]. Popular visualization tools, such as UV-CDAT (Ultrascale Visualization Climate Data Analysis Tool) [18], provide great support for data regridding, exploratory data analysis, and parallel processing of memory-greedy operations [19]. However, these tools suffer great limitations in the context of cyberinfrastructure and big data science [20]. For instance, users need to download and setup the software in order to get access to the tools. To visualize a scene, users also need to write corresponding (python) code, which often requires a long learning curve. As climate simulation data is becoming multi-dimensional, the lack of support in multi-dimensional data analysis poses many challenges for visualizing these data, especially those with time-series stamps. Moreover, in most tools, data is visualized as an individual piece without being integrated with terrain and morphology data to enhance understanding.

Overcoming these limitations has become a significant research thread of virtual globe visualization. Inspired by the vision of "Digital Earth" (by former US vice president Al Gore) [21], a number of virtual globe tools have been developed to digitally depict our living planet. Popular ones include Google Earth [22], NASA WorldWind (National Aeronautics and Space Administration; [23], and Microsoft Virtual Earth [24]. Using these tools, climate data visualization can be integrated into the actual terrain and Earth scene. Sun et al. (2012) developed a geovisual analytical system to support the visualization of climate model output using Google Earth [25]. Varun et al. [26] developed iGlobe, an interactive visualization system that integrates remote sensing data, climate data, and other environmental data to understand weather and climate change impacts. Helbig et al. [27] presents a workflow for 3D visualization of atmospheric data in a virtual reality environment. HurricaneVis [28] is a desktop visualization platform that focuses on scalar data from numerical weather model simulations of tropical cyclones. Leveraging the power of graphics cards, multivariate real-time 4D visualization can also be achieved [29]. These works demonstrate a great advantage in data visualization over traditional approaches that rely solely on 2D maps and scatter plots. However, most of these applications are desktop-based or require pre-installation and configuration, limiting their widespread use and adoption by Internet users.

As cyberinfrastructure evolves, research that develops web-based visual analytical tools has gradually increased. For instance, Fetchclimate [30] and the USGS (United States Geological Survey) National Climate Change Viewer (UCCV) [31] provide solutions for environment information retrieval and mapping. Similar online visualization applications have also been applied to other geoscience disciplines such as hydrology [32–34], oceanography [35], and polar [20,29], etc. Open source packages, such as Cesium [36] or NASA's new WebWorldWind [37], are also exploited to construct web-based environmental applications [38,39]. Though these existing studies provide a satisfying solution to 2D spatiotemporal data visualization, they have very limited capability at visualizing high-dimensional spatiotemporal climate data.

2.2. Key Techniques in Multidimensional Visualization of Spatial Data

Spatiotemporal multidimensional data visualization is a hotspot in the field of scientific visualization. The existing work varies from organizing and visualizing time-varying big data to applying multiple visual analytic techniques for planning, predicting, and decision-making. There are two key issues in developing an efficient web-based visual analytic tool.

The first is efficient management and transmission of big data from the server to client end. With the popularity of 'big data' in both academia and industry, increasing research focuses on managing big data for scientific visualization [40,41]. Others address big data usage on visualization in emerging environments [42,43]. In climate study, Li and Wang [39] proposed a video encoding and compression technique to efficiently organize and transmit time-varying big data over successive timestamps.

The second is exploiting visualization techniques to provide real-time realistic visualization. Wong et al. [44] assembled multiple information visualization and scientific visualization techniques to explore large-scale climate data captured after a natural phenomenon. Li et al. [45] implemented a volume rendering technique for visualizing large-scale geosciences phenomena, i.e., dust storms. An octree data structure, in combination with a view-dependent LOD (Level of Detail) strategy, is used to index 3D spatial data to improve rendering efficiency. Liang et al. [46] further improved this method to introduce a volumetric ray-casting algorithm to avoid loss of accuracy. The algorithm avoids over- or under-sampling when converting geospatial data from a spherical coordinate system to a Cartesian coordinate system for visualization. To boost rendering performance, GPU is always employed for parallel rendering [47,48]. In order to present volumetric data from multiple facets, these techniques need to be extended to include more novel visual effects for a comprehensive visual exploration.

In the next section we describe, in detail, our proposed techniques, including an enhanced octree model to support efficient visualization and analysis of climate data in a cyberinfrastructure environment.

3. Methodology

3.1. Three-Dimensional Data Volume Rendering

The goal of this section is to explain the creation of a panoramic view of climate variables on and above the Earth's surface. To accommodate the multivariate characteristics of climate data (horizontal and vertical), we developed a volume rendering technique to present the variances in the north polar region and its upper air [49]. To take full advantage of highly detailed datasets such as reanalysis data, point clouds visualization is adopted. To support this visualization strategy, a value filter and a region filter were also developed to provide more perspectives and enable a better understanding of various climate phenomena. Challenges in efficiently transferring large datasets between client and server, and rendering big data in the client browser, were also addressed. For instance, there are almost 4 million points at a single timestamp in the climate simulation data we adopted [50]. To overcome these obstacles, an enhanced octree-based Level of Detail (LOD) method is utilized in the point cloud visualization.

The LOD strategy is adopted, because it is a widely-used technique in multi-dimensional visualization that decreases total data volume while boosting rendering and data transfer speed at the same time [51]. Based on the principle that the further the data is observed, the fewer details will need to be shown, the loading and rendering of 3D volume data can be greatly improved. When the octree-based LOD is applied to 3D volume visualization, the data size is reduced by a power of eight (2^3). Because our goal is to realize Web-scale visualization for multi-dimensional data such that people from any place of the world can access the system, the octree is implemented on both the backend (Web server side) and frontend (client browser side). The server side is responsible for preparing the LOD data to respond to clients' requests. The frontend deals with acquiring the proper LOD data and rendering them in the browser using WebGL (Web Graphics Language). We describe the implementation of these two parts in detail below.

3.1.1. Data Preparation at the Server Side

Though the octree can be built from original data and kept in the memory of the server for responding to requests from clients from time to time, this is not only time consuming, which will keep the user waiting longer to acquire the LOD data, but it is also a great burden on the server's memory, especially when the data volume is big or multiple data sources are in use. To overcome these limitations, we adopted a strategy that pre-processes the original data and produces different LOD data as files. In order to prepare the LOD data, the first task is to decompose the original data volume (which consists of numerous points distributed in a 3D grid) into multiple special cubes. These cubes are considered special because their length, measured by the number of points along one side, should be a power of 2. This length will be evenly cut along each side of the cube by 2. Such decomposition ensures an evenly increasing ratio of data size along with the increasing LOD. The iterative decomposition process continues until every cube contains only one point—the highest LOD. The following equation can be used to determine the length L of the initial cube:

$$L = 2^{(floor(log_2^M)+1)} \tag{1}$$

where M is the minimum size amongst the length, width, and height of the original data volume, and *floor()* is a function to receive the integer part of a floating number, given any positive input.

In Equation (1), L is the minimum power of a power of 2 number that is no smaller than M. After identifying the size of the initial cube, the decomposition process starts by dividing the cube into eight equally sized sub-cubes. The sub-cubes generated after the first decomposition are called the roots of the octrees. For convenience, let us state that the roots are at level 0. Because each cube is represented by one data value, a generalization process should be invoked to derive this value. A typical approach is to average the values on all the grid points falling in the cube. This process requires extra storage in order to store the generalized value for the cubes that share the same root.

To address this data challenge, we propose a new octree multi-level storage strategy, or differential storage, to reduce data redundancy across octree layers. The idea comes from differential backup that saves only the changes made in a computer system since the last backup. For our purposes, each higher layer of the octree only stores the differential or incremental dataset(s) from its precedent layers. These values are a sample from the original data points rather than averaged from them. Figure 1 demonstrates a design of the proposed octree with differential storage support. This data structure saves substantial storage space. For example, we estimate saving approximately 14% of the storage space for a three-layer octree using this data structure.

Once the octree model is established, all cubes and sub-cubes are indexed. The root contains eight initial cubes with index from 0 to 7. As the decomposition continues, every cube is split into eight sub-cubes of the same size until the LOD reaches the highest level n. The naming of each sub-cube follows the patterns of (1) the name of its parent cube+; and (2) its index (from 0 to 7). Hence, the sub-cubes with the same root have the same prefix in its name. The strategy used to record data is what distinguishes levels 1 and higher from level 0. At level n, all the generalized data derived from the same roots are written in a single file whose filename contains the level number and the index of the root recorded at level $n-1$. Those sub-cubes with no grid points are ignored. Using this strategy, the web server can rapidly locate the LOD data with a specified region and level. It repeats the process applied to level 1 until level n is reached. It then splits every cube produced in level $n-1$ to get 8 n sub-cubes where no more than one point in a cube exists. The generalized data values, as well as some necessary metadata (such as the range of value variation and the maximum LOD), are recorded in a single file for data initialization at the client side.

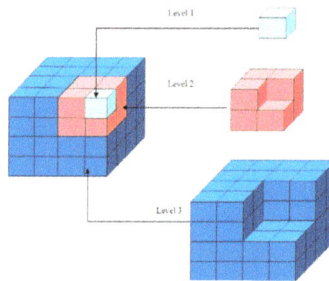

Figure 1. Differential-storage-enhanced octree model.

3.1.2. Data Rendering on the Client Side

Thus far, LOD data is prepared at all levels. The server will return the data by specifying the root index and the LOD as parameters in the request. The client side is now able to access any data it needs. The workflow at the client side starts with retrieving the data at level 0 from the server for initialization. Using the indices recorded in level 0 data, the whole data volume is split up into multiple cubes (differed by their indices) for rendering, management, and further operation. If the cubes have been initialized or a user changes the view perspective, the rendered data will be refreshed by updating the LOD information in each cube. Since the location of each cube is known, the distances from the viewport to all cubes are a fixed number at all levels. For a given cube, the data at which LOD should be rendered is decided by the below equation:

$$LOD = \sqrt{\frac{C}{dist}} \tag{2}$$

where *dist* denotes the view distance, and C is a constant value which needs to be adjusted by user experience.

Applying this equation, the data with higher details are loaded as the view distance decreases. Giving the calculated level and its own index, the cube sends a request to the server and retrieves the desired LOD data. Once a request is sent, the level in the unit is marked and temporarily cached in memory for future rendering purposes. This way, repeated requests for the same data are avoided.

Data loading and rendering performance are greatly enhanced with the proposed LOD strategy, especially when the data is observed from a long distance. If the view distance becomes very short or the observer dives inside the data cubes, however, a large portion or even all of the data is loaded at the highest LOD. This may keep the user waiting a long time and can greatly impede the efficiency of visualization. To resolve this issue, we developed a perspective-based data clipper to eliminate the invisible parts of the data.

The data clipping strategy adopts popular interactive clipping techniques in volume visualization [52] and the WebGL implementation of clipping pixels, in order to remove the data not visible in the current viewport. Benefitting from the decomposition of the whole data volume, the clipping process can be achieved by checking whether the centers of the decomposed cube are within the viewport by applying the following equation:

$$cp = VP \cdot ep \tag{3}$$

where *ep* denotes the vector consisting of the visual coordinates of the position at the center of a decomposed cube with a number 1 appended (e.g., [x, y, z, 1]). *VP* is the view-projection matrix with a 4 × 4 dimension, produced by how a user views the scene and the view projection setting. *cp* is the desired vector, from which whether a point is within the viewport or not can be determined.

For convenience, the elements of the vector *cp* are sequentially marked as x_{clip}, y_{clip}, z_{clip}, and w_{clip}, and the following calculation is performed:

$$x_{viewport} = \frac{x_{clip}}{w_{clip}} \tag{4}$$

$$y_{viewport} = \frac{y_{clip}}{w_{clip}} \tag{5}$$

$$z_{viewport} = \frac{y_{clip}}{w_{clip}} \tag{6}$$

If $x_{viewport}$, $y_{viewport}$, and $z_{viewport}$ both range from −1 to 1 (boundaries excluded), the decomposed units are determined to be visible. In this case, the data point is reserved and rendered. The data points that do not fall in the above range are disregarded.

3.2. Data Filtering

This point cloud visualization provides an overview of the entire multi-dimensional data volume. By adjusting the size or transparency of the points, the internal variation of the data may also be observed. However, when millions of points are being simultaneously rendered in the viewport, the information that a user needs may still be hidden. In order to uncover important details of the data, we suggest three strategies to filter the data points and reveal only the data variables of interests to end-users in both space and time.

First, a value-based filter is introduced in the system. This approach has a great advantage for 3D visualization. Specifically, filtering the value range requires that only part of the data points need to be rendered. For example, high temperature values can be filtered out to allow scientists to focus on analyzing cold spots. These cold spots or regions will form a particular shape and temperature variance will be differed by colors. This function allows users to investigate data variation by colors as well as by shape.

Second, a filter of data by regions is developed to regional studies as well as global studies. For instance, geographical coordinates of the boundary information can filter data within a country or state. To accomplish this task, point-in-polygon operations are performed for each data point in the original dataset after orthogonal projection. However, this may present a problem when the boundary polygons are complex, as computing time for determining the relationship of a polygon and a point substantially increases as the number of the vertices in the polygon increases. In order to reduce the computing burden, a generalization process is applied to the polygon before the point-in-polygon calculation.

In our work, the Douglas-Peucker line simplification algorithm is adopted [53]. The algorithm performs as follows: (1) for every two vertices in the polygon that share the same edge, if the distance between them is under a threshold, they will be merged into a new one—the midpoint of the shared edge; (2) Step 1 is repeated until there are no pairs of points whose interval is less than the threshold. It should be noted that while this simplification process helps accelerate the filtering speed, it could also affect result accuracy. Our preliminary experiments show that with the proper threshold, the time cost of filtering regional points is greatly reduced with little impact on filtering precision.

The last filter targets the vertical variation. Drawing lines on the surface of Earth, the values on the vertical slices of the data volume along those lines are extracted and rendered in a special way. This process is explained in the following section.

3.3. Vertical Profile Visualization

A vertical profile image provides a cross-sectional view of a cubical data. x dimension of the cross-sectional view is a user-defined path along the Earth's surface, and y dimension is normally elevation for spatial data. Interactive vertical profile visualization has the capability to intuitively demonstrate the inner structure and information inside a 3D volume, and reveal the variation of a specific factor in a vertical direction (perpendicular to the Earth's surface), which is very helpful in climate research [54]. Generally, this function is developed in three steps. The first is a response to user requests that involve drawing a polyline on the Earth's surface, indicating a trajectory of interest, e.g., along the east coastline of Greenland. The second is generation of an image according to the original three-dimensional data. Each pixel in the image displays the climate variable values at coordinates (x, z), where the x-axis is the location along the trajectory, and the z-axis shows the location changes on the 29 vertical pressure levels of the raw data. The third step is displaying the image on the ground along the user-defined polyline. Three software modules are developed to handle the above tasks.

The Interactive Module (Figure 2) handles user input. The Image Generator Module acquires the data defined from the user input and converts it to a smooth image, showing the variation from ground to space as well as along the drawn polyline. The 3D Primitive Module projects the image from the Image Generator Module on a blank wall created on the surface of the virtual Earth.

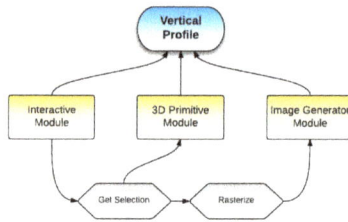

Figure 2. The workflow for creating a vertical profile.

We built the Interactive and 3D Primitive Modules on the open source library Cesium—a very popular virtual globe platform that supports the rendering of spatial data in a 3D virtual space. The following algorithm was developed to generate the image demonstrating the change in some variables along a specific path inside the data cube.

Given that 3D gridded data is being visualized, a horizontal slice of the data means that the vertical dimension is the value distribution in a given study area, identified by latitude and longitude, at a certain altitude. Note that some data might use pressure levels as the vertical dimension, but can be converted to altitude during preprocessing, if preferred. When we look at the data cube, it would look like columns of data points neatly standing on the Earth's surface. When a path of interest is drawn on the Earth's surface, we first need to determine all the nearby columns as the gridded data is not continuous on a 2D surface. The key idea is to rasterize the polyline representing the path. Figure 3 shows an example. Taking a horizontal data slice as a reference, the user-input vector is rasterized on a grid. The highlighted data points falling on the path of interest are extracted for all horizontal data slices to generate the vertical profile image.

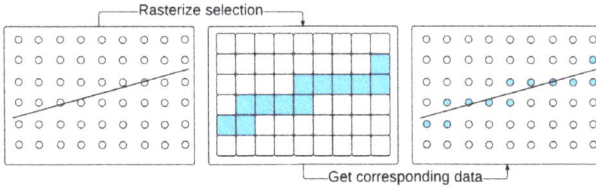

Figure 3. Rasterization of a user-input trajectory of interest on the Earth's Surface.

Rasterization is valid only when all the columns are evenly distributed in a grid, which means there is a specific map projection applied to the data. However, the projection of the input polyline may not always be the same as the one adopted in the original data source. For example, the polyline retrieved using the API was provided by Cesium is WGS (World Geodetic System) 84 [55], while the data used for our study is an azimuthal equidistant projection [56]. In this case, a re-projection process is required before rasterization. The projection information of a data source can usually be found in its metadata.

The next task is to sequentially project the selected columns on a planar coordinate system, whose axes stand for elevation and distance. In this way, a 2D grid can be built. This step unfolds the wall and lays it on a plane (if a user draws a complex path rather than a straight line). Figure 4a shows an example of the selected data points along a user chosen path. Only one image is produced and projected on 3D primitives, rather than multiple images that correspond to all the facets of the folded wall. In this new planar coordinate system, the distance between every two neighboring columns should be the same as the one in the 3D space. The great circle distance between the locations where the columns stand determines this. A final image (Figure 4b) is then generated by interpolating all the data points falling in the image region to generate the raw image matrix. Finally, this matrix is converted to a RGB image with proper color mapping.

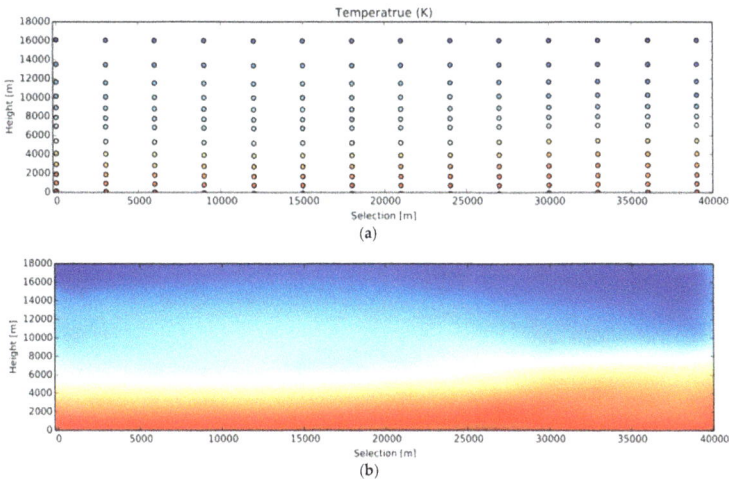

Figure 4. An example of generating a vertical profile image through interpolation: (**a**) An illustration of selected data points (clustered as columns) along a path on the Earth's surface; (**b**) Color-coded image after 2D interpolation (temperature increases from blue to red).

4. Graphic User Interface

Figure 5 demonstrates the PolarGlobe GUI. The climate data used in the visualization is the air temperature output from the Polar version of the WRF Model. This data covers 20 degree latitude and north, and contains 29 vertical layers differed by pressure levels. The spatial resolution is 30 km. By applying a predefined color map, this panoramic view of data clearly shows the change in temperature with a cold spot on top of the Greenland ice sheet.

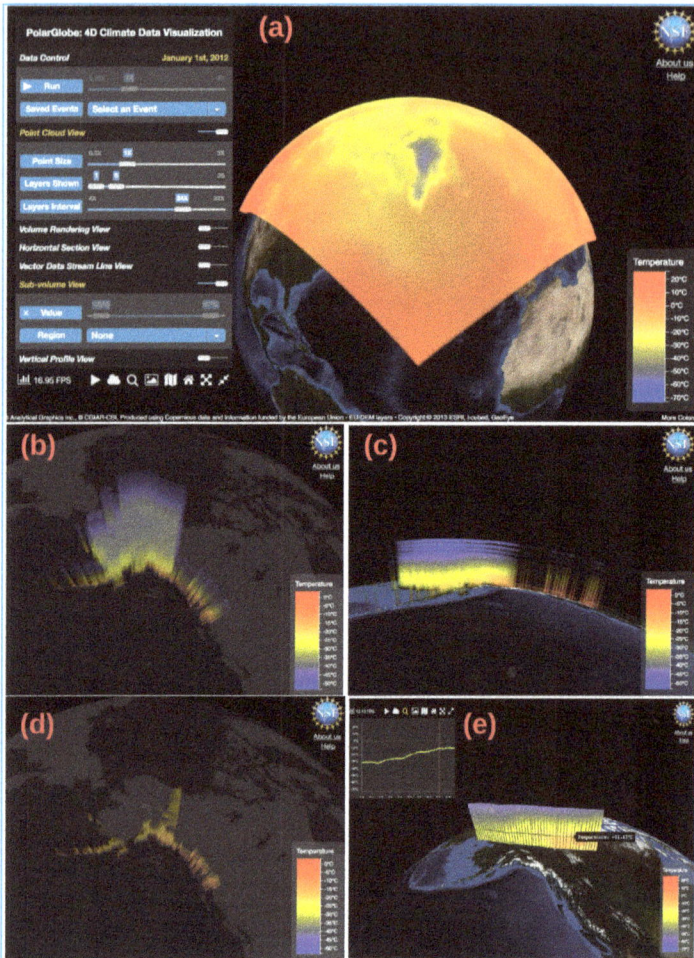

Figure 5. Graphic User Interface of PolarGlobe (http://cici.lab.asu.edu/polarglobe2): (**a**) a screenshot of temperature visualization in the whole study area; (**b**) a close look at the distribution of air temperature in Alaska, US on 1 January 2012; (**c**) a view of temperature of Alaska from the side; (**d**) the value filtering effect (Air temperature higher than −20 degree Celsius for data presented is shown here); (**e**) the vertical profile view and statistics.

A spatial filter can be applied to make further exploration of the air temperature data at an area of interest, such as Alaska, US. Once "Alaska" is selected from the dropdown menu of the spatial region, the original data cube (temperature data at all pressure layers) is cut by the geographical boundary of Alaska. Figure 5b,c demonstrates the temperature data in the winter (1 January 2012), from different angles. It can be seen that it is much warmer along the coast of the Gulf of Alaska and inland (near the south) than in other parts. This observation can be verified by conducting a value filtering in the PolarGlobe system (see results in Figure 5d).

When an inner structure of the data volume needs to be examined, our vertical profile visualization will serve this purpose. Figure 5e shows the temperature change along 65 degree North near Alaska and northwest Canada. It can be observed that the near surface temperature in Canada along the trajectory of interest is higher than that in Alaska, and the value keeps increasing when moving toward the east.

5. Experiments and Results

This section provides quantitative analysis of the performance improved by the proposed methods for accomplishing real-time and interactive visualization of the voluminous dataset. The experiments were conducted on a Dell workstation with 8 cores at 3.4 GHz and 8 gigabytes memory size.

5.1. Performance on Data Loading and Rendering

To accelerate data loading speed, we introduced the enhanced octree-based LOD and viewport clip to filter out the invisible part of the data volume to reduce data size to be loaded. This experiment provides a comparison in terms of data loading time, using: (1) the non-optimized approach, in which the entire dataset will be loading for visualization; (2) the approach that adopts only the LOD; and (3) the one with both the LOD and viewport clip, applied to reveal the advantages of the optimization strategy. In the experiment, we assume a user is viewing the Earth from space and his eye sight falls on the location of 75° W, 60° N on the Earth's surface. As he looks closer or further, the angle remains the same and only the view distance changes. The loading efficiency under the three scenarios is compared and results are shown in Figure 6.

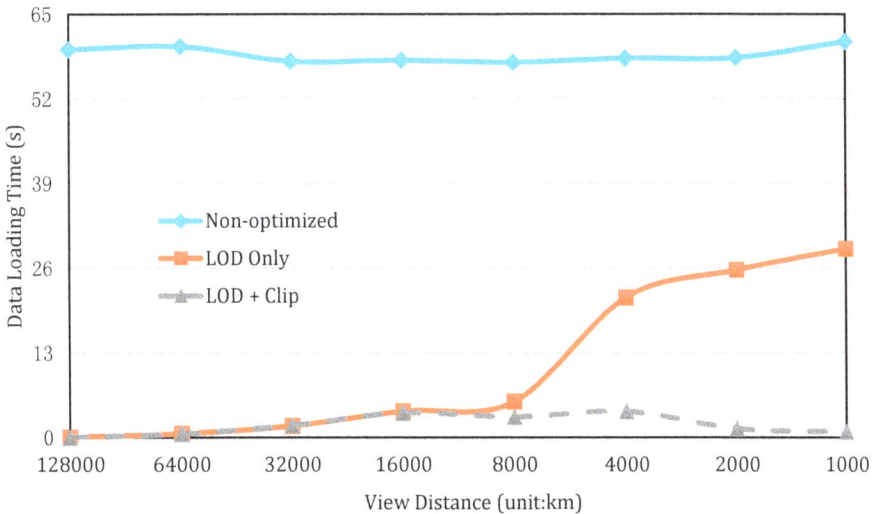

Figure 6. Comparison of data loading time before and after applying optimization strategies as the view distance becomes shorter. (Error bars are smaller than the plot markers).

It can be observed from Figure 6 that data loading with no optimization takes much more time than in the other two situations, no matter how far away a user views the data volume. When the view distance is below 8 million meters, the loading time with LOD-adopted is comparable to the LOD + Clip. Beyond that view distance, the LOD with data clipping presents a conspicuous advantage over the LOD alone. As expected when data volume is observed in shorter distances, a greater amount of points are filtered out because of their invisibility (the smaller the data volume being loaded, the less computation burden on the graphic processor). Therefore, a higher rendering performance (in terms of frame rate) can be achieved. We also compare the rendering performances by their frame rates when refreshing the data volume. Similar results are presented in Figure 7. In Figure 7, an interesting bell-shaped curve is shown for our proposed LOD + Clip method. At the distances of 6000 km and 4000 km, the frame rate drops to about half of that at other distances. This is due to the reason that these distances are close enough to load the data at a higher level of detail, but not close enough to clip the data volume, since most of the data volume remains in the viewport. This fact can be cross validated with the results in Figure 6 (at the given two distances); there is, in fact, an increase in the data loading time. The data-loading curve starts to drop when the view distances move below 4000 km for the LOD + clip method (dashed line in Figure 6).

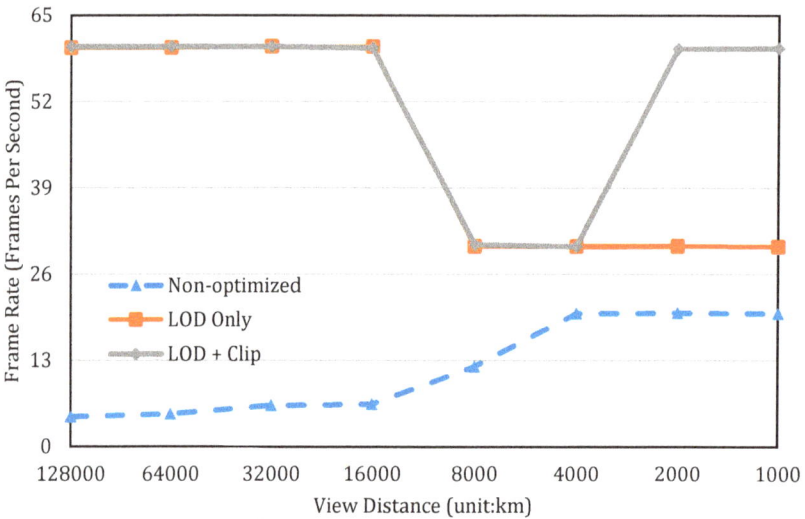

Figure 7. Comparison of data rendering performance, before and after applying optimization strategies.

5.2. Experiment on Accuracy vs. Efficiency in Spatial Filtering and Generalization

As mentioned, we introduced a simplification process for determining spatial boundaries. This clips the original data volume to support the spatial filtering operation. Knowing that an increasing level of simplification introduces a larger error in the boundary data, and thus affects the accuracy of the results, we implemented the filtered grid points. Here the level of simplification was determined by a distance tolerance, used to determine which neighboring vertices should be kept or deleted. The resulting accuracy was measured by the ratio of correctly filtered grid points, using the simplified boundary versus the results obtained by using the original boundary.

We used the boundary of Wisconsin, US as the test data (125,612 vertices in total). Figure 8 illustrates the results. The results reveal a rapid decline of time cost when distance tolerance begins to increase in the simplification process. Here, the unit of this distance tolerance is a degree of latitude/longitude. Accuracy, on the other hand, remains at 100% until the tolerance value is set higher than 0.0064. Hence, the threshold 0.0064 is an ideal parameter setting for spatial boundary simplification that maximizes both filtering efficiency and data accuracy.

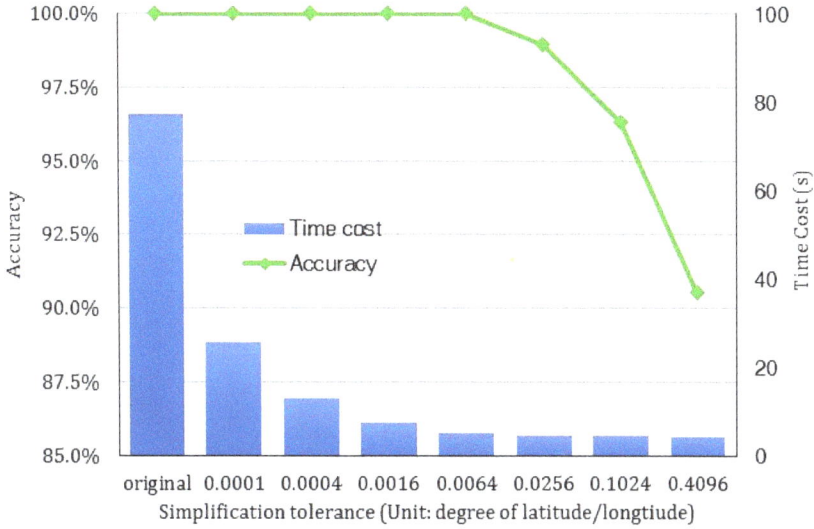

Figure 8. A comparison between time cost and data accuracy in spatial filtering after applying simplification to the spatial boundary.

5.3. Impact of Interpolation on the Efficiency of Vertical Profile Generation and Visualization

Although interpolation is only a small part of implementing vertical profile visualization, it has the greatest impact on efficiency in generating a profile. This step is needed, since our original test data has a relatively coarse resolution—30 km, and the data points are not evenly distributed within space (see illustration in Figure 4). The time cost of the interpolation is affected by three factors: (1) the number of input points, namely, how many data values will be selected as the reference for interpolation; and (2) the number of output points. This is a measure of the resolution of the interpolated image by the total number of pixels, and (3) which interpolation method is used. We designed two experiments to reveal the impact of these three factors on the interpolation performance.

In the first one, we controlled the number of output points (the total pixel numbers of the output image) at 160,000 and compared the performance of different types of interpolation methods by altering the number of input points. Here, the comparison is applied to three types of interpolation which all meet the demand in our case. They are nearest neighbor interpolation [57], linear interpolation [58], and cubic precision Clough-Tocher interpolation [59]. In the second experiment, we controlled the number of input points at 5000 and changed the number of output points. The results of the two experiments are presented in Figures 9 and 10, respectively.

Both figures show that, when either the number of input points or output points increases, an increase in interpolation time can be observed. In addition, it is obvious that linear interpolation achieves the highest efficiency, while the nearest neighbor performs the worst. This is because the nearest neighbor interpolation requires the construction of a KD-tree during the processing, which costs more time. However, the speed is not the only indicator used to evaluate an interpolation method. It is more important to figure out how well an interpolation method emulates our data. Therefore, a third experiment was conducted to test the precision of interpolation.

We choose 10,353 points as the data in this experiment. These 10,353 points consist of 357 columns. In other words, 357 points in the original data grid were selected along a user drawn path on the Earth's surface. This number is related to the density of the raw data points. On each column, there exists 29 data points (since the raw data's z dimension is 29, representing 29 pressure levels). To evaluate the accuracy in the interpolation, part of all the data points were selected and served as the input for interpolation (we call it train data). The rest is used as test/ground truth data.

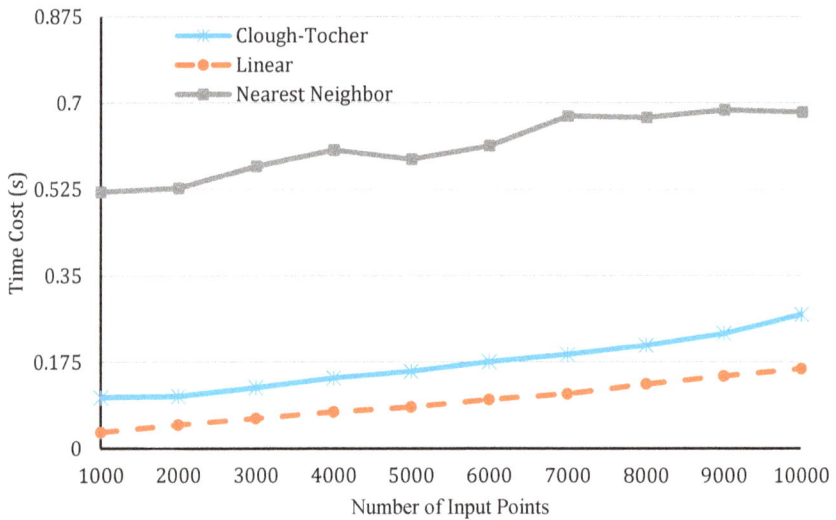

Figure 9. A comparison of efficiency by different interpolation methods, as the number of input points increases.

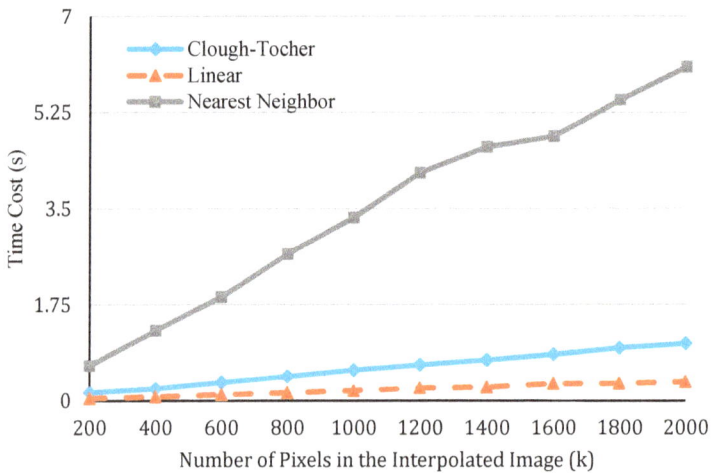

Figure 10. Efficiency of different interpolation methods, as the required resolution of the interpolated image increases.

For the simulated output points, there are two values associated with them: s_i (the interpolated values); and the real values (r_i). The accuracy δ is calculated using the Normalized Euclidean Distance (NED) between the two vectors, composed respectively by the interpolated and real values:

$$\delta = \frac{1}{NED} \qquad (7)$$

$$NED = \sum_{i=1}^{n} \sqrt{\frac{(s_i - r_i)^2}{n}} \tag{8}$$

where n denotes the number of the points in the test data. When a higher NED is observed, a lower accuracy value δ will be obtained. Figure 11 demonstrates the accuracy of each interpolation approach by changing the sampling ratio between train and test data.

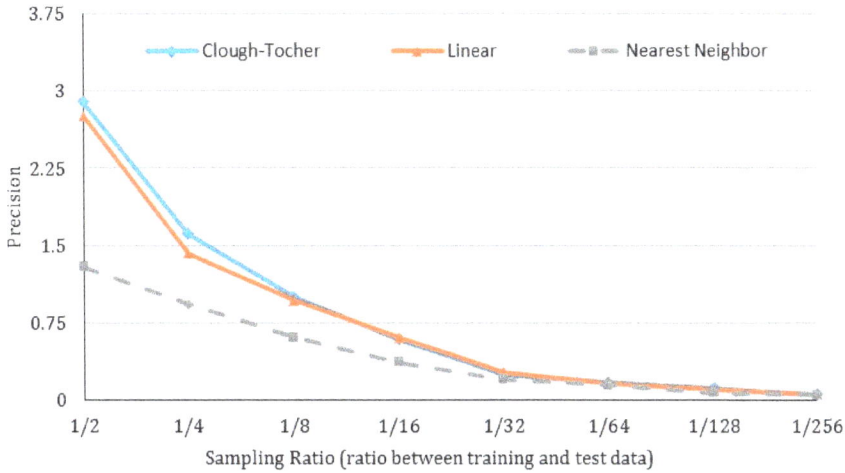

Figure 11. A comparison of accuracy across different interpolation methods, as the number of training data used for interpolation decreases.

As shown, the accuracy of interpolation rapidly declines as fewer training data are selected, as reflected by a decreasing sampling ratio. This is especially true for linear and Clough-Tocher interpolations. On the other hand, these two methods still perform better than the nearest neighbor interpolation at each sampling rate. Clough-Tocher interpolation performs a bit better than linear interpolation when the sampling ratio is controlled above 1/8. When the ratio is below 1/8, we observe very similar resultant accuracy values. Synthesizing results from all three experiments, linear interpolation is the best fit in our visualization system, with real-time requirement due to its fast speed and high accuracy.

6. Conclusions

This paper introduces PolarGlobe, a Web-based virtual globe system to allow Web-scale access of big climate simulation data. Different from previous work, our proposed platform does not require installation of plugins. This substantially reduces the learning curve of the software tool. Technically, the major contributions of this work include: (1) a server-client architecture powered up by a new differential-storage-enhanced octree model to support efficient spatial indexing, transmission, and rendering of big climate data; (2) a combined value and spatial filter to enable perception-based visualization and interactive data exploration; and (3) vertical profile visualization to allow examination of variations in climate variables on a cross-section inside the data cube. Although primarily tested on climate simulation data, visualization techniques can be widely applied to other Earth science domains, such as oceanography, hydrology, and geology. We believe this platform will provide strong support to scientists for testing models and validating hypotheses, as well as for the general public to understand different components of the Earth system and its interactions.

In the future, we will enhance the PolarGlobe system in the following directions: first, methods will be developed to effectively present multivariate geoscientific data for an integrated analysis; second, strategies for visualizing vector data on the globe will also be exploited; and third, we will extend the current visualization capability with advanced data mining or spatial analysis capability, to equip PolarGlobe as not only a system for visualization but also for knowledge discovery.

Acknowledgments: This project is supported by the National Science Foundation (PLR-1349259; BCS-1455349; and PLR-1504432).

Author Contributions: Sizhe Wang and Wenwen Li originated the idea of the paper. Sizhe Wang developed the algorithm and the PolarGlobe system, with assistance from Feng Wang. Sizhe Wang, Wenwen Li wrote the paper. Feng Wang assists with some graphics and references. All authors discussed the implementation details of the work.

Conflicts of Interest: The authors declare no conflict of interest.

References

1. Dasgupta, S.; Mody, A.; Roy, S.; Wheeler, D. Environmental regulation and development: A cross-country empirical analysis. *Oxf. Dev. Stud.* **2001**, *29*, 173–187. [CrossRef]
2. Grimm, N.B.; Faeth, S.H.; Golubiewski, N.E.; Redman, C.L.; Wu, J.; Bai, X.; Briggs, J.M. Global change and the ecology of cities. *Science* **2008**, *319*, 756–760. [CrossRef] [PubMed]
3. Tacoli, C. Crisis or adaptation? Migration and climate change in a context of high mobility. *Environ. Urban.* **2009**, *21*, 513–525. [CrossRef]
4. Li, Y.; Li, Y.; Zhou, Y.; Shi, Y.; Zhu, X. Investigation of a coupling model of coordination between urbanization and the environment. *J. Environ. Manag.* **2012**, *98*, 127–133. [CrossRef] [PubMed]
5. Emanuel, K. Increasing destructiveness of tropical cyclones over the past 30 years. *Nature* **2005**, *436*, 686–688. [CrossRef] [PubMed]
6. Goudie, A.S. Dust storms: Recent developments. *J. Environ. Manag.* **2009**, *90*, 89–94. [CrossRef] [PubMed]
7. Meehl, G.A.; Zwiers, F.; Evans, J.; Knutson, T. Trends in extreme weather and climate events: Issues related to modeling extremes in projections of future climate change. *Bull. Am. Meteorol. Soc.* **2000**, *81*, 427–436. [CrossRef]
8. Mitrovica, J.X.; Tamisiea, M.E.; Davis, J.L.; Milne, G.A. Recent mass balance of polar ice sheets inferred from patterns of global sea-level change. *Nature* **2001**, *409*, 1026–1029. [CrossRef] [PubMed]
9. Cook, A.J.; Fox, A.J.; Vaughan, D.G.; Ferrigno, J.G. Retreating glacier fronts on the Antarctic Peninsula over the past half-century. *Science* **2005**, *308*, 541–544. [CrossRef] [PubMed]
10. Sheppard, S.R. *Visualizing Climate Change: A Guide to Visual Communication of Climate Change and Developing Local Solutions*; Routledge: Florence, KY, USA, 2012; p. 511.
11. Giorgi, F.; Mearns, L.O. Approaches to the simulation of regional climate change: A review. *Rev. Geophys.* **1991**, *29*, 191–216. [CrossRef]
12. Chervenak, A.; Deelman, E.; Kesselman, C.; Allcock, B.; Foster, I.; Nefedova, V.; Lee, J.; Sim, A.; Shoshani, A.; Drach, B. High-performance remote access to climate simulation data: A challenge problem for data grid technologies. *Parallel Comput.* **2003**, *29*, 1335–1356. [CrossRef]
13. Overpeck, J.T.; Meehl, G.A.; Bony, S.; Easterling, D.R. Climate data challenges in the 21st century. *Science* **2011**, *331*, 700–702. [CrossRef] [PubMed]
14. Gordin, D.N.; Polman, J.L.; Pea, R.D. The Climate Visualizer: Sense-making through scientific visualization. *J. Sci. Educ. Technol.* **1994**, *3*, 203–226. [CrossRef]
15. Van Wijk, J.J. The value of visualization. In Proceedings of the IEEE Visualization VIS 05, Minneapolis, MN, USA, 23–28 October 2005; pp. 79–86.
16. Galton, F. *Meteographics, or, Methods of Mapping the Weather, Macmillan, London*; British Library: London, UK, 1863.
17. Nocke, T.; Heyder, U.; Petri, S.; Vohland, K.; Wrobel, M.; Lucht, W. Visualization of Biosphere Changes in the Context of Climate Change. In Proceedings of the Conference on Information Technology and Climate Change, Berlin, Germany, 25–26 September 2008.

18. Santos, E.; Poco, J.; Wei, Y.; Liu, S.; Cook, B.; Williams, D.N.; Silva, C.T. UV-CDAT: Analyzing Climate Datasets from a User's Perspective. *Comput. Sci. Eng.* **2013**, *15*, 94–103. [CrossRef]

19. Williams, D. The ultra-scale visualization climate data analysis tools (UV-CDAT): Data analysis and visualization for geoscience data. *IEEE Comp.* **2013**, *46*, 68–76. [CrossRef]

20. Li, W.; Wu, S.; Song, M.; Zhou, X. A scalable cyberinfrastructure solution to support big data management and multivariate visualization of time-series sensor observation data. *Earth Sci. Inform.* **2016**, *9*, 449–464. [CrossRef]

21. Gore, A. The Digital Earth: Understanding our planet in the 21st century. *Aust. Surv.* **1998**, *43*, 89–91. [CrossRef]

22. Sheppard, S.R.; Cizek, P. The ethics of Google Earth: Crossing thresholds from spatial data to landscape visualisation. *J. Environ. Manag.* **2009**, *90*, 2102–2117. [CrossRef] [PubMed]

23. Boschetti, L.; Roy, D.P.; Justice, C.O. Using NASA's World Wind virtual globe for interactive internet visualization of the global MODIS burned area product. *Int. J. Remote Sens.* **2008**, *29*, 3067–3072. [CrossRef]

24. Boulos, M.N. Web GIS in practice III: Creating a simple interactive map of England's strategic Health Authorities using Google Maps API, Google Earth KML, and MSN Virtual Earth Map Control. *Int. J. Health Geogr.* **2005**, *4*, 22. [CrossRef] [PubMed]

25. Sun, X.; Shen, S.; Leptoukh, G.G.; Wang, P.; Di, L.; Lu, M. Development of a Web-based visualization platform for climate research using Google Earth. *Comput. Geosci.* **2012**, *47*, 160–168. [CrossRef]

26. Varun, C.; Vatsavai, R.; Bhaduri, B. iGlobe: An interactive visualization and analysis framework for geospatial data. In Proceedings of the 2nd International Conference on Computing for Geospatial Research & Applications, Washington, DC, USA, 23–25 May 2011; p. 21.

27. Helbig, C.; Bauer, H.S.; Rink, K.; Wulfmeyer, V.; Frank, M.; Kolditz, O. Concept and workflow for 3D visualization of atmospheric data in a virtual reality environment for analytical approaches. *Environ. Earth Sci.* **2014**, *72*, 3767–3780. [CrossRef]

28. Berberich, M.; Amburn, P.; Dyer, J.; Moorhead, R.; Brill, M. HurricaneVis: Real Time Volume Visualization of Hurricane Data. In Proceedings of the Eurographics/IEEE Symposium on Visualization, Berlin, Germany, 10–12 June 2009.

29. Wang, F.; Li, W.; Wang, S. Polar Cyclone Identification from 4D Climate Data in a Knowledge-Driven Visualization System. *Climate* **2016**, *4*, 43. [CrossRef]

30. Li, Z.; Yang, C.; Sun, M.; Li, J.; Xu, C.; Huang, Q.; Liu, K. A high performance web-based system for analyzing and visualizing spatiotemporal data for climate studies. In Proceedings of the International Symposium on Web and Wireless Geographical Information Systems, Banff, AB, Canada, 4–5 April 2013; Springer: Berlin/Heidelberg, Germany, 2013; pp. 190–198.

31. Alder, J.R.; Hostetler, S.W. Web based visualization of large climate data sets. *Environ. Model. Softw.* **2015**, *68*, 175–180. [CrossRef]

32. Liu, Z.; Ostrenga, D.; Teng, W.; Kempler, S. Developing online visualization and analysis services for NASA satellite-derived global precipitation products during the Big Geospatial Data era. In *Big Data: Techniques and Technologies in Geoinformatics*; CRC Press: Boca Raton, FL, USA, 2014; pp. 91–116.

33. Van Meersbergen, M.; Drost, N.; Blower, J.; Griffiths, G.; Hut, R.; van de Giesen, N. Remote web-based 3D visualization of hydrological forecasting datasets. In Proceedings of the EGU General Assembly Conference, Vienna, Austria, 12–17 April 2015; Volume 17, p. 4865.

34. Hunter, J.; Brooking, C.; Reading, L.; Vink, S. A Web-based system enabling the integration, analysis, and 3D sub-surface visualization of groundwater monitoring data and geological models. *Int. J. Digit. Earth* **2016**, *9*, 197–214. [CrossRef]

35. Moroni, D.F.; Armstrong, E.; Tsontos, V.; Hausman, J.; Jiang, Y. Managing and servicing physical oceanographic data at a NASA Distributed Active Archive Center. In Proceedings of the OCEANS 2016 MTS/IEEE Monterey, Monterey, CA, USA, 19–23 September 2016; pp. 1–6.

36. AGI. Cesium-WebGL Virtual Globe and Map Engine. Available online: https://cesiumjs.org/ (accessed on 17 January 2015).

37. Brovelli, M.A.; Hogan, P.; Prestifilippo, G.; Zamboni, G. NASA Webworldwind: Multidimensional Virtual Globe for Geo Big Data Visualization. In Proceedings of the ISPRS-International Archives of the Photogrammetry, Remote Sensing and Spatial Information Sciences 2016, Prague, Czech Republic, 12–19 July 2016; pp. 563–566.

38. Voumard, Y.; Sacramento, P.; Marchetti, P.G.; Hogan, P. WebWorldWind, Achievements and Future of the ESA-NASA Partnership (No. e2134v1). Available online: https://peerj.com/preprints/2134.pdf (accessed on 18 April 2017).

39. Li, W.; Wang, S. PolarGlobe: A web-wide virtual globe system for visualizing multidimensional, time-varying, big climate data. *Int. J. Geogr. Inf. Sci.* **2017**, *31*, 1562–1582. [CrossRef]

40. Cox, M.; Ellsworth, D. *Managing Big Data for Scientific Visualization*; ACM Siggraph: Los Angeles, CA, USA, 1997; Volume 97, pp. 146–162.

41. Demchenko, Y.; Grosso, P.; De Laat, C.; Membrey, P. Addressing big data issues in scientific data infrastructure. In Proceedings of the IEEE 2013 International Conference on Collaboration Technologies and Systems (CTS), San Diego, CA, USA, 20–24 May 2013; pp. 48–55.

42. Baccarelli, E.; Cordeschi, N.; Mei, A.; Panella, M.; Shojafar, M.; Stefa, J. Energy-efficient dynamic traffic offloading and reconfiguration of networked data centers for big data stream mobile computing: Review, challenges, and a case study. *IEEE Netw.* **2016**, *30*, 54–61. [CrossRef]

43. Cordeschi, N.; Shojafar, M.; Amendola, D.; Baccarelli, E. Energy-efficient adaptive networked datacenters for the QoS support of real-time applications. *J. Supercomput.* **2015**, *71*, 448–478. [CrossRef]

44. Wong, P.C.; Shen, H.W.; Leung, R.; Hagos, S.; Lee, T.Y.; Tong, X.; Lu, K. Visual analytics of large-scale climate model data. In Proceedings of the 2014 IEEE 4th Symposium on Large Data Analysis and Visualization (LDAV), Paris, France, 9–10 November 2014; pp. 85–92.

45. Li, J.; Wu, H.; Yang, C.; Wong, D.W.; Xie, J. Visualizing dynamic geosciences phenomena using an octree-based view-dependent LOD strategy within virtual globes. *Comput. Geosci.* **2011**, *37*, 1295–1302. [CrossRef]

46. Liang, J.; Gong, J.; Li, W.; Ibrahim, A.N. Visualizing 3D atmospheric data with spherical volume texture on virtual globes. *Comput. Geosci.* **2014**, *68*, 81–91. [CrossRef]

47. Kruger, J.; Westermann, R. Acceleration techniques for GPU-based volume rendering. In Proceedings of the 14th IEEE Visualization 2003 (VIS'03), Seattle, WA, USA, 13–24 October 2003; p. 38.

48. Wang, F.; Wang, G.; Pan, D.; Liu, Y.; Yang, Y.; Wang, H. A parallel algorithm for viewshed analysis in three-dimensional Digital Earth. *Comput. Geosci.* **2015**, *75*, 57–65.

49. Drebin, R.A.; Carpenter, L.; Hanrahan, P. Volume rendering. *Comput. Graph.* **1988**, *22*, 65–74. [CrossRef]

50. National Center for Atmospheric Research Staff. The Climate Data Guide: Arctic System Reanalysis (ASR). Available online: https://climatedataguide.ucar.edu/climate-data/arctic-system-reanalysis-asr (accessed on 22 June 2017).

51. Luebke, D.P. *Level of Detail for 3D Graphics*; Morgan Kaufmann: San Francisco, CA, USA, 2003.

52. Weiskopf, D.; Engel, K.; Ertl, T. Interactive clipping techniques for texture-based volume visualization and volume shading. *IEEE Trans. Vis. Comput. Graph.* **2003**, *9*, 298–312. [CrossRef]

53. Douglas, D.H.; Peucker, T.K. Algorithms for the reduction of the number of points required to represent a digitized line or its caricature. *Cartogr. Int. J. Geogr. Inf. Geovis.* **1973**, *10*, 112–122. [CrossRef]

54. Menon, S.; Hansen, J.; Nazarenko, L.; Luo, Y. Climate effects of black carbon aerosols in China and India. *Science* **2002**, *297*, 2250–2253. [CrossRef] [PubMed]

55. Malys, S. The WGS84 Reference Frame. National Imagery and Mapping Agency. 1996. Available online: http://earth-info.nga.mil/GandG/publications/tr8350.2/wgs84fin.pdf (accessed on 10 March 2017).

56. Hinks, A.R. A retro-azimuthal equidistant projection of the whole sphere. *Geogr. J.* **1929**, *73*, 245–247. [CrossRef]

57. Parker, J.A.; Kenyon, R.V.; Troxel, D.E. Comparison of interpolating methods for image resampling. *IEEE Trans. Med. Imaging* **1983**, *2*, 31–39. [CrossRef] [PubMed]

58. De Boor, C.; De Boor, C.; Mathématicien, E.U.; De Boor, C.; De Boor, C. *A Practical Guide to Splines*; Springer: New York, NY, USA, 1978; Volume 27, p. 325.

59. Mann, S. Cubic precision Clough-Tocher interpolation. *Comput. Aided Geom. Des.* **1999**, *16*, 85–88. [CrossRef]

informatics

MDPI

Article

Constructing Interactive Visual Classification, Clustering and Dimension Reduction Models for n-D Data

Boris Kovalerchuk * and Dmytro Dovhalets

Department of Computer Science, Central Washington University, Ellensburg, WA 98926, USA; Dmytro.Dovhalets@cwu.edu
* Correspondence: Boris.Kovalerchuk@cwu.edu

Academic Editors: Achim Ebert and Gunther H. Weber
Received: 31 May 2017; Accepted: 19 July 2017; Published: 25 July 2017

Abstract: The exploration of multidimensional datasets of all possible sizes and dimensions is a long-standing challenge in knowledge discovery, machine learning, and visualization. While multiple efficient visualization methods for n-D data analysis exist, the loss of information, occlusion, and clutter continue to be a challenge. This paper proposes and explores a new interactive method for visual discovery of n-D relations for supervised learning. The method includes automatic, interactive, and combined algorithms for discovering linear relations, dimension reduction, and generalization for non-linear relations. This method is a special category of reversible General Line Coordinates (GLC). It produces graphs in 2-D that represent n-D points losslessly, i.e., allowing the restoration of n-D data from the graphs. The projections of graphs are used for classification. The method is illustrated by solving machine-learning classification and dimension-reduction tasks from the domains of image processing, computer-aided medical diagnostics, and finance. Experiments conducted on several datasets show that this visual interactive method can compete in accuracy with analytical machine learning algorithms.

Keywords: interactive visualization; classification; clustering; dimension reduction; multidimensional visual analytics; machine learning; knowledge discovery; linear relations

1. Introduction

Many procedures for n-D data analysis, knowledge discovery and visualization have demonstrated efficiency for different datasets [1–5]. However, the *loss of information, occlusion*, and *clutter* in visualizations of n-D data continues to be a challenge for knowledge discovery [1,2]. The *dimension scalability* challenge for visualization of n-D data is present at a low dimension of $n = 4$. Since only 2-D and 3-D data can be directly visualized in the physical 3-D world, visualization of n-D data becomes more difficult with higher dimensions as there is greater loss of information, occlusion and clutter, Further progress in data science will require greater involvement of end users in constructing machine learning models, along with more scalable, intuitive and efficient visual discovery methods and tools [6].

A representative software system for the interactive visual exploration of multivariate datasets is XmdvTool [7]. It implements well-established algorithms such as parallel coordinates, radial coordinates, and scatter plots with hierarchical organization of attributes [8]. For a long time, its functionality was concentrated on exploratory manipulation of records in these visualizations. Recently, its focus has been extended to support data mining (version 9.0, 2015), including interactive parameter space exploration for association rules [9], interactive pattern exploration in streaming [10], and time series [11].

The goal of this article is to develop a new *interactive visual machine learning system for solving supervised learning classification tasks* based on a new algorithm called GLC-L [12]. This study expands the base GLC-L algorithm to new interactive and automatic algorithms GLC-IL, GLC-AL and GLC-DRL for discovery of linear and non-linear relations and dimension reduction.

Classification and dimension reduction tasks from three domains: image processing, computer-aided medical diagnostics, and finance (stock market) are used to illustrate the method. This method belongs to a class of General Line Coordinates (GLC) [12–15] where the review of the state of the art is provided. The applications of GLC in finance are presented in [16]. The rest of this paper is organized as follows. Section 2 presents the approach that includes the base algorithm GLC-L (Section 2.1) the interactive version of the base algorithm (Section 2.2), the algorithm for automatic discovery of relations combined with interactions (Section 2.3), visual structure analysis of classes (Section 2.4), and generalization of algorithms for non-linear relations (Section 2.5). Section 3 presents the results for five case studies using the algorithms presented in Section 2. Section 4 discusses and analyses the results in comparison with prior results and software implementation. The conclusion section presents the advantages and benefits of proposed algorithms for multiple domains.

2. Methods: Linear Dependencies for Classification with Visual Interactive Means

Consider a task of visualizing an n-D linear function $F(x) = y$ where $x = (x_1, x_2, \ldots, x_n)$ is an n-D point and y is a scalar, $y = c_1x_1 + c_2x_2 + c_3x_3 + \ldots + c_nx_n + c_{n+1}$. Such functions play important roles in classification, regression and multi-objective optimization tasks. In regression, $F(x)$ directly serves as a regression function. In classification, $F(x)$ serves as a discriminant function to separate the two classes with a classification rule with a threshold T: if $y < T$ then x belongs to class 1, else x belongs to class 2. In multi-objective optimization, $F(x)$ serves as a tradeoff to reconcile n contradictory objective functions with c_i serving as weights for objectives.

2.1. Base GLC-L Algorithm

This section presents the visualization *algorithm* called *GLC-L* for a linear function [12]. It is used as a base for other algorithms presented in this paper.

Let $K = (k_1, k_2, \ldots, k_{n+1})$, $k_i = c_i/c_{max}$, where $c_{max} = |\max_{i=1:n+1}(c_i)|$, and $G(x) = k_1x_1 + k_2x_2 + \ldots + k_nx_n + k_{n+1}$. Here all k_i are normalized to be in $[-1,1]$ interval. The following property is true for F and G: $F(x) < T$ if and only if $G(x) < T/c_{max}$. Thus, F and G are equivalent linear classification functions. Below we present steps of *GLC-L* algorithm for a given linear function $F(x)$ with coefficients $C = (c_1, c_2, \ldots, c_{n+1})$.

Step 1: *Normalize* $C = (c_1, c_2, \ldots, c_{n+1})$ by creating as set of normalized parameters $K = (k_1, k_2, \ldots, k_{n+1})$: $k_i = c_i/c_{max}$. The resulting normalized equation $y_n = k_1x_1 + k_2x_2 + \ldots + k_nx_n + k_{n+1}$ with normalized rule: if $y_n < T/c_{max}$ then x belongs to class 1, else x belongs to class 2, where y_n is a normalized value, $y_n = F(x)/c_{max}$. Note that for the classification task we can assume $c_{n+1} = 0$ with the same task generality. For regression, we also deal with all data normalized. If actual y_{act} is known, then it is normalized by C_{max} for comparison with y_n, y_{act}/c_{max}.

Step 2: *Compute all angles* $Q_i = arccos(|k_i|)$ of absolute values of k_i and locate coordinates $X_1 - X_n$ in accordance with these angles as shown in Figure 1 relative to the horizontal lines. If $k_i < 0$, then coordinate X_i is oriented to the left, otherwise X_i is oriented to the right (see Figure 1). For a given n-D point $x = (x_1, x_2, \ldots, x_n)$, draw its values as *vectors* $x_1, x_2, \ldots, x_n$ in respective coordinates $X_1 - X_n$ (see Figure 1).

Step 3. *Draw vectors* $x_1, x_2, \ldots, x_n$ *one after another*, as shown on the left side of Figure 1. Then *project* the last point for x_n onto the horizontal axis U (see a red dotted line in Figure 1). To simplify, visualization axis U can be collocated with the horizontal lines that define the angles Q_i as shown in Figure 2.

Step 4.

Step 4a. For regression and linear optimization tasks, repeat step 3 for all n-D points as shown in the upper part of Figure 2a,b.

Step 4b. For the two-class classification task, repeat step 3 for all n-D points of classes 1 and 2 drawn in different colors. Move points of class 2 by mirroring them to the bottom with axis U doubled as shown in Figure 2. For more than two classes, Figure 1 is created for each class and m parallel axis U_j are generated next to each other similar to Figure 2. Each axis U_j corresponds to a given class j, where m is the number of classes.

Step 4c. For multi-class classification tasks, conduct step 4b for all n-D points of each pair of classes i and j drawn in different colors, or draw each class against all other classes together.

This algorithm uses the property that $cos(arccos\ k) = k$ for $k \in [-1,1]$, i.e., projection of vectors $\mathbf{x}_i$ to axis U will be $k_i x_i$ and with consecutive location of vectors $\mathbf{x}_i$, the projection from the end of the last vector $\mathbf{x}_n$ gives a sum $k_1 x_1 + k_2 x_2 + \ldots + k_n x_n$ on axis U. It does not include k_{n+1}. To add k_{n+1}, it is sufficient to shift the start point of $\mathbf{x}_1$ on axis U (in Figure 1) by k_{n+1}. Alternatively, for the visual classification task, k_{n+1} can be omitted by subtracting k_{n+1} from the threshold.

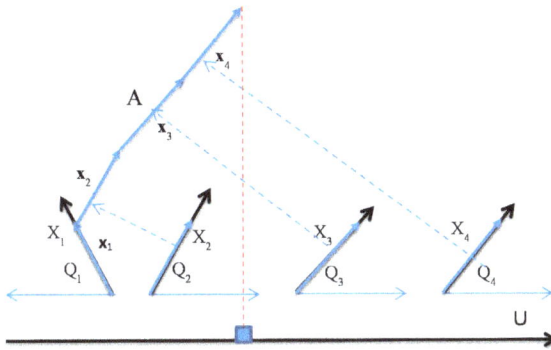

Figure 1. 4-D point $A = (1,1,1,1)$ in GLC-L coordinates $X_1 - X_4$ with angles (Q_1,Q_2,Q_3,Q_4) with vectors $\mathbf{x}_i$ shifted to be connected one after another and the end of last vector projected to the black line. X_1 is directed to the left due to negative k_1. Coordinates for negative k_i are always directed to the left.

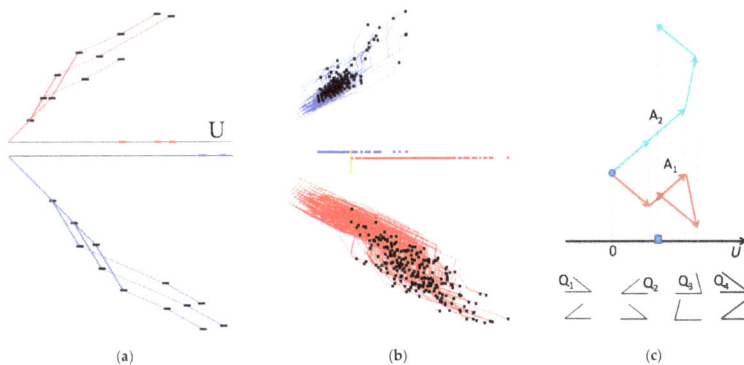

(a)   (b)   (c)

Figure 2. GLC-L algorithm on real and simulated data. (**a**) Result with axis X_1 starting at axis U and repeated for the second class below it; (**b**) Visualized data subset from two classes of Wisconsin breast cancer data from UCI Machine Learning Repository [17]; (**c**) 4-D point $A = (-1,1,-1,1)$ in two representations A_1 and A_2 in GLC-L coordinates $X_1 - X_4$ with angles $Q_1 - Q_4$.

Steps 2 and 3 of the algorithm for negative coefficients k_i and negative values x_i can be implemented in two ways. The first way represents a negative value x_i, e.g., $x_i = -1$ as a vector x_i that is directed backward relative to the vector that represent $x_i = 1$ on coordinate X_i. As a result, such vectors x_i go down and to the right. See representation A_1 in Figure 2c for point $A = (-1,1,-1,1)$ that is also self-crossing. The alternative representation A_2 (also shown in Figure 2c) uses the property that $k_i x_i > 0$ when both k_i and x_i are negative. Such $k_i x_i$ increases the linear function F by the same value as positive k_i and x_i. Therefore, A_2 uses the positive x_i, k_i and the "positive" angle associated with positive k_i. This angle is shown below angle Q_1 in Figure 2c. Thus, for instance, we can use $x_i = 1$, $k_i = 0.5$ instead of $x_i = -1$ and $k_i = -0.5$. An important advantage of A_2 is that it is perceptually simpler than A_1. The visualizations presented in this article use A_2 representation.

A linear function of n variables, where all coefficients c_i have similar values, is visualized in GLC-L by a line (graph, path) that is similar to a straight line. In this situation, all attributes bring similar contributions to the discriminant function and all samples of a class form a "strip" that is a simple form GLC-L representation.

In general, the term c_{n+1} is included in F due to both mathematical and the application reasons. It allows the coverage of the most general linear relations. If a user has a function with a non-zero c_{n+1}, the algorithm will visualize it. Similarly, if an analytical machine learning method produced such a function, the algorithm will visualize it too. Whether c_{n+1} is a meaningful bias or not in the user's task does not change the classification result. For regression problems, the situation is different; to get the exact meaningful result, c_{n+1} must be added and interpreted by a user. In terms of visualization, it only leads to the scale shift.

2.2. Interactive GLC-L Algorithm

For the data *classification* task, the interactive algorithm **GLC-IL** is as follows:

- It starts from the results of GLC-L such as shown in Figure 2b.
- Next, a user can interactively slide a yellow bar in Figure 2b to change a classification threshold. The algorithm updates the confusion matrix and the accuracy of classification, and pops it up for the user.
- An appropriate threshold found by a user can be interactively recorded. Then, a user can request an analytical form of the linear discrimination rule be produced and also be recorded.
- A user sets up two new thresholds if the accuracy is too low with any threshold (see Figure 3a with two green bars). The algorithm retrieves all n-points with projections that end in the interval between these bars. Next, only these n-D points are visualized (see Figure 3b).
- At this stage of the exploration the user has three options:

  (a)   modify interactively the coefficients by rotating the ends of the selected arrows (see Figure 4),
  (b)   run an automatic coefficient optimization algorithm GLC-AL described in Section 2.3,
  (c)   apply a visual structure analysis of classes presented in the visualization described in section.

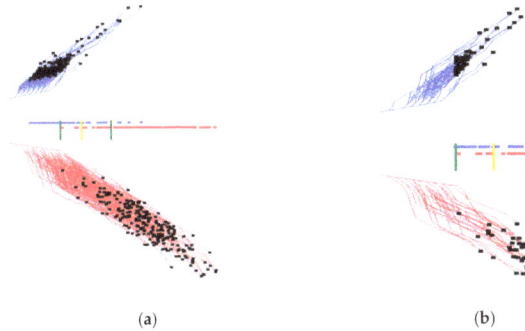

(a)   (b)

Figure 3. Interactive GLC-L setting with sliding green bars to define the overlap area of two classes for further exploration.(**a**) Interactive defining of the overlap area of two classes; (**b**) Selected overlapped n-D points.

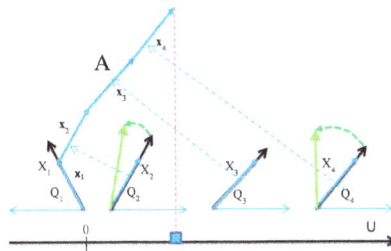

Figure 4. Modifying interactively the coefficients by rotating the ends of selected arrows, X_2 and X_4 are rotated.

For *clustering*, the interactive algorithm *GLC-IL* is as follows. A user interactively selects an n-D point of interest P by clicking on its 2-D graph (path) P^*. The system will find all graphs H^* that are close to it according to the rule below.

Let $P^* = (p_1, p_2, \ldots, p_n)$ and $H^* = (h_1, h_2, \ldots, h_n)$, where $p_i = (p_{i1}, p_{i2})$ and $h_i = (h_{i1}, h_{i2})$ are 2-D points (nodes of graphs),

T be a threshold that a user can change interactively,

$L(P, T)$ be a set of n-D points that are close to point P with threshold T (i.e., a cluster for P with T),

$L(P, T) = \{H: D(P^*, H^*) \leq T\}$, where $D(P^*, H^*) \leq T \Leftrightarrow \forall i \ | \ | p_i - h_i | \ | < T$, and

$| \ | p_i - h_i | \ |$ be the Euclidian distance between 2-D points p_i and h_i.

The automatic version of this algorithm searches for the largest T, such that only n-D points of the class, which contains point P, are in $L(P, T)$ assuming that the class labels are known,

$$max \ T: \{H \in L(P, T) \Rightarrow H \in \text{Class}(P)\},$$

where Class(P) is a class that includes n-D point P.

2.3. Algorithm GLC-AL for Automatic Discovery of Relation Combined with Interactions

The GLC-AL algorithm differs from the Fisher Linear Discrimination Analysis (FDA), Linear SVM, and Logistic Regression algorithms in the criterion used for optimization. The GLC-AL algorithm directly maximizes some value computed from the confusion matrix (typically accuracy), $A = (TP + TN)/(TP + TN + FP + FN)$, which is equivalent to the optimization criterion used in the linear perceptron [18] and Neural Networks in general. In contrast, the Logistic Regression minimizes the Log-likelihood [19]. The GLC-AL algorithm also allows maximization of the truth positive (TP). Fisher Linear Discrimination Analysis maximizes the ratio of between-class to within-class scatter [20]. The Linear SVM algorithm searches for a hyperplane with a large margin of classification, using the regularization and quadratic programming [21].

The automatic algorithm GLC-AL is combined with interactive capabilities as described below. The progress in accuracy is shown after every m iterations of optimization, and the user can stop the optimization at any moment to analyze the current result. It also allows interactive change of optimization criterion, say from maximization of accuracy to minimization of False Negatives (FN), which is important in computer-aided cancer diagnostic tasks.

There are several common computation strategies to maximize accuracy A in $[-1,1]^{n+1}$ space of coefficients k_i Gradient-based search, random search, genetic and other algorithms are commonly used to make the search feasible.

For the practical implementation, in this study, we used a simple random search algorithm that starts from a randomly generated set of coefficients k_i, computes the accuracy A for this set, then generates another set of coefficients k_i again randomly, computes A for this set, and repeats this process m times. Then the highest value of A is shown to the user to decide if it is satisfactory. This is Step 1 of the algorithm shown below. It is implemented in C++ and linked with OpenGL visualization and interaction program that implements Steps 2–4. A user runs the process m times more if it is not satisfactory. In Section 4.1, we show that this automatic step 1 is computationally feasible.

```
Step 1:
best_coefficients = []
while n > 0
        coefficients <- random(−1, 1)
        all_lines = 0
        for i data_samples:
            line = 0
            for x data_dimensions:
                if coefficients[x] < 0:
                    line = line – data_dimensions[x]*cos(acos(coefficients[x]))
                else:
                    line = line + data_dimensions[x]*cos(acos(coefficients[x]))
            all_lines.append(line)
            //update best_coefficients
        n−
```

Step 2: Projects the end points for the set of coefficients that correspond to the highest A value (in the same way as in Figure 4) and prints off the confusion matrix, i.e., for the best separation of the two classes.

Step 3:

Step 3a:

1: User moves around the class separation line.
2: A new confusion matrix is calculated.

Step 3b:

1: User picks the two thresholds to project a subset of the dataset.
2: n-D points of this subset (between the two thresholds) are projected.
3: A new confusion matrix is calculated.
4: User visually discovers patterns from the projection.

Step 4: User can repeat Step 3a or Step 3b to further zoom in on a subset of the projection or go back to Step 1.

Validation process. Typical 10-fold cross validation with 90–10% splits produces *10 different 90–10% splits* of data on the training and validation data. In this study, we used *10 different 70–30% splits* with 70% for the training set and 30% for the validation set in each split. Thus, we have the same 10 tests of accuracy as in the typical cross validation. Note that supervised learning tasks with 70–30% splits are more challenging than the tasks with 90–10% splits.

These 70–30% splits were selected by using permutation of data. The *splitting process* is as follows:

(1) indexing all *m* given samples from 1 to *m*, $w = (1,2, \dots ,m)$
(2) randomly permuting these indexes, and getting a new order of indexes, $\pi(w)$
(3) picking up first 70% of indexes from $\pi(w)$
(4) assigning samples with these indexes to be training data
(5) assigning remaining 30% of samples to be validation data.

This splitting process also can be used for a 90–10% split or other splits.
The total validation process for each set of coefficients k includes:

(i) applying data splitting process.
(ii) computing accuracy *A* of classification for this **k**.
(iii) repeating (i) and (ii) *t* times (each times with different data split).
(iv) computing average of accuracies found in all these runs.

2.4. Visual Structure Analysis of Classes

For the visual structure analysis, a user can interactively:

- Select border points of each class, coloring them in different colors.
- Outline classes by constructing an envelope in the form of a convex or a non-convex hull.
- Select most important coordinates by coloring them differently from other coordinates.
- Selecting misclassified and overlapped cases by coloring them differently from other cases.
- Drawing the prevailing direction of the envelope and computing its location and angle.
- Contrasting envelopes of difference classes to find the separating features.

2.5. Algorithm GLC-DRL for Dimension Reduction

A user can apply the automatic algorithm for dimension reduction anytime a projection is made to remove dimensions that don't contribute much to the overall line in the x direction (angles close to 90°). The contribution of each dimension to the line in the horizontal direction is calculated each time the GLC-AL finds coefficients. The algorithm for automatic dimension reduction is as follows:

Step 1: Setting up a threshold for the dimensions, which did not contribute to the line significantly in the horizontal projection.
Step 2: Based on the threshold from Step 1, dimensions are removed from the data, and the threshold is incremented by a constant.
Step 3: A new projection is made from the reduced data.
Step 4: A new confusion matrix is calculated.

The interactive algorithm for dimension reduction allows a user to pick up any coordinate arrow X_i and remove it by clicking on it, which leads to the zeroing of its projection. See coordinates X_2 and X_7 (in red) in Figure 5b. The computational algorithm for dimension reduction is as follows.

Step 1: The user visually examines the angles for each dimension, and determines which one is not contributing much to the overall line.

Step 2: The user selects and clicks on the angle from Step 1.

Step 3: The dimension, which has been selected, is removed from the dataset and a new projection is made along with a new confusion matrix. The dimension, which has been removed, is highlighted.

Step 4:

> Step 4a: The user goes back to Step 1 to further reduce the dimensions.
>
> Step 4b: The user selects to find other coefficients with the remaining dimensions for a better projection using the automatic algorithm GLC-AL described above.

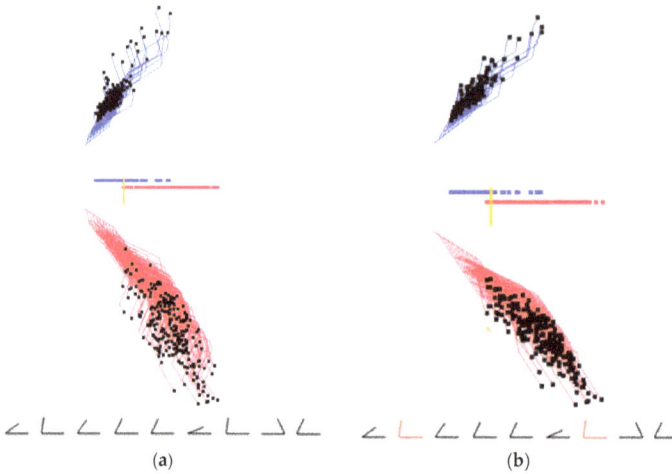

(a)   (b)

Figure 5. Interactive dimension reduction, angles for each dimension are shown on the bottom. (a) Initial visualization of two classes optimized by GLC-AL algorithm; (b) Visualization of two classes after 2nd and 7th dimensions (red) with low contribution (angle about 90°) have been removed.

2.6. Generalization of the Algorithms for Discovering Non-Linear Functions and Multiple Classes

Consider a goal of visualizing a function $F(x) = c_{11}x_1 + c_{12}x_1^2 + c_{21}x_2 + c_{22}x_2^2 + c_3x_3 + \ldots + c_nx_n + c_{n+1}$ with quadratic components. For this F, the algorithm treats x_i and x_i^2 as two different variables X_{i1} and X_{i2} with the separate coordinate arrows similar to Figure 1. Polynomials of higher order will have more than two such arrows. For a non-polynomial function $F(x) = c_1f_1(x_1) + c_2f_2(x_2) + \ldots + c_nf_n(x_n) + c_{n+1}$, which is a linear combination of non-linear functions f_i, the only modification in GLC-L is the substitution of x_i by $f_i(x_i)$ in the multiplication with angles still defined by the coefficients c_i. The rest of the algorithm is the same. For the multiple classes the algorithm follows the method used in the multinomial logistic regression by discrimination of one class against all other k-1 classes together. Repeating this process k times for each class will give k discrimination functions that allow the discrimination of all classes.

3. Results: Case Studies

Below we present the results of five case studies. In the selection of data for these studies, we followed a common practice in the evaluation of new methods—using benchmark data from the repositories with the published accuracy results for alternative methods as a more objective and less biased way than executing alternative methods by ourselves. We used two repositories: Machine Learning Repository at the University of California Irvine [17,22], and the Modified National Institute of Standards and Technology (MNIST) set of images of digits [23]. In addition, we used S&P 500 data for the period that includes the highly volatile time of Brexit.

3.1. Case Study 1

For the first study, Wisconsin Breast Cancer Diagnostic (WBC) data set was used [17]. It has 11 attributes. The first attribute is the id number which was removed and the last attribute is the class label which was used for classification. These data were donated to the repository in 1992. The samples with missing values were removed, resulting in 444 benign cases and 239 malignant cases. Figure 6 shows samples of screenshots where these data are interactively visualized and classified in GLC-L for different linear discrimination functions, providing accuracy over 95% of these data. The malignant cases are drawn in red and benign in blue.

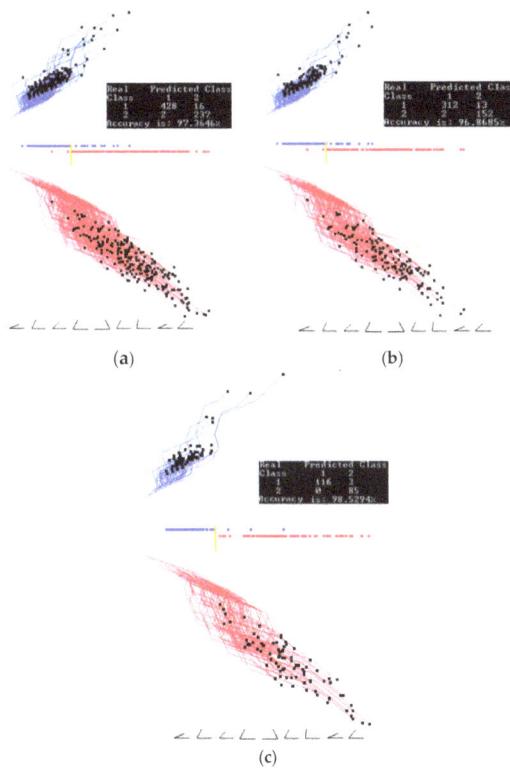

Figure 6. Results for the Wisconsin breast cancer data showing the training, validation, and the entire data set when trained on the entire data set. (**a**) Entire training and validation data set. Best projections of one of the first runs of GLC-AL. Coefficients found on the entire data set; (**b**) Data split into 70/30 (training and validation) showing only 70% of the data, using coefficients and the separation line found on the entire data set in (a); (**c**) Showing the 30% (validation set). Using the coefficients and the separation line same as in (a). Accuracy goes up.

Figures 6 and 7 show examples of how splitting the data into training and validation affects the accuracy. Figure 6 shows results of training on the entire data set, while Figure 7 shows results of training on 70% of the data randomly selected. The visual analysis of Figure 7 shows that 70% of data used for training are representative for the testing data too. This is also reflected in similar accuracies of 97.07% and 96.56% on these training and validation data. The next case studies are shown first on the entire data set to understand the whole dataset. Accuracy on the training and validation data can be found in Section 3.4, where a 70/30 split was also used.

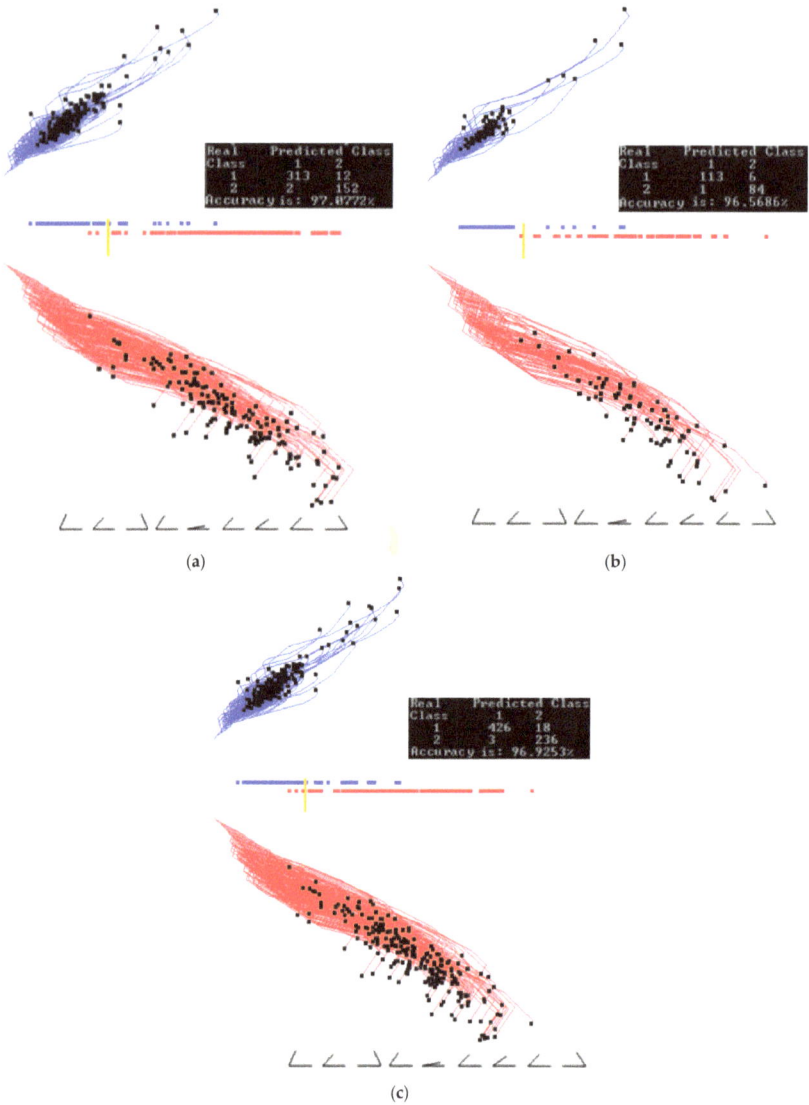

Figure 7. Results for the Wisconsin breast cancer data showing the training, validation, and the entire data set when trained on the training set. (**a**) Data are split using 70% (training set) to the find coefficients with the projecting training set. Best result from the first runs of GLC-AL; (**b**) Using the coefficients found by the training set in (**a**) and projecting the validation data set (30% of the data); (**c**) Projecting the entire data set using the coefficients found by the training set in (**a**).

Figure 8a shows the results for the best linear discrimination function obtained in the first 20 runs of the random search algorithm GLC-AL. The threshold found by this algorithm automatically is shown as a yellow bar. Results for the alternative discriminant functions from multiple runs of the random search by algorithm GLC-AL are shown in Figure 8b,c and Figure 9.

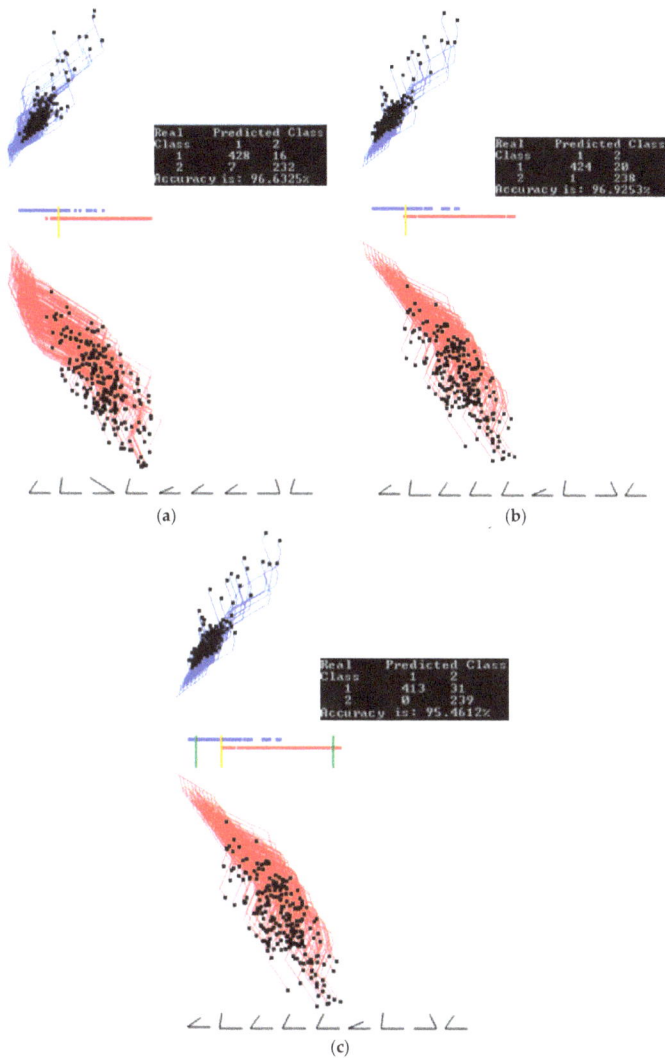

Figure 8. Wisconsin breast cancer data interactively visualized and classified in GLC-L for different linear discrimination functions. (**a**) Data visualized and classified using the best function of the first 20 runs of the random search with a threshold found automatically shown as a yellow bar; (**b**) Data visualized and classified using an alternative function from the first 20 runs with the threshold (yellow bar) at the positions having only one malignant (red case) on the wrong side and higher overall accuracy than in (a). (**c**) Visualization (b) where the separation threshold is moved to have all malignant (red cases) on the correct side with the tradeoff in the accuracy.

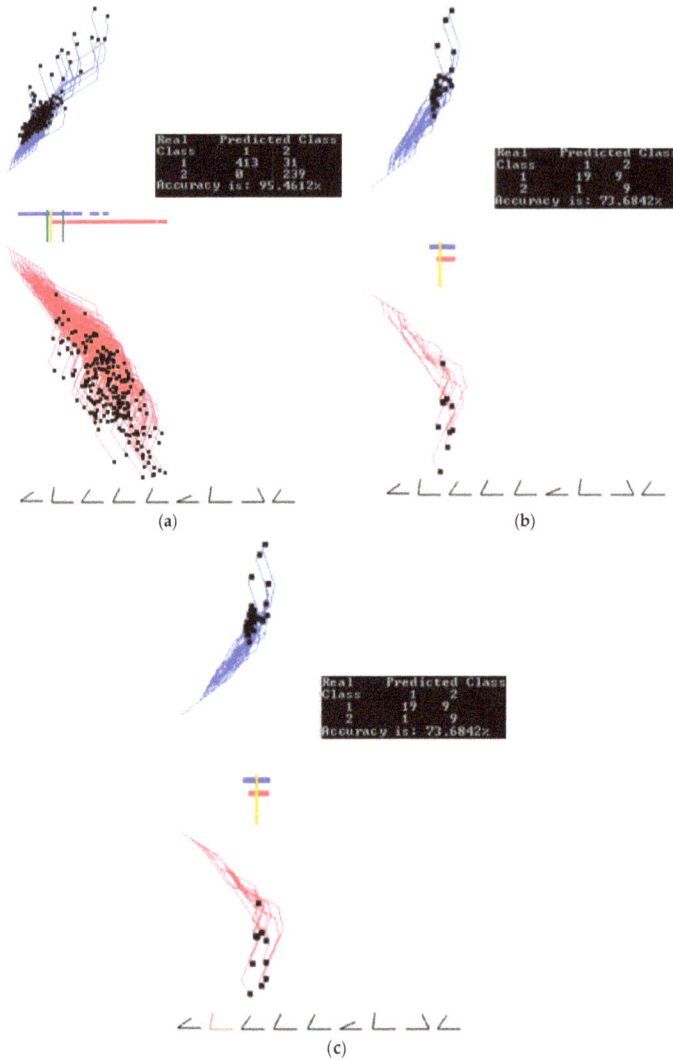

Figure 9. Wisconsin breast cancer data interactively projecting a selected subset. (**a**) Two thresholds are set from Figure 8c for selecting overlapping cases. (**b**) Overlapping cases from the interval between two thresholds from (a). (**c**) Overlapping cases from the interval between two thresholds, with the 2nd dimension with low contribution removed without decreasing accuracy.

In these examples the threshold (yellow bar) is located at the different positions, including the situations, where all malignant cases (red cases) are on the correct side of the threshold, i.e., no misclassification of the malignant cases.

Figure 9a–c shows the process and results of interactive selecting subsets of cases using two thresholds. This tight threshold interval selects heavily overlapping cases for further detailed analysis and classification. This analysis removed interactively the 2nd dimension with low contribution without decreasing accuracy (see Figure 9c).

3.2. Case Study 2

In this study the Parkinson's data set from UCI Machine Learning Repository [22] was used. This data set, known as Oxford Parkinson's Disease Detection Dataset, was donated to the repository in 2008. The Parkinson's data set has 23 attributes, one of them being status if the person has Parkinson's disease.

The dataset has 195 voice recordings from 31 people of which 23 have Parkinson's disease. There are several recordings from each person. Samples with Parkinson's disease present are colored red in this study. In the data preparation step of this case study, each column was normalized between 0 and 1 separately.

Figures 10 and 11 show examples of how splitting the data into training and validation sets affects the accuracy. Figure 10 shows results of training on the entire dataset, while Figure 11 shows results of training on 70% of the data randomly selected. The visual analysis of Figure 11 shows that 70% data used for training is also representative for the validation data. This is also reflected in similar accuracies of 91.24% and 88.71% respectively. The rest of illustration for this case study is for the entire dataset to understand the dataset as a whole. Accuracy on the training and validation can be found later in Section 3.4, where a 70/30 split was also used.

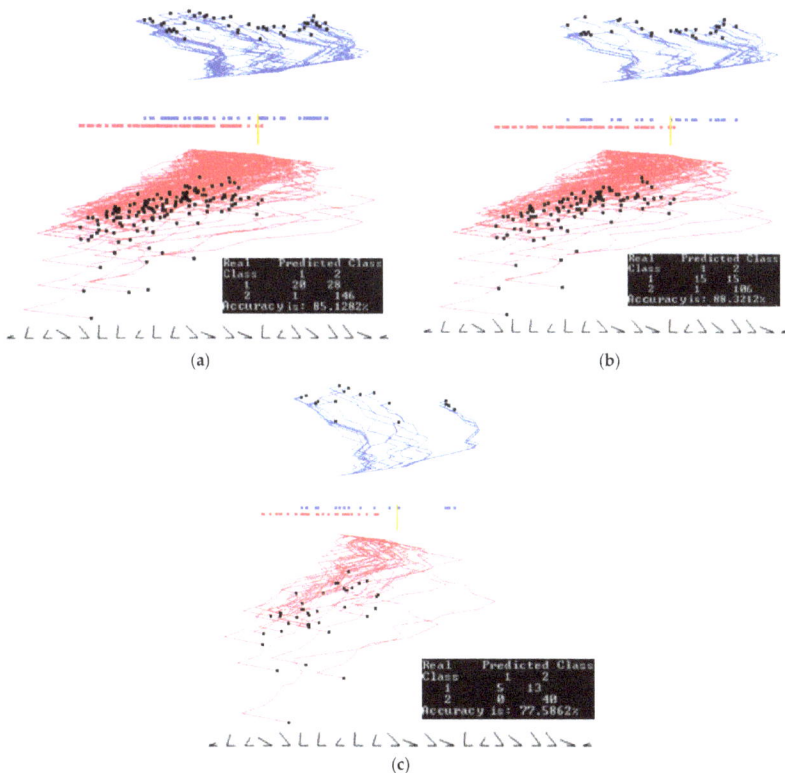

Figure 10. Results with Parkinson's disease data set showing the training, validation and the entire data set when trained on the entire data set. (**a**) Training and validation on the entire dataset. Best projections of one of the first runs of GLC-AL. Coefficients found on the entire data set; (**b**) Data split into 70/30 (training/validation) showing only 70% of the data, using the coefficients and the separation line found on the entire data set in (a). (**c**) Showing the 30% (validation set). Using coefficients and the separation line the same as in (a). Accuracy goes down.

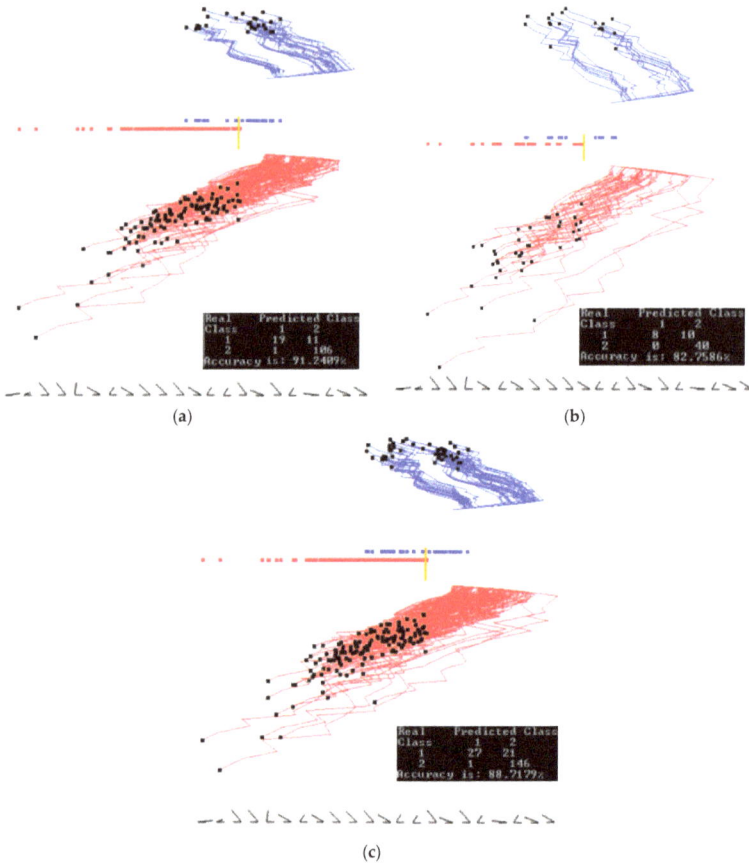

Figure 11. Results with Parkinson's disease data set showing the training, validation and the entire data set when trained on the training set. (**a**) Data is split. Using 70% (training set) to find the coefficients. Projecting training set, best from the first runs of GLC-A; (**b**) Using coefficients found by the training set in (a) and projecting the validation dataset (30% of the data); (**c**) Projecting the entire data set using coefficients found by the training set in (a).

The result for the best discrimination function found from the second run of 20 epochs is shown in Figure 12a. In Figure 12b, five dimensions are removed, some of them are with angles close to 90°, and the separation line threshold is also moved relative to Figure 12a. In Figure 12c, the two limits for a subinterval are set to zoom in on the overlapping samples.

In Figure 12d, where the subregion is projected and 42 samples are removed, the accuracy only decreases by 4% from 86.15% to 82.35%. Out of those 42 cases, 40 of them are samples of Parkinson's disease (red cases), and only 2 cases are not Parkinson's disease.

With such line separation as in Figure 12a–c, it is very easy to classify cases with Parkinson's disease from this dataset (high True Positive rate, TP); however, a significant number of cases with no Parkinson's disease are classified incorrectly (high False Positive rate, FP).

This indicates the need for improving FP more exploration, such as preliminary clustering of the data, more iterations to find coefficients, or using non-linear discriminant functions. The first attempt can be a quadratic function that is done by adding a squared coordinate X_i^2 to the list of coordinates without changing the GLC-L algorithm (see Section 2.5).

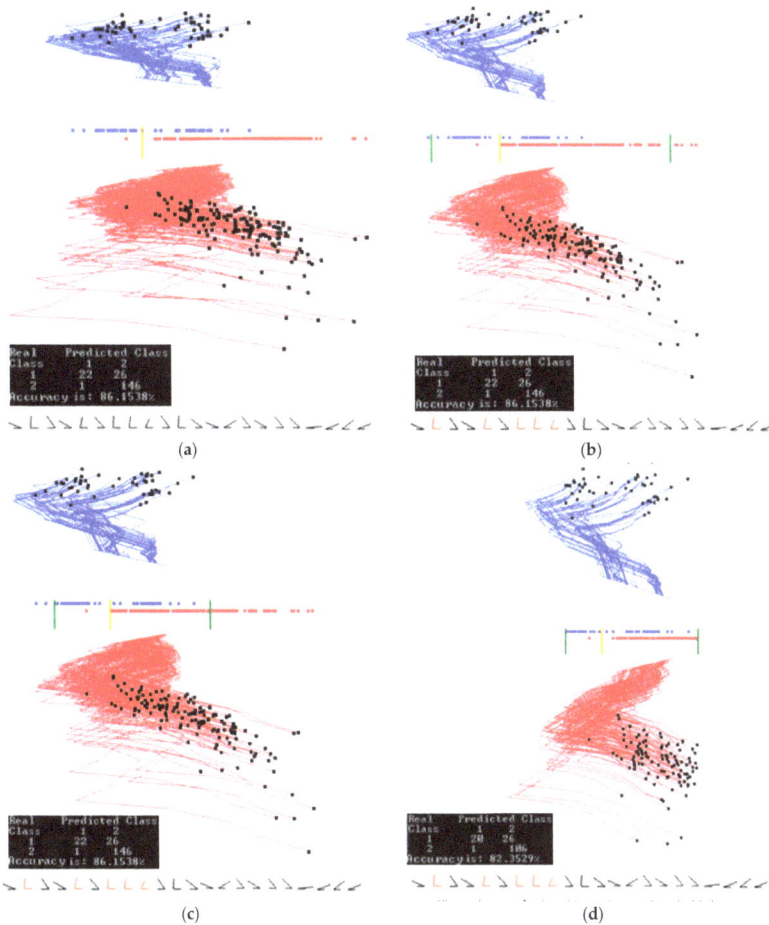

Figure 12. Additional Parkinson's disease experiments. (**a**) Best projection from the second run of 20 epochs. (**b**) Projection with 5 dimensions removed. Separation line threshold is also moved. Accuracy stays the same; (**c**) Two limits for a subinterval are set; (**d**) Only cases for the subinterval are projected with the separation line moved. Accuracy drops.

3.3. Case Study 3

In this study, a subset of the Modified National Institute of Standards and Technology (MNIST) database [23] was used. Images of digit 0 (red) and digit 1 (blue) were used for projection with 900 samples for each digit. In the preprocessing step, each image is cropped to remove the border. The images after cropping were 22 × 22, which is 484 dimensions.

Figures 13 and 14 show examples of how splitting the data into training and validation changes the accuracy. Figure 13 shows the results of training on the entire data set, while Figure 14 shows the results of training on 70% of the data, which are randomly selected. The visual analysis of Figure 14 shows that 70% of the data used for training are also representative for the validation data. This is also reflected in similar accuracies of 91.58% and 91.44% respectively. The rest of this case study is illustrated on the entire data set to understand the dataset as a whole. Accuracy on the training and validation can be found later in Section 4.1, where a 70/30 split was also used.

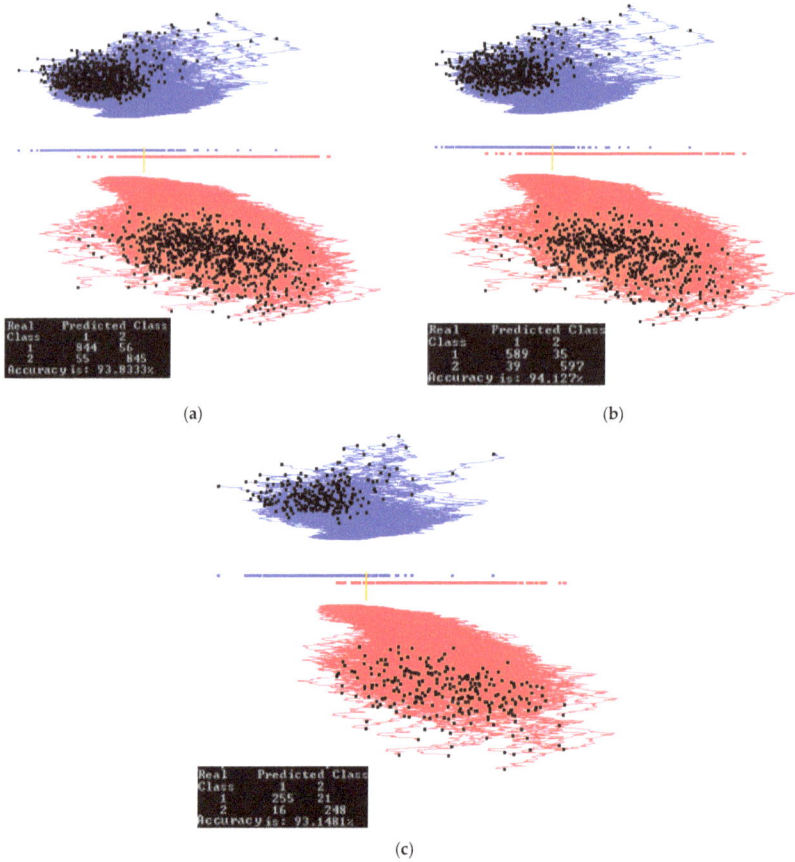

(a)

(b)

(c)

Figure 13. MNIST subset for digits 0 and 1 dataset showing training, validation, and the entire data set when trained on the entire data set. (**a**) Training and validation on the entire data set. Best projections of one of the first runs of GLC-A. Coefficients found on the entire data set; (**b**) Data split into 70/30% (training/validation) showing only 70% of the data, using coefficients and separation line found on the entire data set in (a); (**c**) Showing the 30% (validation set). Using the coefficients and the separation line same as in (a). Accuracy goes down.

(a)　　　　　　　　　　　　　　　　(b)

(c)

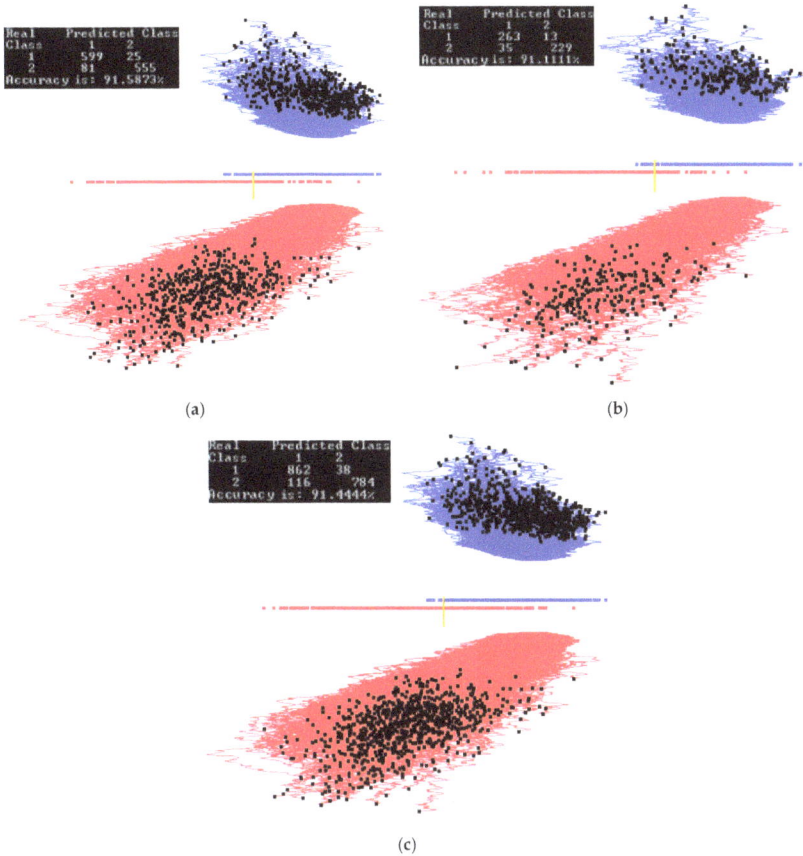

Figure 14. MNIST subset for digits 0 and 1 data set showing training, validation, and the entire data set when trained on the training set. (**a**) Data is split. Using 70% (training set) to find the coefficients. Projecting training set, best from the first runs of GLC-A; (**b**) Using the coefficients found by the training set in (a) and projecting the validation dataset (30% of the data); (**c**) Projecting the entire data set using the coefficients found by the training set in (a).

Figure 15a shows the results of applying the algorithm GLC-AL to these MNIST images. It is the best discriminant function of the first run of 20 epochs with the accuracy of 95.16%. Figure 15b shows the result of applying the automatic algorithm GLC-DRL to these data and the discriminant function. It displays 249 dimensions and removes 235 dimensions, dropping the accuracy only slightly by 0.28%. Figure 15c shows the result when a user decided to run the algorithm GLC-DRL a few more times. It removed a total of 393 dimensions, kept and projected the remaining 91 dimensions with the accuracy dropping to 83.77% from 93.84% as shown in Figure 15b.

Figure 15d shows the result of user interaction with the system by setting up the interval (using two bar thresholds) to select a subset of the data of interest in the overlap of the projections. The selected data are shown in Figure 16a. Figure 16b shows the results of running GLC-AL algorithm on the subinterval to find a better discriminant function and projection. Accuracy goes up by 5.6% in this subinterval.

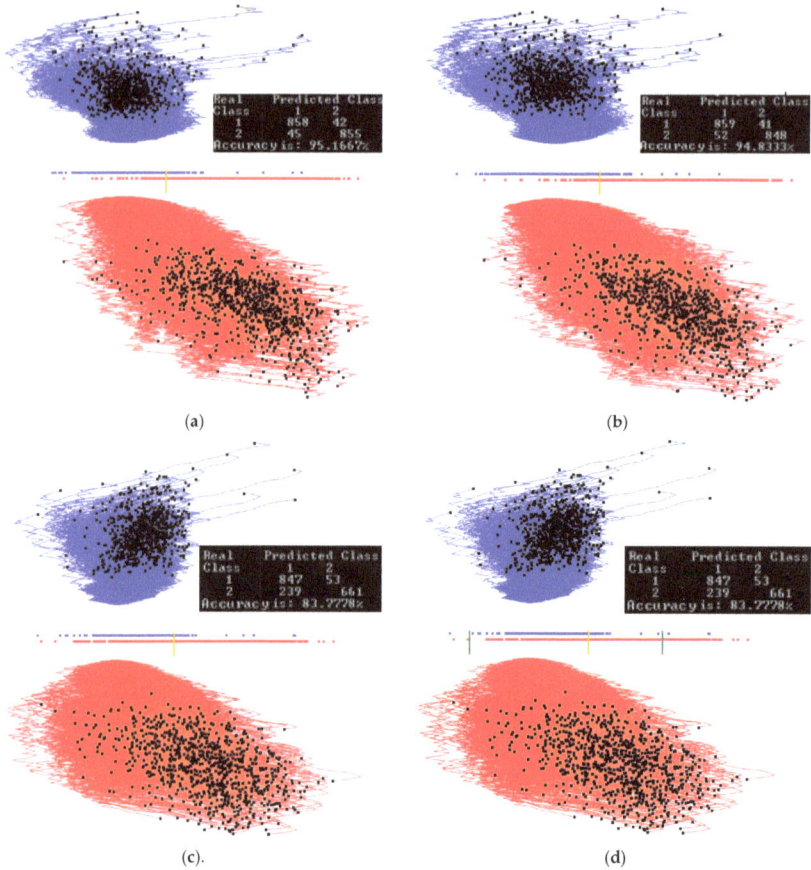

Figure 15. Experiments with 900 samples of MNIST dataset for digits 0 and 1. (**a**) Results for the best linear discriminant function of the first run of 20 epochs; (**b**) Results of the automatic dimension reduction displaying 249 dimensions with 235 dimensions removed with the accuracy dropped by 0.28%; (**c**) Automatic dimension reduction, which is run a few more times removing a total of 393 dimensions and keeping 91 dimensions with dropped accuracy; (**d**) Thresholds for a subinterval are set (green bars).

Next, the automatic dimension reduction algorithm GLC-DRL is run on these subinterval data, removing the 46 dimensions and keeping and projecting the 45 dimensions with the accuracy going up by 1% (see Figure 16c). Figure 16d shows the result when a user decided to run the algorithm GLC-DRL a few more times on these data, removing 7 more dimensions, and keeping 38 dimensions, with the accuracy gaining 6.8%, and reaching 95.76%.

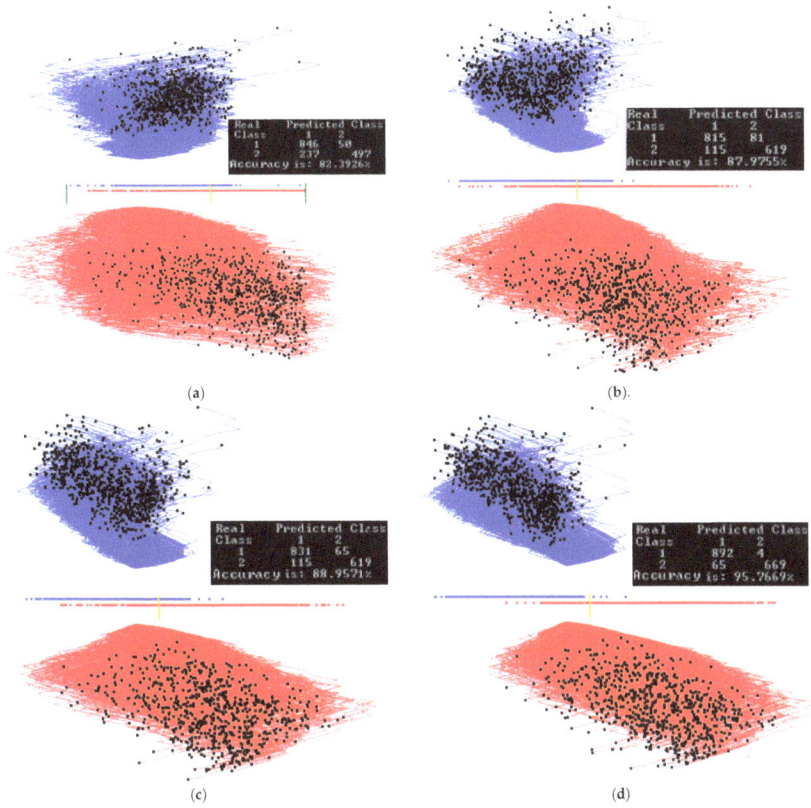

Figure 16. Experiments with a 900 samples of MNIST dataset for digits 0 and 1 doing automatic dimension reduction. (**a**) Data between the two green thresholds are visualized and projected; (**b**) GLC-AL algorithm on the subinterval to find a better projection. Accuracy goes up by 5.6% in the subregion; (**c**) Result of automatic dimension reduction running a few more times that removes 46 dimensions and keeps 45 dimensions with accuracy going up by 1%; (**d**) Result of automatic dimension reduction running a few more times that removes 7 more dimensions and keeps 38 dimensions with accuracy going up 6.8% more.

3.4. Case Study 4

Another experiment was done on a different subset of the MNIST database to see if any visual information could be extracted on encoded images. For this experiment, the training set consisted of all samples of digit 0 and digit 1 from the training set of MNIST (60,000 images). There was 12,665 samples of digit 0 and digit 1 combined in the training set. The validation set consisted of all the samples of digit 0 and digit 1 from the validation set of MNIST (10,000 images). There was 2115 samples of digit 0 and digit 1 combined in the validation set. The preprocessing step for the data was the same as for case study 3, where pixel padding was removed resulting in 22×22 images.

A Neural Network Autoencoder which was constructed using Python library Keras [24] encoded the images from 484 (22×22) dimensions to 24 dimensions. The Keras library originally was developed by François Chollet [25]. We used Keras version 1.0.2 running with python version 2.7.11. The Autoencoder had one hidden layer and was trained on the training set (12,665 samples). Examples of decoded images can be seen in Figure 17. The validation set (2115 images) was passed through the encoder to get its representation in 24 dimensions. The encoded validation set is what has been used in this case study images) was passed through the encoder to get its representation in 24 dimensions. The encoded validation set is what has been used in this case study.

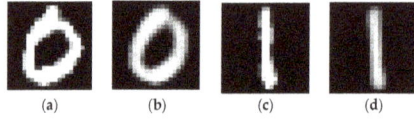

(a) (b) (c) (d)

Figure 17. Examples of original and encoded images. (**a**) Example of a digit after preprocessing; (**b**) Decoded image from 24 values into 484. Same image as in (a); (**c**) Another example of a digit after preprocessing; (**d**) Decoded image from 24 values into 484. Same image as in (c).

The goal of this case study is to compare side-by-side GLC-L visualization with parallel coordinates. Figure 18 shows the comparison of these two visualizations using 24 dimensions found by the Autoencoder among the original 484 dimensions. Figure 18a,b shows the difference between the two classes more clearly than Parallel coordinates (PC) in Figure 18c,d. The shapes of the clouds in GLC-L are very different. The red class is elongated and the blue one is more rounded and shifted to the right relative to the red cloud. It also shows the separation threshold between classes and the accuracy of classification.

Parallel coordinates in Figure 18c,d do not show a separation between classes and accuracy of classification. Only a visual comparison of Figure 18c,d can be used for finding the features that can help in building a classifier. In particular, the intervals on features 10, 8, 23, 24 are very different, but the overlap is not, allowing the simple visual separation of the classes. Thus there is no clear way to classify these data using PCs. PCs give only some informal hints for building a classifier using these features.

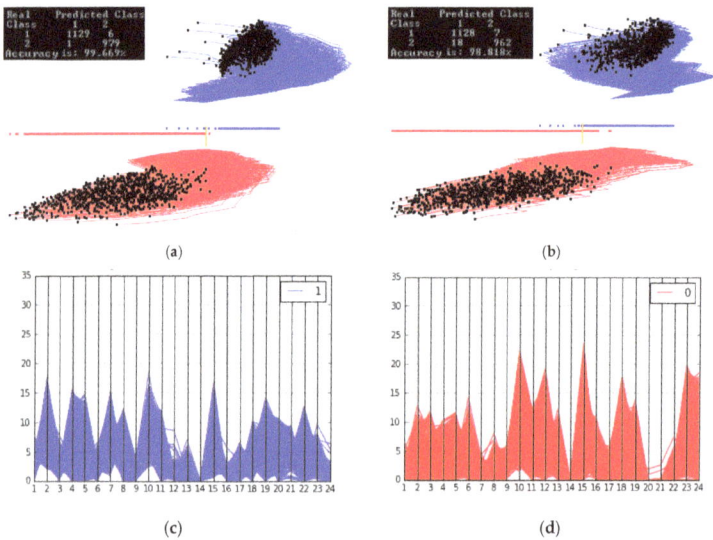

(a) (b)

(c) (d)

Figure 18. Comparing encoded digit 0 and digit 1 on the parallel coordinates and GLC-L, using 24 dimensions found by the Autoencoder among 484 dimensions. Each vertical line is one of the 24 features scaled in the [0,35] interval. (**a**) Results for the best linear discriminant function of the first run of 20 epochs; (**b**) Another run of 20 epochs, best linear discriminant function from this run. Accuracy drops 1%; (**c**) Digit 1 is visualized on the parallel coordinates; (**d**) Digit 0 is visualized on the parallel coordinates

3.5. Case Study 5

This study uses S & P 500 data for the first half of 2016 that include highly volatile S&P 500 data at the time of the Brexit vote. S & P 500 lost 75.91 points in one day (from 2113.32 to 2037.41) from 23 June to 24 June 2016 and continued dropping to 27 June to 2000.54 with total loss of 112.78 points since 23 June. The loss of value in these days is 3.59% and 5.34%, respectively. The goal of this case study is predicting S&P 500 up/down changes on Fridays knowing S&P 500 values for the previous four days of the week. The day after the Brexit vote is also a Friday. Thus, it is also of interest to see if the method will be able to predict that S&P 500 will go down on that day.

Below we use the following notation to describe the construction of features used for prediction:

$S_1(w)$, $S_2(w)$, $S_3(w)$, $S_4(w)$, and $S_5(w)$ are S&P 500 values for Monday-Friday, respectively, of week w;

$D_i(w) = S_{i+1}(w) - S_i(w)$ are differences in S&P 500 values on adjacent days, $i = 1:4$;

Class(w) = 1 (down) if $D_4(w) < 0$, Class(w) = 2 (up) if $D_4(w) > 0$, Class(w) = 0 (no change) if $D_4(w) = 0$.

We computed the attributes $D_1(w)$–$D_4(w)$ and Class(w) from the original S & P 500 time series for the trading weeks of the first half of 2016. Attributes D_1–D_3 were used to predict Class (up/down). We excluded incomplete trading weeks, getting 10 "Friday down" weeks, and 12 "Friday up" weeks available for training and validation. We used multiple 60%:40% splits of these data on training and validation due to a small size of these data. The Brexit week was a part of the validation data set and was never included in the training datasets. Figures 19 and 20 show the best results on the training and validation data, which are: 76.92% on the training data, and 77.77% on the validation data. We attribute the greater accuracy on the validation data to a small dataset. These accuracies are obtained for two different sets of coefficients. The accuracy of one of the runs was 84.81% on training data, but its accuracy on validation was only 55.3%. The average accuracy on all 10 runs was 77.78% on all training data, and 61.12% on the validation data. While these accuracies are lower than in the case studies 1–4, they are quite common in such market predictions [16]. The accuracy of the down prediction for 24 June (after Brexit) in those 10 runs was correct in 80% of the runs, including the runs shown in Figures 19 and 20 as green lines.

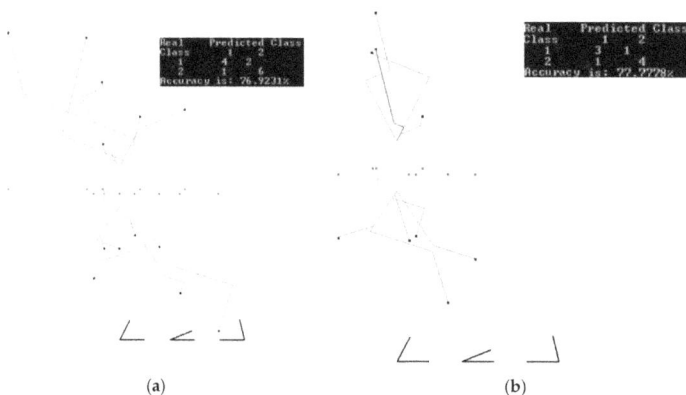

Figure 19. Run 3: data split 60%/40% ((training/validation) for the coefficients K = (0.3, 0.6, −0.6). (**a**) Results on training data (60%) for the coefficients found on these training data; (**b**) Results on validation data (40%) for coefficients found on the training data. A green line (24 June after Brexit) is correctly classified as S&P 500 down.

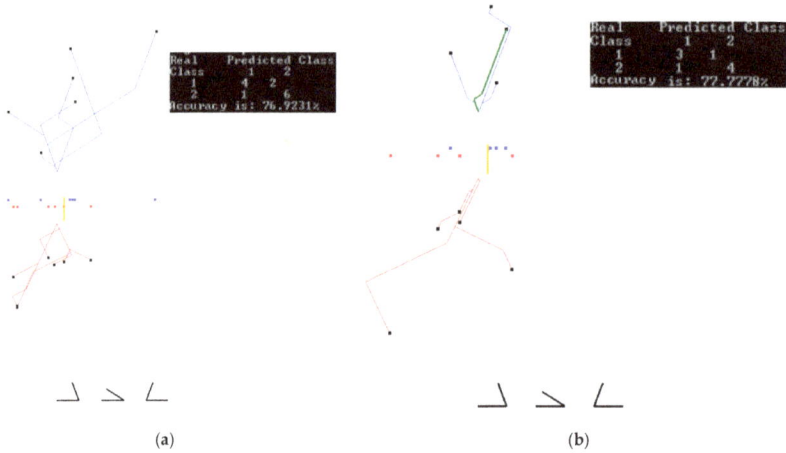

Figure 20. Run 7: data split 60%/40% ((training/validation) for coefficients K = (−0.3, −0.8, 0.3). (a) Results on training data (60%) for coefficients found on these training data; (b) Results on validation data (40%) for a coefficient found on training data. A green line (24 June after Brexit) is correctly classified as S&P 500 down.

4. Discussion and Analysis

4.1. Software Implementation, Time and Accuracy

All algorithms, including visualization and GUI, were implemented in OpenGL and C++ in the Windows Operating System. A Neural Network Autoencoder was implemented using the Python library Keras [24,25]). Later, we expect to make the programs publicly available.

Experiments presented in Section 3 show that 50 iterations produce a nearly optimal solution in terms of accuracy. Increasing the number of iterations to a few hundred did not show a significant improvement in the accuracy after 50 epochs. In the cases where better coefficients were found past 50 epochs, the accuracy increase was less than 1%.

The automatic step 1 of the algorithm GLC-AL in the case studies above had been computationally feasible. For $m = 50$ (number of iterations) in the case study, it took 3.3 s to get 96.56% accuracy on a PC with 3.2 GHz quad core processor. Running under the same conditions, case study 2 took 14.18 s to get 87.17% accuracy, and for case study 3, it took 234.28 s to get 93.15% accuracy. These accuracies are on the validation sets for the 70/30 split of data into the training and validation sets. In the future, work on a more structured random search can be implemented to decrease the computation time in high dimensional studies.

Table 1 shows the accuracy on training and validation data for the case studies 1–3. We conducted 10 different runs of 50 epochs each with the automatic step 1 of the algorithm GLC-AL. Each run had freshly permutated data, which were split into 70% training and 30% validation as described in Section 2.3. In some runs for case study 1, the validation accuracy was higher than the training one, which is because the data set is fairly small and the split heavily influences the accuracy.

The training accuracies in Table 1 are the highest ones obtained in each run of the 50 epochs. Validation accuracies in this table are computed for the discrimination functions found on respective training data. The variability of accuracy on training data among 50 epochs in each run is quite high. For instance, the accuracy varied from 63.25% to 97.91% with the average of 82.46% in case study 1.

Table 1. Best accuracies of classification on training and validation data with 70–30% data splits.

Case Study 1 Results. Wisconsin Breast Cancer Data			Case Study 2 Results. Parkinson's Disease			Case Study 3 Results. MNIST-Subset		
Run	Training Accuracy, %	Validation Accuracy, %	Run	Training Accuracy, %	Validation Accuracy, %	Run	Training Accuracy, %	Validation Accuracy, %
1	96.86	96.56	1	89.05	74.13	1	98.17	98.33
2	96.65	97.05	2	85.4	84.48	2	94.28	94.62
3	97.91	96.56	3	84.67	94.83	3	95.07	94.81
4	96.45	96.56	4	84.67	84.48	4	97.22	96.67
5	97.07	96.57	5	85.4	77.58	5	94.52	93.19
6	97.91	96.07	6	84.67	93.1	6	92.85	91.48
7	97.07	96.56	7	84.67	86.2	7	96.03	95.55
8	97.49	98.04	8	87.59	87.93	8	94.76	94.62
9	97.28	98.03	9	86.13	82.76	9	96.11	95.56
10	96.87	97.55	10	83.94	87.93	10	95.43	95.17
Average	97.16	96.95	Average	85.62	85.34	Average	95.44	95.00

4.2. Comparison with Published Results

Case study 1. For the Wisconsin breast cancer (WBC) data, the best accuracy of *96.995%* is reported in [26] for the 10 fold cross-validation tests using SVM. This number is slightly above our average accuracy of *96.955* on the validation data in Table 1, using the GLC-L algorithm. We use the 70:30 split, which is more challenging, for getting the highly accurate training, than the 10-fold 90:10 split used in that study. The best result on validation is *98.04%*, obtained for the run 8 with 97.49% accuracy on training (see Table 1) that is *better than 96.995%* in [26].

The best results from previous studies collected in [26] include *96.84%* for SVM-RBF kernel [27] and *96.99%* for SVM in [28]. In [26] the best result by a combination of major machine learning algorithms implemented in Weka [29] is *97.28%*. The combination includes SVM, C4.5 decision tree, naïve Bayesian classifier and k-Nearest Neighbors algorithms. This result is below the *98.04%* we obtained with GLC-L. It is more important in the cancer studies to control the number of misclassified cancer cases than the number of misclassified benign cases. The total accuracy of classification reported [26] does not show these actual false positive and false negative values, while figures in Section 3.2 for GLC-L show them along with GLC-L interactive tools to improve the number of misclassified cancer cases.

This comparison shows that the GLC-L algorithm can compete with the major machine learning algorithms in accuracy. In addition, the GLC-L algorithm has the following important advantages: it is (i) simpler than the major machine learning algorithms, (ii) visual, (iii) interactive, and (iv) understandable by a user without advanced machine learning skills. These features are important for *increasing user confidence* in the learning algorithm outcome.

Case study 2. In [30] 13 Machine Learning algorithms have been applied to these Parkinson's data. The accuracy ranges from 70.77 for Partial Least Square Regression algorithm, 75.38% to ID3 decision tree, to 88.72 for Support Vector Machine, 97.44 for k-Nearest Neighbors and 100% for Random decision tree forest, and average accuracy equal to *89.82%* for all 13 methods.

The best results that we obtained with GLC-L for the same 195 cases are *88.71%* (Figure 11c) and *91.24 %* (Figure 11a). Note that the result in Figure 11a was obtained by using only 70% of 195 cases for training, not all of them. The split for training and validation is not reported in [30], making the direct comparison with results from GLC-L difficult. Just by using more training data, the commonly used split 90:10 can produce higher accuracy than a 70:30 split. Therefore, the only conclusion that can be made is that the accuracy results of GLC-L are comparable with average accuracy provided by common analytical machine learning algorithms.

Case study 3. While we did not conduct the full exploration of MNIST database for all digits, the accuracy with GLC-AL is comparable and higher than the accuracy of other linear classifiers reported in the literature for all digits and whole dataset. Those errors are 7.6% (accuracy 92.4%), for a pairwise linear classifier, and 12% and 8.4% (accuracy 88% and 91.6%), for two linear classifiers

(1-layer NN) [31]. From 1998, dramatic progress was reached in this dataset with non-linear SVM and deep learning algorithms with over 99% accuracy [23]. Therefore, future application of the non-linear GLC-L (as outlined in Section 2.5) also promises higher accuracy.

For **Case study 4**, the comparison with parallel coordinates is presented in Section 3.4. For **case study 5,** we are not aware of similar experiments, but the accuracy in this case study is comparable with the accuracy of some published stock market predictions in the literature.

5. Conclusions

This paper presented GLC-L, GLC-IL, GLC-AL, and GLC-DRL algorithms, and their use for knowledge discovery in solving the machine learning classification tasks in n-D interactively. We conducted the five case studies to evaluate these algorithms using the data from three domains: computer-aided medical diagnostics, image processing, and finance (stock market). The utility of our algorithms was illustrated by these empirical studies.

The main advantages and benefits of these algorithms are: (1) lossless (reversible) visualization of n-D data in 2-D as graphs, (2) absence of the self-crossing of graphs (planar graphs), (3) dimension scalability (the case studies had shown the success with 484 dimensions), (4) scalability to the number of cases (interactive data clustering increases the number of data cases that can be handled), (5) integration of automatic and interactive visual means, (6) reached the accuracy on the level of analytical methods, (7) opportunities to justify the linear and non-linear models vs. guessing a class of predictive models, (8) simple to understand for a non-expert in Machine Learning (can be done by a subject matter expert with minimal support from the data scientist), (9) supports multi-class classification, (10) easy visual metaphor of a linear function, and (11) applicable (scalable) for discovering patterns, selecting the data subsets, classifying data, clustering, and dimension reduction.

In all experiments in this article, 50 simulation epochs were sufficient to get acceptable results. These 50 epochs correspond to computing 500 values of the objective function (accuracy), due to ten versions of training and validation data in each epoch. The likely contribution to rapid convergence of data that we used is the possible existence of many "good" discrimination functions that can separate classes in these datasets accurately enough. This includes the situations with a wide margin. In such situations a "small" number of simulations can find "good" functions. The indirect confirmation of multiplicity of discriminant functions is a variety of machine learning methods that produced quite accurate discrimination on these data that can be found in the literature. These methods include SVM, C4.5 decision tree, naïve Bayesian classifier, k-Nearest Neighbors, and Random forest. The likely contribution of the GLC-AL algorithm is in random generation of vectors of coefficients {K}. This can quickly cover a *wide range* of K in the hypercube $[-1,1]^{n+1}$ and capture K that quickly gives high accuracy. Next, a "small" number of simulations is a typical fuzzy set with multiple values. In our experiments, this number was about 50 iterations. Building a full membership function of this fuzzy set is a subject of future studies on multiple different datasets.

The proposed visual analytics developments can help to improve the control of underfitting (overgeneralization) and overfitting in learning algorithms known as bias-variance dilemma [32], and will help in more rigorous selection of a class of machine learning models. Overfitting rejects relevant cases by considering them irrelevant. In contrast, underfitting accepts irrelevant cases, considering them as relevant.

A common way to deal with overfitting is adding a *regularization term* to the cost function that penalizes the complexity of the discriminant function, such as requiring its smoothness, certain prior distributions of model parameters, limiting the number of layers, certain group structure, and others [32]. Moreover, such cost functions as the least-squares can be considered as a form of regularization for the regression task. The regularization approach makes ill-posed problems mathematically *solvable*. However, those extra requirements often are *external* to the user task. For instance, the least-square method may not optimize the error for most important samples because it is not weighted. Thus, it is difficult to justify a regularizer including λ parameter, which controls the importance of the regularization term.

The linear discriminants considered in this article are among the simplest discriminants with *lower variance* predictions outside training data and do not need to be penalized for complexity. Further simplification of linear functions commonly is done by dimension reduction that is explored in Section 2.5. Linear discriminants suffer much more from overgeneralization (*higher bias*).

The expected contribution of the visual analytics discussed in the article to deal with the bias-variance dilemma is not in a direct justification of a regularization term to the cost function. It is in the introduction of another form of a *regularizer outside of the cost function.* In general, it is not required for a regularizer to be a part of the cost function to fulfil its role of *controlling underfitting and overfitting.*

The opportunity for such an outside control can be illustrated by multiple figures in his article. For instance, Figure 7a shows that all cases of class 1 form an elongated shape, and all cases of class 2 form another elongated shape on the training data. The assumption that the training data are representative for the whole data leads to the expectation that all new data of each class must be within these elongated shapes of the training data with some margin. Figure 7b shows that this is the case for validation data from Figure 7b. The analysis shows that only a few cases of validation data are outside of the convex hulls of the respective elongated shapes of classes 1 and 2 of the training data. This confirms that these training data are representative for the validation data. The linear discriminant in Figure 7a,b, shown as a yellow bar, classifies any case on the left to class 1 and any case on the right to class 2, i.e., significantly underfits (overgeneralizes) elongated shapes of the classes.

Thus, elongated shapes and their convex hull can serve as alternative *regularizers*. The idea of convex hulls is outlined in Section 2.4. Additional requirements may include snootiness or simplicity of an envelope at the margin distance μ from the elongated shapes. Here μ serves as a generalization parameter that plays a similar role that λ parameter plays to control the importance of the regularization term within a cost function.

Two arguments for simpler linear classifiers vs. non-linear classifiers are presented in [33]: (1) non-linear classifiers not always provide a significant advantage in performance, and (2) the relationship between features and the prediction can be harder to interpret for non-linear classifiers. Our approach, based on a linear classifier with non-linear constraints in the form of envelopes, takes a middle ground, and thus provides an opportunity to combine advantages from both linear and non-linear classifiers.

While the results are positive, these algorithms can be improved in multiple ways. Future studies will expand the proposed approach to knowledge discovery in datasets of larger dimensions with the larger number of instances and with heterogeneous attributes. Other opportunities include using GLC-L and related algorithms: (a) as a visual interface to larger repositories: not only data, but models, metadata, images, 3-D scenes and analyses, (b) as a conduit to combine visual and analytical methods to gain more insight.

Acknowledgments: We are very grateful to the anonymous reviewers for their valuable feedback and helpful comments.

Author Contributions: Boris Kovalerchuk and Dmytro Dovhalets conceived and designed the experiments; Dmytro Dovhalets performed the experiments; Boris Kovalerchuk and Dmytro Dovhalets analyzed the data; Boris Kovalerchuk contributed materials/analysis tools; Boris Kovalerchuk wrote the paper.

Conflicts of Interest: The authors declare no conflict of interest.

References

1. Bertini, E.; Tatu, A.; Keim, D. Quality metrics in high-dimensional data visualization: An overview and systematization. *IEEE Trans. Vis. Comput. Gr.* **2011**, *17*, 2203–2212. [CrossRef] [PubMed]
2. Ward, M.; Grinstein, G.; Keim, D. *Interactive Data Visualization: Foundations, Techniques, and Applications*; A K Peters/CRC Press: Natick, MA, USA, 2010.

3. Rübel, O.; Ahern, S.; Bethel, E.W.; Biggin, M.D.; Childs, H.; Cormier-Michel, E.; DePace, A.; Eisen, M.B.; Fowlkes, C.C.; Geddes, C.G.; et al. Coupling visualization and data analysis for knowledge discovery from multi-dimensional scientific data. *Procedia Comput. Sci.* **2010**, *1*, 1757–1764. [CrossRef] [PubMed]
4. Inselberg, A. *Parallel Coordinates: Visual Multidimensional Geometry and Its Applications*; Springer: New York, NY, USA, 2009.
5. Wong, P.; Bergeron, R. 30 years of multidimensional multivariate visualization. In *Scientific Visualization—Overviews, Methodologies and Techniques*; Nielson, G.M., Hagan, H., Muller, H., Eds.; IEEE Computer Society Press: Washington, DC, USA, 1997; pp. 3–33.
6. Kovalerchuk, B.; Kovalerchuk, M. Toward virtual data scientist. In Proceedings of the 2017 International Joint Conference On Neural Networks, Anchorage, AK, USA, 14–19 May 2017; pp. 3073–3080.
7. XmdvTool Software Package for the Interactive Visual Exploration of Multivariate Data Sets. Version 9.0 Released 31 October 2015. Available online: http://davis.wpi.edu/~xmdv/ (accessed on 24 June 2017).
8. Yang, J.; Peng, W.; Ward, M.O.; Rundensteiner, E.A. Interactive hierarchical dimension ordering, spacing and filtering for exploration of high dimensional datasets. In Proceedings of the 9th Annual IEEE Conference on Information Visualization, Washington, DC, USA, 19–21 October 2003; pp. 105–112.
9. Lin, X.; Mukherji, A.; Rundensteiner, E.A.; Ward, M.O. SPIRE: Supporting parameter-driven interactive rule mining and exploration. *Proc. VLDB Endow.* **2014**, *7*, 1653–1656. [CrossRef]
10. Yang, D.; Zhao, K.; Hasan, M.; Lu, H.; Rundensteiner, E.; Ward, M. Mining and linking patterns across live data streams and stream archives. *Proc. VLDB Endow.* **2013**, *6*, 1346–1349. [CrossRef]
11. Zhao, K.; Ward, M.; Rundensteiner, E.; Higgins, H. MaVis: Machine Learning Aided Multi-Model Framework for Time Series Visual Analytics. *Electron. Imaging* **2016**, 1–10. [CrossRef]
12. Kovalerchuk, B.; Grishin, V. Adjustable general line coordinates for visual knowledge discovery in n-D data. *Inf. Vis.* **2017**. [CrossRef]
13. Grishin, V.; Kovalerchuk, B. Multidimensional collaborative lossless visualization: Experimental study. In Proceedings of the International Conference on Cooperative Design, Visualization and Engineering (CDVE 2014), Seattle, WA, USA, 14–17 September 2014; Luo, Y., Ed.; Springer: Basel, Switzerland, 2014; pp. 27–35.
14. Kovalerchuk, B. Super-intelligence challenges and lossless visual representation of high-dimensional data. In Proceedings of the 2016 International Joint Conference on Neural Networks (IJCNN), Vancouver, BC, Canada, 24–29 July 2016; pp. 1803–1810.
15. Kovalerchuk, B. Visualization of multidimensional data with collocated paired coordinates and general line coordinates. *Proc. SPIE* **2014**, *9017*. [CrossRef]
16. Wilinski, A.; Kovalerchuk, B. Visual knowledge discovery and machine learning for investment strategy. *Cogn. Syst. Res.* **2017**, *44*, 100–114. [CrossRef]
17. UCI Machine Learning Repository. Breast Cancer Wisconsin (Original) Data Set. Available online: https://archive.ics.uci.edu/ml/datasets/breast+cancer+wisconsin+(original) (accessed on 15 June 2017).
18. Freund, Y.; Schapire, R. Large margin classification using the perceptron algorithm. *Mach. Learn.* **1999**, *37*, 277–296. [CrossRef]
19. Freedman, D. *Statistical Models: Theory and Practice*; Cambridge University Press: Cambridge, UK, 2009.
20. Maszczyk, T.; Duch, W. Support vector machines for visualization and dimensionality reduction. In *International Conference on Artificial Neural Networks*; Springer: Berlin/Heidelberg, Germany, 2008; pp. 346–356.
21. Cristianini, N.; Shawe-Taylor, J. *An Introduction to Support Vector Machines and Other Kernel-Based Learning Methods*; Cambridge University Press: Cambridge, UK, 2000.
22. UCI Machine Learning Repository. Parkinsons Data Set. Available online: https://archive.ics.uci.edu/ml/datasets/parkinsons (accessed on 15 June 2017).
23. LeCun, Y.; Cortes, C.; Burges, C. MNIST Handwritten Digit Database, 2013. Available online: http://yann.lecun.com/exdb/mnist/ (accessed on 12 March 2017).
24. Keras: The Python Deep Learning Library. Available online: http://keras.io (accessed on 14 June 2017).
25. Chollet, F. Keras. 2015. Available online: https://github.com/fchollet/keras (accessed on 14 June 2017).
26. Salama, G.I.; Abdelhalim, M.; Zeid, M.A. Breast cancer diagnosis on three different datasets using multi-classifiers. *Breast Cancer (WDBC)* **2012**, *32*, 2.

27. Aruna, S.; Rajagopalan, D.S.; Nandakishore, L.V. Knowledge based analysis of various statistical tools in detecting breast cancer. *Comput. Sci. Inf. Technol.* **2011**, *2*, 37–45.

28. Christobel, A.; Sivaprakasam, Y. An empirical comparison of data mining classification methods. *Int. J. Comput. Inf. Syst.* **2011**, *3*, 24–28.

29. Weka 3: Data Mining Software in Java. Available online: http://www.cs.waikato.ac.nz/ml/weka/ (accessed on 14 June 2017).

30. Ramani, R.G.; Sivagami, G. Parkinson disease classification using data mining algorithms. *Int. J. Comput. Appl.* **2011**, *32*, 17–22.

31. LeCun, Y.; Bottou, L.; Bengio, Y.; Haffner, P. Gradient-Based Learning Applied to Document Recognition. *Proc. IEEE* **1998**, *86*, 2278–2324. [CrossRef]

32. Domingos, P. A unified bias-variance decomposition. In Proceedings of the 17th International Conference on Machine Learning, Stanford, CA, USA, 29 June–2 July 2000; Morgan Kaufmann: Burlington, MA, USA, 2000; pp. 231–238.

33. Pereira, F.; Mitchell, T.; Botvinick, M. Machine learning classifiers and fMRI: A tutorial overview. *NeuroImage* **2009**, *45* (Suppl. 1), S199–S209. [CrossRef] [PubMed]

informatics

MDPI

Article

Visual Analysis of Relationships between Heterogeneous Networks and Texts: An Application on the IEEE VIS Publication Dataset

Björn Zimmer [1], Magnus Sahlgren [2] and Andreas Kerren [1,*]

[1]   Department of Computer Science, Linnaeus University, 351 95 Växjö, Sweden; bjorn.zimmer@lnu.se
[2]   Swedish Research Institute (RISE SICS), Box 1263, 164 29 Kista, Sweden; magnus.sahlgren@ri.se
*   Correspondence: kerren@acm.org; Tel.: +46-470-767-502

Academic Editors: Achim Ebert and Gunther H. Weber
Received: 6 April 2017; Accepted: 8 May 2017; Published: 11 May 2017

Abstract: The visual exploration of large and complex network structures remains a challenge for many application fields. Moreover, a growing number of real-world networks is multivariate and often interconnected with each other. Entities in a network may have relationships with elements of other related datasets, which do not necessarily have to be networks themselves, and these relationships may be defined by attributes that can vary greatly. In this work, we propose a comprehensive visual analytics approach that supports researchers to specify and subsequently explore attribute-based relationships across networks, text documents and derived secondary data. Our approach provides an individual search functionality based on keywords and semantically similar terms over the entire text corpus to find related network nodes. For examining these nodes in the interconnected network views, we introduce a new interaction technique, called Hub2Go, which facilitates the navigation by guiding the user to the information of interest. To showcase our system, we use a large text corpus collected from research papers listed in the visualization publication dataset that consists of 2752 documents over a period of 25 years. Here, we analyze relationships between various heterogeneous networks, a bag-of-words index and a word similarity matrix, all derived from the initial corpus and metadata.

Keywords: heterogeneous networks; interaction; graph drawing; multivariate datasets; NLP; text analysis; visualization; visual analytics

1. Introduction

The combination of different heterogeneous networks and related textual data is crucial for various application domains. Libraries, for example, are nowadays interested in analyzing (known or hidden) relationships among various collections of books, which might be related to each other even though they do not share the same author or topic. Based on an initial book search, analysts want to find out what terms were used in a specific book, find related ones that might use the same or similar terms and also visualize the direct neighborhood network of those books, which could, for instance, consist of other books written by the same authors.

Similar text corpora may be derived from conference proceeding publications, such as the IEEE Visualization Conference [1] proceedings, which we use as an application example and use case in this paper. Here, a researcher could be interested in finding out more about a specific topic and therefore wants to explore publications that use a number of specific keywords or terms. However, related publications might use different terms, but still talk about the same idea; or those publications mention certain terms that do not appear in the title or their keyword list. For less experienced researchers, it might help to alleviate the initial process of getting an overview of their new field if they would have access to tools that are able to find these concealed connections.

Integrating different network visualizations together with related multivariate data and helping analysts to discover relevant information, in the context of this paper by providing them with interactive possibilities to view and explore the relations between networks and large text corpora, is still an open challenge and has not yet been sufficiently addressed to the best of our knowledge. This approach is also important for other application domains such as biology, where the correlation of large metabolic pathways, protein-protein-interaction networks or regulatory networks is an important issue and the interconnections between various networks are not always obvious. Different networks come with numerous multivariate attributes (e.g., coming from experimental data), and exploring just one network and its related multivariate data is already a challenge in itself. This has been addressed by various techniques and tools; see the state-of-the-art survey of Kerren et al. [2]. Another survey for the visual analysis of graphs and open challenges in this area is provided by von Landesberger et al. [3]. In a collection of datasets (see the illustration in Figure 1), an attribute in one network might be related to other attributes in various other networks, even though they might not have the same identifier or name. To make interesting assessments about the data, analysts need to search for specific keywords, terms, authors or other metadata across different networks that might or might not be related.

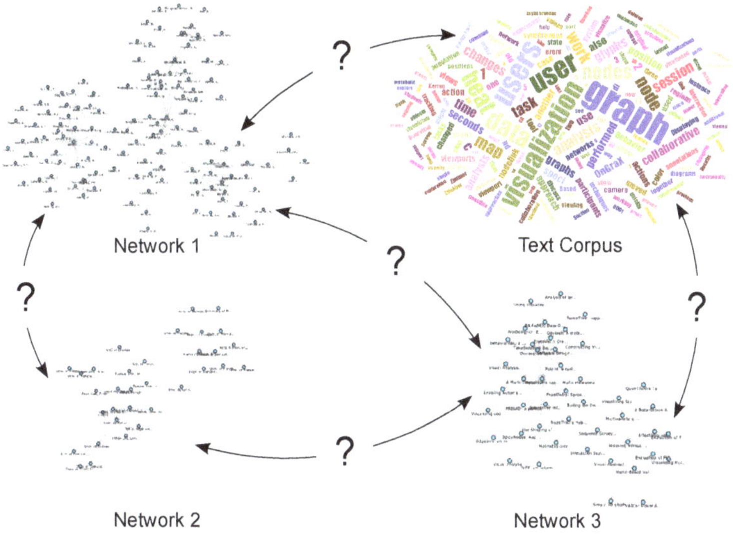

Figure 1. Different networks and datasets can be interconnected with each other, but these connections may not be immediately apparent. How can users visualize and explore them interactively?

If they also have text corpora available in addition to the network data, it can be interesting to see the relationships between documents that result from analyzing the text corpora. Such relations could be derived from the usage of specific terms in the documents, from a list of papers written by a specific author or be based on texts written at the same institution. It could also be worth investigating whether authors who use similar terms throughout their publications have a connection to each other. Examples of such a connection could be co-authoring a paper, performing research in the same field or having the same affiliation. Maybe the authors never really worked together, but one author was citing another author's work and was influenced by the choice of words in the original document. In this article, we address these challenges and analysis tasks. Our main contributions in the field of heterogeneous network exploration in the context of visual text analytics are:

- we present a scalable system that gives analysts the possibility to interactively specify and explore mappings across interconnected heterogeneous networks with thousands of nodes and related data derived from text corpora;
- we provide a way to perform a search based on main keywords and semantically similar terms over the entire text corpus to find related nodes, which represent, for instance, papers, authors or affiliations;
- we introduce the technique "Hub2Go", which enables users to quickly add and examine these nodes in the interconnected networks views; and
- we give the option to directly compare the usage of specific terms across a selection of nodes.

The remainder of this article is organized as follows. In the next section, we discuss related work in network visualization and document analysis. The compilation of our example data is described thoroughly in Section 3. We discuss our visualization together with a method for exploring mappings between different networks and datasets in Sections 4 and 5. Then, we showcase our approach with a use case in Section 6 followed by a discussion on limitations and performance aspects in Section 7. We conclude this article in Section 8.

2. Related Work

The visualization of links across related networks has so far mostly been addressed for biological pathways. Due to the sheer size of these networks, biologists usually create a number of smaller subsets that are then analyzed using different techniques. Caleydo [4], for instance, shows networks side by side in a 2.5D environment. Entourage [5] uses a focus+context approach to explore relationships in biological pathways. While analysts explore a pathway and focus on specific nodes, parts of related pathways are shown in additional views as contextual subsets. Our application uses dynamic network views similar to this idea; based on a focus node and configured attribute matching settings between networks and additional datasets, our system allows one to explore nodes and their direct neighborhood in related network views. Exploring the connections in just one larger network is often already a challenging task: navigating across links to adjacent nodes is difficult if the nodes are far away from each other and not in the current view. This usually requires much panning and also zooming of the graph view and puts a lot of cognitive load on the user. To tackle this problem, May et al. [6] proposed signposts to provide cues about off-screen regions in a graph visualization. Henry et al. [7] combine traditional node-link diagrams with adjacency matrices for a compact view on a network. The links between the matrix representations can be considered as connections between subnetworks within one large network. Another approach from Moscovich et al. [8] introduced (1) link sliding to automatically move the camera across links to adjacent nodes and (2) bring and go, which moves adjacent nodes into the current view, close to the selected node. Our Hub2Go technique uses a similar approach, but supports automatic camera movements in multiple network views to ease the navigation from and to nodes across interconnected networks and relations to other data. There are also tools to assist users in creating and exploring networks from raw tabular data, such as Orion [9] or Ploceus [10]. These tools use single node-link and matrix views to visualize the generated networks. Our approach differs from these techniques because we expect already existing network data, visualize related networks in multiple node-link views and give options to dynamically define and visualize links between those views.

Tools for the visual analysis of documents and literature publications are related to our use case application. A general overview of text visualization techniques is available in the interactive TextVis Browser [11]. Furthermore, Federico et al. [12] provide a broad survey about various techniques and open challenges for analyzing the scientific literature. More specifically, Görg et al. [13] use Jigsaw for the exploration and sensemaking of document collections. They use a list view to show the relationships among co-authors, keywords and other attributes, together with a clustered view of related papers. Shen et al. [14] use an interactive table together with multiple stacked network planes in a 2D or 3D view to visualize interconnections between various paper, author and publication networks.

PivotSlice [15] uses a more general approach to visualize relationships in datasets. It supports dynamic queries on attributes and divides datasets into facets and connected views on the data. Phrase Nets [16] offers techniques to depict relationships in a text and map it into a network-based overview. Refinery [17] shows combined heterogeneous networks with subgraphs generated from user queries in one network view. Chen et al. [18] analyze the metadata of document collections and visualize identified topics together with co-authors in a single network visualization to show collaborations over time. Instead of only using titles, abstracts, keywords and other metadata as the previous tools, our system also supports exploring the complete text corpus of a document collection, which gives analysts the possibility to find documents based on terms that might otherwise not be visible. There are various other systems that use the complete text corpora of document collections. For instance, TextPioneer [19] uses on automatic topic extraction to build a hierarchical overview of interesting topics across multiple text corpora, whereas our approach focuses more on the exploration of specific terms of a document corpus. CiteRivers [20] aims at visualizing citation patterns of scientific document collections over time, but does not feature the exploration of author collaborations as we do in our network visualizations.

3. Data Sources and Preprocessing

Our visualization runs in a web browser and uses a client/server-based architecture, which is based on our Online Graph Exploration System (OnGraX) [21,22]. Generally, OnGraX can be used to specify and explore mappings across any networks and related quantitative data. For our use case employed throughout this article, we focus on analyzing the complete text corpus, which we derived from all papers in the IEEE VIS conference proceedings with the help of the visualization publication dataset [23]. We created three networks from the initial data, which help us to explore relations within the data. Nodes in the first network (2752 nodes and 10,021 edges) represent all papers from 1990–2015, and edges represent paper citations. Node shapes in the resulting network visualization encode the VIS conferences: rectangles for the Visual Analytics Science and Technology (VAST) conference, circles for the Information Visualization (InfoVis) conference, triangles for the Scientific Visualization (SciVis) conference and diamonds for older Vis papers. Nodes in the second network (4890 nodes and 14,023 edges) represent all authors and co-authors from the dataset. Edges are added between authors if they wrote a paper together at least once. Nodes in the third network (1539 nodes and 5773 edges) represent all author affiliations, and edges are added if authors from different affiliations published a paper together. As we did not prune the initial VIS dataset, the co-authorship and affiliation networks may contain duplicates with slightly different wording.

Moreover, we did a pdf-to-text conversion for all papers in the IEEE VIS conference proceedings and used the resulting text corpus to create a word occurrence index (a so-called bag-of-words (BoW) index) and a word importance index (a so-called term frequency–inverse document frequency (TF-IDF) index) to find the most frequent and most important terms. The resulting indices contain 12,346 words/terms each and can also be used to uncover the main representative words/terms in a specific paper or to find terms that were used by a specific author or even in an affiliation. The indices contain single words, as well as bigrams, which are discussed in Section 3.1. Additionally, we use a word similarity matrix to find and analyze the relationships between terms that are used in the same context between various papers.

Figure 2 shows the structure and attributes of the networks and datasets in detail. The resulting list and matrix datasets are stored as plain, space-delimited text files on the server. At runtime, the server loads and keeps a copy of all required datasets in memory to provide faster query times for mapping requests during the exploration of the networks.

Note that the difference between the BoW and TF-IDF indices is that the former only quantifies the frequency of occurrence of terms in the data (i.e., has a term occurred or not?), whereas the latter tries to qualify the importance of occurrences (i.e., is a term important or not?). Having identified the main keywords in a paper (or author/affiliation) of interest, it is also possible to find other papers (authors/affiliations) that have used the same terms. As an example, we may be interested in a specific paper about "sentiment analysis". By using the BoW index, we can find all other papers that have mentioned this term, and we can subsequently analyze the co-authorship networks of this group of papers (see Section 5 for details on this process). If we use the TF-IDF index instead, we find other papers for which the term is a useful index term (i.e., for which it is important). Finally, the word similarity matrix enables us to find other terms that have been used near-synonymously (or, more generally, in the same way) as some keyword. Searching for the term "sentiment analysis" again, we can use the word similarity matrix to find other terms that might be of interest to us; perhaps terms such as "opinion mining" and "topic detection", which could then be included in the current analysis. In the following, we provide details on how we produced the BoW, TF-IDF indices and word similarity matrices.

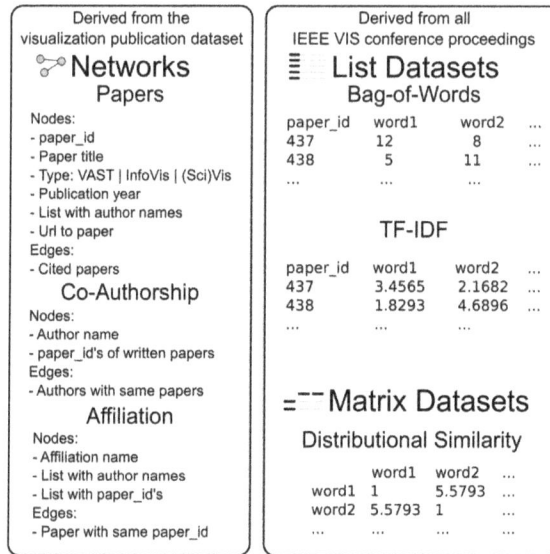

Figure 2. Overview of our dataset: all networks, node attributes and edge types were taken from the visualization publication dataset, with additional textual data derived from all papers in the IEEE Visualization Conference proceedings.

3.1. Text Preprocessing

After the pdf-to-text conversion, we removed the punctuation from the resulting text and also lower-cased and tokenized it. We even identified salient bigrams in the data, such as the term "traffic_flows" from our use case in Section 6, using a variant of mutual information that is commonly used for bigram detection in natural language processing:

$$p(a, b) = \frac{f_{a,b} - \delta}{f_a f_b} \tag{1}$$

where $f_{a,b}$ is the co-occurrence frequency of a and b, f_a and f_b are the individual frequencies of a and b and δ is a discounting factor that counteracts the tendency of mutual information to promote infrequent items, which we set to $\delta = 5$. The resulting mutual information scores are sorted, and only the highest-scoring bigrams are included in the data. The threshold for the mutual information scores is a trade-off between precision and recall: a lower threshold leads to more, but less precise bigrams, while a higher threshold leads to fewer, but more precise bigrams. We opt for a more conservative threshold in this paper.

3.2. Bag-of-Words

Bag-of-words (BoW) is a standard text representation formalism in natural language processing (NLP), which represents text simply as a bag containing frequency counts of the words that occur in the text. Formally, the BoW representation of a text is a vector:

$$\vec{w} = [t_1, \ldots, t_n] \tag{2}$$

whose dimensionality n is the size of the vocabulary (i.e., each word in the vocabulary is represented by one separate dimension) and t_i is some weight that quantifies the importance of the word in the text. There are many ways to quantify the importance of a word in a document; the arguably most common term weighting scheme is TF-IDF, which in its simplest form is defined as:

$$\text{TF-IDF}_{i,j} = \text{TF}_{i,j} \cdot \log \frac{N}{\text{DF}_i} \tag{3}$$

where $\text{TF}_{i,j}$ is the frequency of word i in document j, DF_i is the number of documents word i occurs in and N is the total number of documents in the data. It should be noted that there are many variations, refinements and alternatives to using TF-IDF to extract useful terms as indicated by Chuang et al. [24], for example. However, we opt for the standard TF-IDF measure in this work, since it is simple to compute and produces useful results. Using BoW representations enables us to find all documents that contain a certain term (i.e., all documents for which the value of the dimension representing the term in question is not zero). The TF-IDF representation also enables us to list the most useful index terms for a document (i.e., the terms with highest TF-IDF weights for that particular document).

3.3. Word Similarity

In order to produce the word similarity matrix, we use distributional semantics, which is the practice of using information about the co-occurrence patterns of terms in order to quantify semantic similarity. In a standard distributional semantic model (DSM), each word is represented by a distributional vector, $\vec{w}_f = [w_1, \ldots, w_m]$, where w_i is a function of the co-occurrence count between the focus word w_f and each context word w_i that has occurred within a window of k tokens around the focus word [25]. Words that have co-occurred with the same other words (i.e., that are interchangeable in context) get similar distributional vectors, which means we can use the resulting model to quantify semantic similarity between terms.

There are many variations of DSMs, ranging from simple count-based methods to the currently more popular, and more complex, models based on factorization techniques or neural networks [26–28]. We opt for an approach based on an incremental random projection of a count-based model, called random indexing (RI) [29], which accumulates distributional vectors $\vec{v}(a)$ by summing sparse random index vectors $\vec{r}(b)$ that act as fingerprints for the context items:

$$\vec{v}(a) \leftarrow \vec{v}(a_i) + \sum_{j=-c, j\neq 0}^{c} w(x^{(i+j)}) \pi^j \vec{r}(x^{(i+j)}) \tag{4}$$

where c is the extension of the context window, $w(b)$ is some weight function that quantifies the importance of context term b and π^j is a permutation that rotates the random index vectors according to the position j of the context items within the context windows, thus enabling the model to take word order into account [30]. We use 2000-dimensional vectors, $c = 2$, and the weight function suggested in Sahlgren et al. [31]. There are several reasons why we use RI instead of one of the currently more popular DSM models. One is the ability of RI to encode word order, which is not possible to do using, e.g., the word2vec library or the GloVe model. Another reason is the scalability and efficiency of RI, which makes it suitable for collections with large amounts of data. It should also be mentioned that the most current DSMs perform similarly, given that parameters have been optimized for the specific data and task at hand.

In order to produce the word similarity matrix, we compute pairwise similarities between all words with frequency $\geq$ 50 in the data using cosine similarity. We utilize a comparatively high frequency cut-off for the word similarity matrix, since low-frequent words have insufficient distributional statistics to produce reliable distributional vectors.

4. Specifying Attribute Mappings

Before exploring the interconnections and relations between networks and additional datasets, analysts have to specify which attributes they want to compare throughout the complete data collection. Instead of having to do this programmatically, our application enables analysts to create and subsequently explore arbitrary mappings between all graphs and additional datasets (e.g., lists or matrices) on the fly by using a small bipartite graph visualization as shown in Figure 3. Nodes either represent a graph or a dataset that was imported into the application. The node colors are selected by the tool from a color list created with ColorBrewer [32]. They are also used as background colors for the hubs during the exploration of a mapping to indicate the target dataset (see Figure 4h). Every dialog with a view on a graph or dataset has a small colored rectangle in the top right corner to indicate its mapped color. Users can add unidirectional and bidirectional mappings by dragging the mouse from a source to a target node. The application will ask which attribute from the source should be mapped against the target.

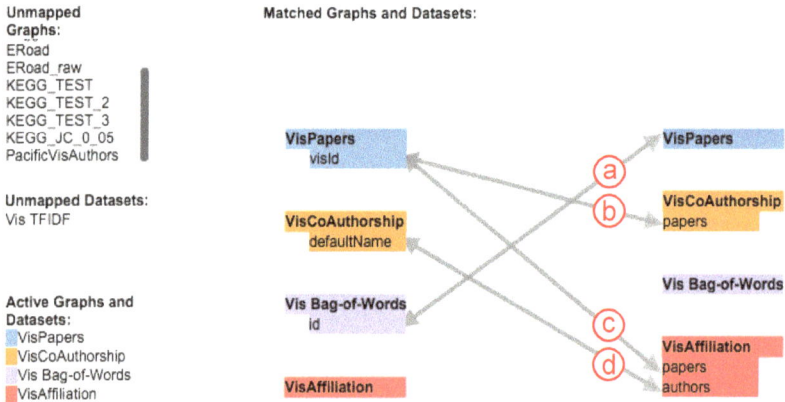

Figure 3. The dialog to configure the mapping between graphs and datasets. This figure shows four configured mappings between three networks and the BoW index.

As performance can be an issue if a large number of graphs and datasets with many entries are loaded, users can add and remove graphs and datasets (that are already imported into the system) to the matcher on the fly. Depending on a priori knowledge, mappings can be defined from and to specific attributes or be more general, to iterate over all possible attributes of a target graph or dataset. For the data from our use case, all matches across the datasets are calculated and visualized in less than a second, but to provide better scalability for bigger datasets with a large number of entities and additional attributes, specific mappings could be helpful to reduce the required time to iterate over all possible matches.

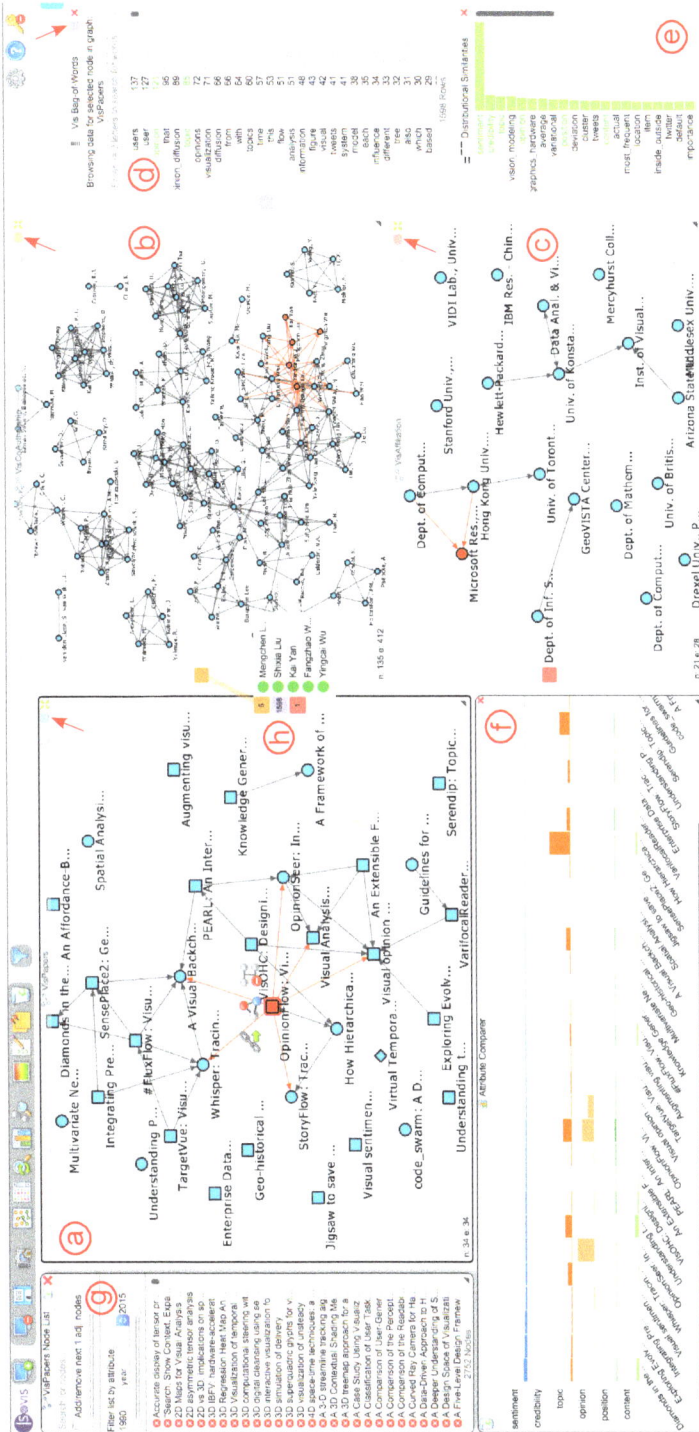

Figure 4. An analysis session in progress. Here, a mapping between a paper citation network (**a**), a co-authorship network (**b**) and an affiliation network (**c**) is explored. The bag-of-words chart (**d**) shows all terms for the currently-selected paper in the paper network, and the chart at (**e**) shows terms used in a similar context as the word "sentiment" throughout the complete text corpus of the IEEE conference proceedings. The attribute comparer (**f**) visualizes the occurrence of six terms over a selection of papers. The dialog at (**g**) contains the complete list of nodes for the active network that the user is currently exploring; in this case, the paper network. The three colored rectangles at (**h**) realize our interactive Hub2Go, which currently indicates matches from the paper "Opinion Flow", which is selected in the paper network, to three other networks or datasets. The small rectangles at the top right of the three network views and the bag-of-words chart (marked with red arrows) represent the colors used for the Hub2Go instances (**h**) to indicate matches across networks and datasets.

An actual mapping is performed dynamically, i.e., if a user clicks on a node in one of the network views or an entry in one of the additional bar charts. In Figure 3, the analyst declared four mappings between three graphs and the BoW index. The first mapping (see Figure 3a) from the BoW index to the papers network is used if a user clicks on a node in the respective network view. The application will then iterate through the BoW index, locate the row with the matching paper ID and fetch all terms used in that paper to show their BoW values in the BoW chart. As this mapping is bidirectional, the mapping will also be performed if a term is clicked in the BoW chart. In this case, the application will fetch all papers that use this specific term. The next two mappings (see Figure 3b,c) from the paper publications network to the co-authorship and affiliation networks are used to find related authors and affiliations for a clicked paper. The application will map the paper to all authors in the co-authorship network and affiliations that have the paper ID in their list of papers. The last mapping (see Figure 3d) maps an author's name from the co-authorship network to the affiliation network, thereby enabling the user to find all affiliations a specific author wrote papers for and also to find all authors of a specific affiliation. Currently, our application only discovers matches between attributes with identical values. For instance, the last mapping (see Figure 3d) only finds completely identical strings between the configured attributes of the co-authorship and affiliation network. To be able to support fuzzy matching, we plan to extend the mapping feature in a future version. String similarities and configurable range values for numerical attributes could be used to find and display similar results.

In our example, the latter three attribute mappings (see Figure 3b–d) are defined to specific attributes in the source and target nodes, as the exact attributes, interconnections and relations are already known for these datasets. Alternatively, if the attributes of a newly-imported graph are unknown, it is possible to specify an attribute mapping that tries to find identical attribute values by iterating over all entities and their attributes of a mapped graph or dataset. The mapping in Figure 3a, for instance, implies the paper network without declaring a specific target attribute. In this case, the application will try to match all possible attributes of the target to the specified source attribute. If an attribute can be mapped depends on the type of the source attribute (e.g., nominal or ordinal). The application will only try to map between attributes of the same type.

5. Visualization Approach

Depending on the use case, it is not always possible to merge several heterogeneous networks into one view. For instance, some metabolic networks from the Kyoto Encyclopedia of Genes and Genomes (KEGG) pathway database [33] have a precomputed layout, which is preferred by biologists, but the networks are still connected with each other. In this case, it would be beneficial to have multiple views on the networks and provide a means to navigate the relationships, similar to the approach facilitated by Entourage [5]. In another scenario, multiple networks could be merged into one single network view. The Apolo system [34] uses this technique to make sense of large network data using machine learning and user interaction to find nodes that might be of interest to the user. Our goal was to design a flexible system that can be used for different application areas. We also wanted to be able to visualize additional related multivariate data across the networks. It is possible to show attributes of target nodes directly in a graph visualization, for instance as information wedges at the borders of the visualization [35], but the number of attributes that can be visualized with this technique is limited and not applicable for all possible use cases. Consequently, we decided to use multiple coordinated views together with brushing and linking techniques [36] for our tool.

5.1. Visual Design

Graphs are visualized as interactive node-link views (see Figure 4a), whereas list-based datasets, such as the bag-of-words index, are shown as horizontal bar charts (see Figure 4d,e) due to space efficiency. The distributional similarity dialog (see Figure 4e) displays the most similar terms of a selected word in the related BoW chart. The attribute comparer (see Figure 4f) helps to compare quantitative attributes, such as the BoW values for instance, among a selection of nodes that can be added from any network view by performing a right-click on a node. It also supports sorting of nodes based on a selected attribute. The concrete processing depends on the network type in our case: if a node is added from the affiliation or the co-authorship network, the accumulated value from all papers for that affiliation or author is visualized, whereas a node from the paper network shows the actual BoW or TF-IDF value for that paper directly. If a node in one of the network views or an entry in the charts is hovered with the mouse, all related nodes or entries in all other views are highlighted, which helps to interactively explore the relationships between the datasets.

Node-link visualizations of networks with many thousands of nodes and edges are usually quite cluttered and difficult to explore without further filtering or clustering. To circumvent this issue, we facilitate a bottom-up approach to explore the datasets: when the application is started, all network views are empty, and the BoW/TF-IDF charts show all terms over the complete dataset. Nodes from networks are added into or removed from the views on demand, either by directly searching for a specific node in the node list (see Figure 4g), finding nodes by keywords from the BoW chart or by requesting connected matches from a node in another node-link view. If a network does not already have precomputed node positions, the network views use a force-directed layout from the yFiles library [37]. The layout is calculated whenever new nodes are added to the view and tries to minimize the position changes of already existing nodes in order to preserve the mental map. If necessary, users can adjust the layout settings and select different layout algorithms such as circular or hierarchical graph layouts.

5.2. Hub2Go

In order to perform a bottom-up exploration across interconnected heterogeneous networks and related data views, it is essential to quickly find nodes of interest. Users should also have the option to seamlessly add nodes into a related target network and swiftly go to their location to explore their local neighborhood. Moreover, it should be possible to remove nodes from a search result, if it did not yield interesting information. Depending on the use case, it is also not always feasible to directly show all interconnections to a target node if the node has a high number of interconnections.

To provide a flexible solution for these problems, we introduce the Hub2Go technique. Whenever a search is requested by clicking on a node in a network view or by selecting a term in the BoW or TF-IDF charts, the system represents the search results as Hub2Go instances at the border of the initial network view or chart (see Figure 4h). The hubs show the connections to other datasets: in the context of Figure 4, the user clicked on the paper "Opinion Flow" (marked in red) in the paper network (see Figure 4a) and initialized a search for connected entities in other networks or datasets based on the configured mappings. Three Hub2Go instances show the result: five nodes in the co-authorship network, 1598 terms in the BoW chart and one node in the affiliation network.

To create the concrete network views given in Figure 4, we started by searching for the term "sentiment" in the BoW chart (see Figure 5a). By clicking on the resulting term, the system matches the term over all configured networks and visualizes the result as three Hub2Go instances (see Figure 5b). In this case, the hubs show 34 papers, 166 authors and 21 affiliations that mention this term at least once. Hovering over the blue hub, which is related to the paper network, shows a list of all 34 found papers. Left-clicking the hub adds all nodes from the search result to the target network (see Figure 5c); or removes them, if they are already in the view. Alternatively, users could browse through the list and manually add or remove nodes of interest by left-clicking them. Going back to Figure 4, we also added all matching nodes to the other node-link views using our Hub2Go technique. Additionally, users can perform a right-click on a hub or a node in a hub-list to add all or specific nodes to the attribute comparer. In this case, we added all 34 papers to the attribute comparer (see Figure 4f) by right-clicking on the blue Hub2Go instance, and to analyze the occurrence of related terms among these papers, we also added "sentiment", "credibility", "topic", "opinion", "position" and "content" to the attribute comparer by right-clicking them in the distributional similarity dialog, which currently shows the most similar terms for the term "sentiment".

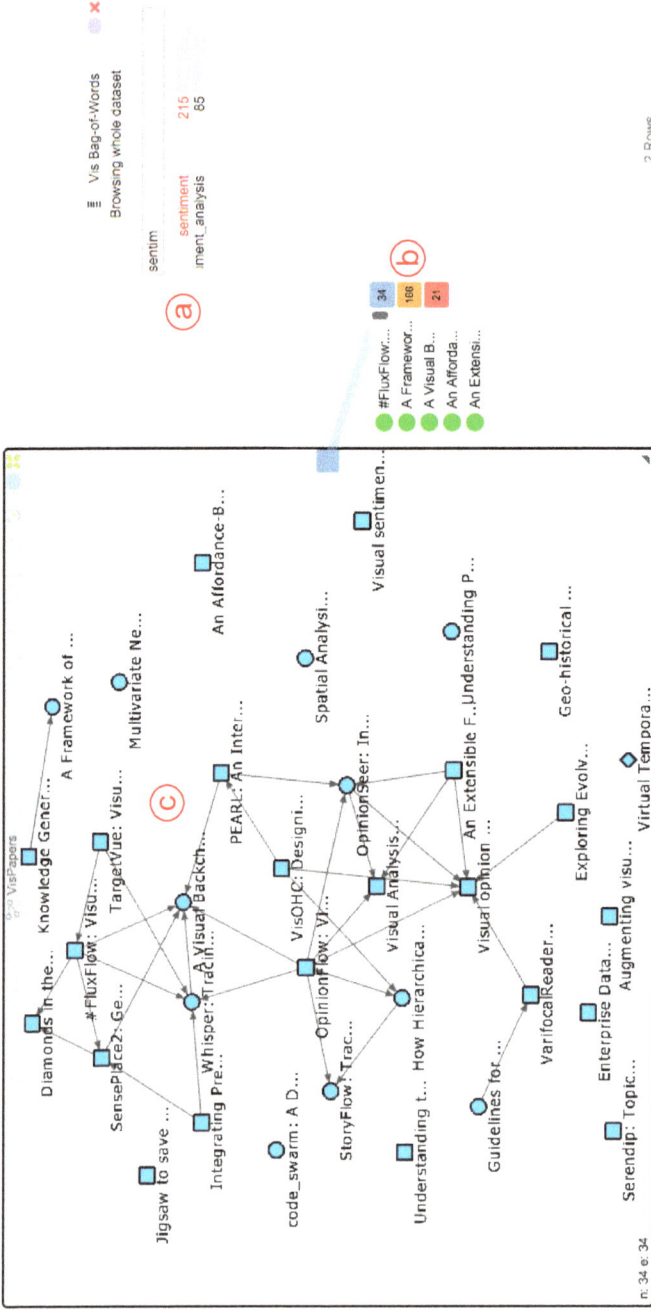

Figure 5. Usage of the Hub2Go instances: The user searched for the term "sentiment" in the BoW chart (a) and clicked on it. The resulting matches across the networks are shown as three Hub2Go instances at the border of the dialog (b). Hovering the blue hub shows a scrollable list with all related papers. Clicking the hub adds all papers to the paper network view (c).

In Figure 4, the user is currently hovering the orange hub, associated with the co-authorship network, with the mouse cursor, which brings up a list with all found authors for the selected paper and also highlights all connected nodes in the other views. Hovering the mouse over a specific item in the hub-list automatically centers the connected target node in the relevant view, enabling the analyst to swiftly explore its local neighborhood. To indicate whether nodes of a search result are already in the view, circles in this list appear green if they are already added to the network view or red if they have not been added to the view. When clicking on a node in a hub list, the neighborhood of this node can optionally also be added to the network view (based on a user setting). This is helpful to quickly find related nodes in the dataset, for instance citations from or to other papers or authors who also wrote other papers together with the initial author of a paper.

6. Use Case

Our analysis goal for the use case described in the following is to discover relationships between papers that cannot be spotted by an analysis of pure metadata, such as the data provided by the visualization publication dataset [23]. For instance, we may want to discover thematic outliers in the paper collection of a specific author and analyze if there are similarities between those outliers and other papers written by the author (or by others). The three network views could also be of variable interest to different users. Students or researchers usually want to explore research results and are interested in the relations between existing papers and authors, whereas libraries or funding organizations might also be interested in co-authorship and the related affiliations to explore the results of an interdisciplinary project.

Let us assume for this use case that a student previously read an interesting paper about "traffic flows" and that the student remembers one of the authors of this paper: "Jarke J. van Wijk". The student wants to find out more about the author and might also have a few questions related to the paper and wants to use our system to answer them:

Q1, author question: Which words/terms does van Wijk use in his papers? Answering this question would provide the student a first impression about van Wijk's most important research areas.

Q2, author question: Does he usually focus on the same terms throughout his papers or does he address various different topics, which would indicate a broad research interest?

Q3, paper question: Are there other papers related to the term "traffic_flows" that might be of interest to the student?

Q4, paper question: Considering the term "traffic_flows": are other terms in other papers used in a similar context, which would make those papers also interesting to the student?

Analysis Step 1:

Aiming to answer Q1, the student first adds the TF-IDF chart to this analysis (see Figure 6g) to get the most important words among selected papers. He/she then activates the co-authorship network view and searches for van Wijk's name in the search dialog (see Figure 6a) and adds van Wijk's corresponding author node together with all other nodes of authors who collaborated with him to the egocentric network view (see Figure 6b). After clicking on van Wijk's node in this view, the system shows four Hub2Go instances (see Figure 6c), which indicate 38 related papers in the paper network (blue), 8 nodes in the affiliation network (terra cotta) and 7245 terms in the BoW (violet) and TF-IDF charts (jade green). The two charts already give the student an overview of all terms used in van Wijk's papers, thereby answering Q1. Left-clicking on the blue hub adds all papers to the paper network view, and the student also puts all of van Wijk's affiliations into the affiliation network view (note that in this case, some of the affiliations are actually duplicates with slightly different wording in the visualization publication dataset).

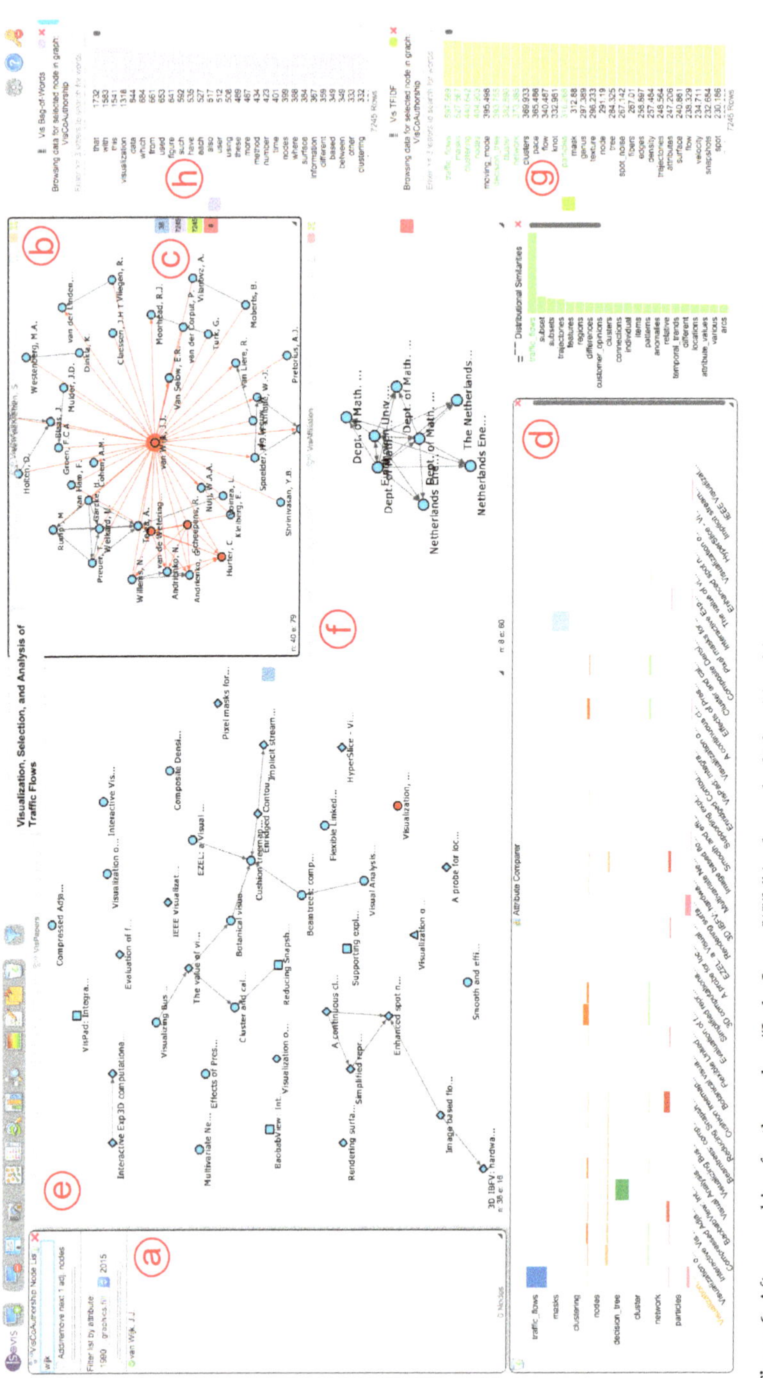

Figure 6. After searching for the author "Jarke J. van Wijk" in the node dialog (**a**), adding his corresponding node and all of his co-authors to the co-authorship network (**b**) and performing a matching over all graphs and datasets (**c**), the system also allows the exploration of all of van Wijk's papers (**e**), affiliations (**f**) and the terms he used throughout his papers with the help of the TF-IDF(**g**) and BoW (**h**) charts. The attribute comparer (**d**) shows TF-IDF values of van Wijk's papers for eight chosen terms, which are marked in green in the TF-IDF chart. Currently, the paper "Visualization, Selection, and Analysis of Traffic Flows" has been selected in the VisPapers and Attribute Comparer views.

The student could now further explore which terms are used in a specific paper by clicking on its node in the paper network. The student could also explore which of his own (or his co-authors') papers van Wijk cites in the VIS publication dataset. For doing so, we use a pop-up menu that appears while an interesting node (i.e., a specific paper) is hovered with the mouse in order to add all nodes that are adjacent to the hovered node into the network view. This way, the student is able to easily identify papers that cite or are cited by one of van Wijk's papers. By using the pop-up menu, he/she could also directly open the IEEE-Xplore link, which is saved in the URL attribute of each node, to take a closer look at the actual papers.

Hovering over a node in the VisPapers view also highlights connected nodes in the other networks, enabling the student to see the authors and affiliations of specific papers. Furthermore, the student could also investigate the co-authorship network and hover over a node to highlight all papers that an author wrote together with van Wijk.

Analysis Step 2:

To answer Q2, the student investigates the TF-IDF chart, which shows the terms for all of van Wijk's papers, as long as his node is selected in the co-authorship network. The student adds eight of van Wijk's most used/important terms "traffic_flows", "masks", "clustering", "nodes", "decision_tree", "cluster", "network" and "particles" to the attribute comparer (see Figure 7) and right clicks the blue hub in the co-authorship view to also add van Wijk's papers to the comparer. The student notices that the terms "clustering", "nodes", "cluster", "network" and "particles" are used quite evenly throughout van Wijk's paper collection. However, the other three terms are interesting outliers: "traffic_flows", "masks" and "decision_tree" are each uniquely used in three different papers. In this case, the student discovers from the chosen terms that van Wijk addresses similar topics in most of his papers (in the case of the terms that are used evenly), but the three outliers could also indicate a broader research interest. The student could now further investigate and use the paper network view to open the links to the papers, and he/she could also use the TF-IDF chart to get an overview of the other terms that were used in those papers.

Analysis Step 3:

The student now decides to address Q3 and to find more papers about the term "traffic_flows". Figure 7 already revealed that van Wijk only has one single paper that mentions this term, and searching for the term in the titles of all papers via the node list of the paper network also only reveals van Wijk's paper. The addition of the BoW and TF-IDF indices makes it possible to search for terms in the data, rather than being limited to searching for keywords in the titles of papers. Therefore, the student can now find all papers that mention some specific term, regardless of whether it is in the title of the paper or not. To see if there are other papers in the dataset that use this term, the student clicks on "traffic_flows" in the TF-IDF chart and gets three Hub2Go instances linking to 12 papers, 42 authors and 9 affiliations (see Figure 8). After clearing the attribute comparer from the previous search queries, the student adds all 12 papers to the paper network and the attribute comparer using the blue hub. Van Wijk's paper has by far the highest frequency of the term, but the student could now also open the other papers to see if they are relevant for his research.

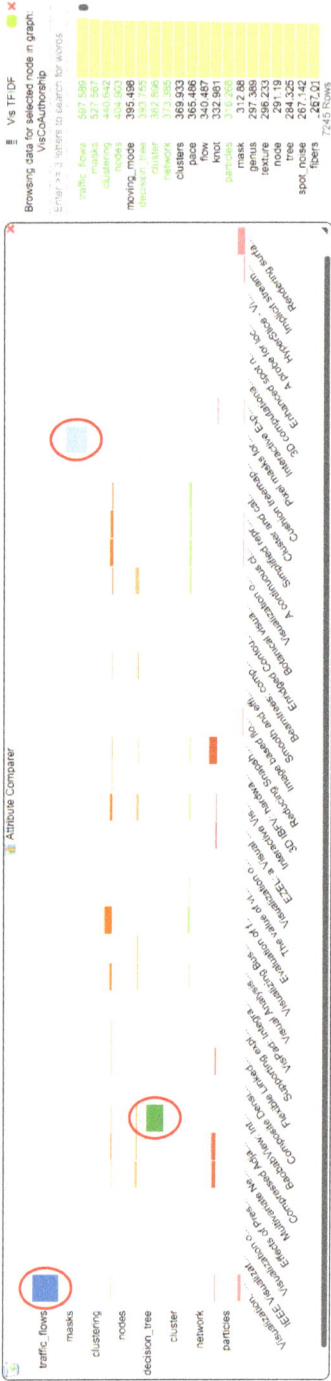

Figure 7. Distribution of eight selected terms in van Wijk's papers. The terms "traffic_flows", "masks" and "decision_tree" (marked by red ellipsoids) are each uniquely used in different papers.

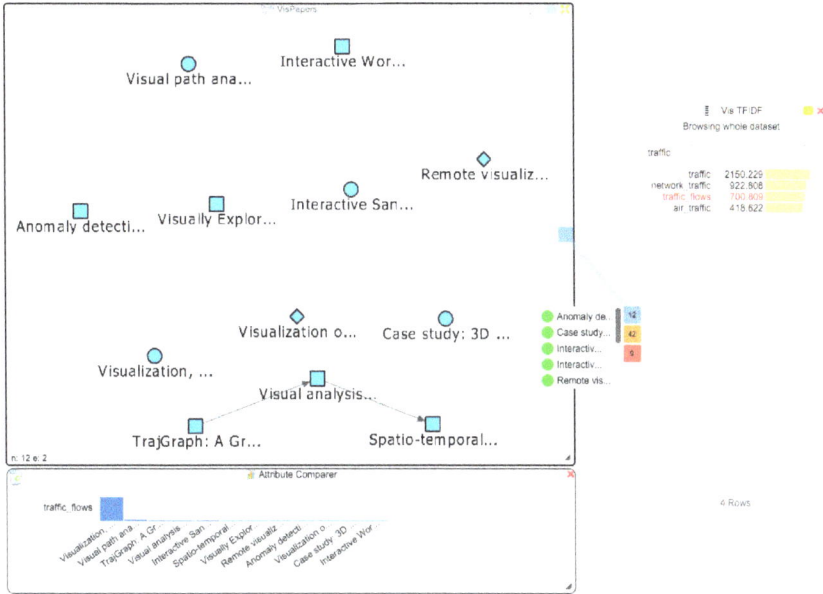

Figure 8. The network view shows all papers mentioning the term "traffic_flows". The attribute comparer shows the distribution of that term among the papers.

Analysis Step 4:

The student can now utilize the TF-IDF (or BoW) values together with the distributional similarities to further investigate these 12 papers and find additional work that might be thematically related to traffic flows, thereby addressing Q4. The student selects van Wijk's paper in the paper network to get the TF-IDF values for that paper and adds the next three terms with the highest values ("aircraft", "traffic", "trajectories") to the comparer. A well-known issue with keyword search is vocabulary variation (sometimes referred to as the synonymy problem), which means that several different terms can be used to refer to the same thing. In the case of "traffic_flows", there could be other terms that are also used by researchers to refer to this type of visualization. The student therefore consults the distributional similarities view (see Figure 9), which shows the most similar terms, and identifies four interesting ones "trajectories", "features", "regions" and "patterns", which might be relevant for investigation besides the term "traffic_flows". The first term from the distributional similarities "trajectories" was already in the TF-IDF chart, which encourages the student to further investigate the papers mentioning this term.

Right-clicking on the term "trajectories" in the attribute comparer orders the papers in the chart based on the TF-IDF values for this term. The student finds five new papers that mention this term (see Figure 9(1–5) at the bottom in the attribute comparer dialog). They could be of interest, and the student decides to add them into the paper network view that still contains all of van Wijk's papers. Right-clicking the papers in the attribute comparer adds them to the paper network view: four of these papers (1–4) are actually connected to each other, and one of van Wijk's papers (see Figure 9(6); "Composite Density Maps for Multivariate Trajectories") is part of this group, as it cites two papers with the term "trajectories". Since it does not mention the term "traffic_flows", it was not added to the attribute comparer in the initial search. To investigate the occurrence of the already selected terms in this paper, the student also adds it to the attribute comparer by right clicking on it in the paper network. The student finds out that the paper also uses the terms "trajectories", "features" and "patterns", which suggests that he/she might also take a closer look at this paper.

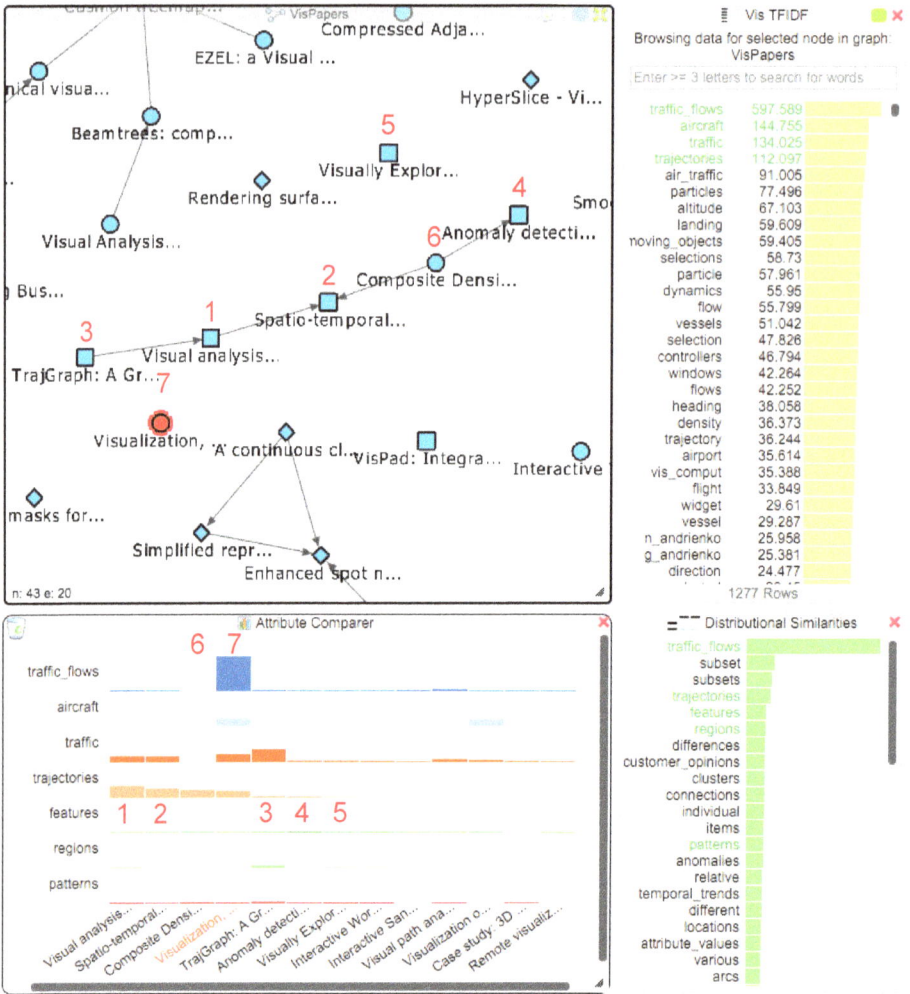

Figure 9. The TF-IDF chart now shows all values for the same focus paper ((7); currently selected in the paper network view) as in Figure 6. Adding all papers mentioning the term "traffic_flows" together with six related terms from the TF-IDF chart and the distributional similarities to the attribute comparer reveals five potentially interesting papers (1–5), which we added to van Wijk's initial paper network view. Four of these papers are connected due to citations, and one of van Wijk's papers (6) is also part of the resulting group.

Note that the distributional similarities encode the similarity of usage in the data rather than some generic (or "objective") notion of semantic similarity. This means that we will find terms that have been used in the same way in this view. Such similarities may sometimes be somewhat surprising and perhaps even counter-intuitive (in this case the terms "subset" and "subsets" are not related to the initial search term), but are a direct effect of word usage in the data at hand.

7. Discussion

The employment of multiple coordinated views has the advantage that we can explore various interconnected networks at once. This is of course limited to the available screen space. On a 2K desktop resolution, three to four networks in addition to the BoW and distributional similarity charts are a realistic usage scenario (Figures 4 and 6 were taken at a 1900 × 1080 resolution), whereas a 4K resolution offers even more space for additional network views. A higher resolution is essential for bigger document collections, for instance, if users plan to explore the interconnections between various different conference proceedings and would have three or four paper citation networks and not just a single one, as we did in our use case.

Analyzing a bigger text corpus should have a big impact on the performance of the system and the interaction between the views, as all word indices and similarities are calculated beforehand and loaded into the main memory on the OnGraX server. Exploring a larger number of different networks also scales well for realistic use cases. Our mapping system on the server uses multiple parallel threads and can iterate through up to 16 datasets or networks at the same time to find matches. As such, only the highest number of nodes (n) and the number of mapped attributes (m) in a graph have an impact on the performance, and a matching usually runs in $\mathcal{O}(nm)$ time. Additionally, all networks are rendered on the client computer's GPU by using WebGL [38]. This approach is faster than using the more common SVG-based node-link visualization approaches, and our system provides good scalability for up to 10,000 nodes and edges (depending on the hardware specifications of the client). As we use a bottom up approach and only load nodes into the views on demand, the number of nodes that have to be visualized at once is usually quite small and as such also scales well with bigger networks.

While users can quickly specify new attribute mappings between imported graphs and datasets, this process usually requires some a priori knowledge about the structure and attributes of the data. OnGraX currently only visualizes attributes of a graph or dataset during the creation of a new mapping in the mapping dialog. Our plan is to improve this view to give users more intuitive access to this information.

8. Conclusions

In this work, we presented a system to explore the structure and relations between heterogeneous interconnected networks and additional metadata. The inclusion of large textual information, in the form of both a BoW/TF-IDF index and a word similarity matrix, together with the possibility to map these data across different networks, constitutes a novel contribution in visual text analytics and allows us to extensively explore relations between related papers, authors or affiliations based on selected keywords. Being able to search for terms and semantically similar used words and visualize their usage throughout the complete text corpus can reveal relations between thematically-related documents, which would otherwise not be apparent if an analyst could only search through the titles and keywords of a document collection. Our Hub2Go technique assists users to quickly add (or remove) interconnected nodes in interactive network views and charts and assists in navigating to the locations of these nodes to explore their local network structure. Our system has been developed in close collaboration with NLP experts, and the next step is to assess the usability of our system by performing a user study with a larger document collection. We also plan to make the OnGraX system publicly available in the near future. Due to copyright reasons, we cannot make the whole full text dataset accessible, but the extracted BoW and TF-IDF files, as well as the word similarity matrix are available at the provided URL in the Supplementary Materials.

Supplementary Materials: The following are available online at http://bit.ly/2qS79rw Video S1: Intro Video, Video S2: Use Case, File S3: BoW, TF-IDF and word similarity files for the VIS publication dataset.

Acknowledgments: This work was funded by the StaViCTA (Stance in discourse using Visual and Computational Text Analytics) project, framework grant "The Digitized Society – Past, Present, and Future" with No. 2012-5659 from the Swedish Research Council (Vetenskapsrådet).

Author Contributions: Björn Zimmer and Andreas Kerren conceived of and designed the solution. Björn Zimmer implemented the software. Björn Zimmer and Magnus Sahlgren analyzed and preprocessed the data. Magnus Sahlgren contributed critical advice for the use case. Björn Zimmer, Magnus Sahlgren and Andreas Kerren wrote the paper.

Conflicts of Interest: The authors declare no conflict of interest.

References

1. IEEE VIS. IEEE Visualization Conference (VIS). Available online: http://ieeevis.org. (accessed on 9 May 2017).
2. Kerren, A.; Purchase, H.; Ward, M.O. *Multivariate Network Visualization*; Lecture Notes in Computer Science; Springer: Berlin, Germany, 2014.
3. von Landesberger, T.; Kuijper, A.; Schreck, T.; Kohlhammer, J.; van Wijk, J.; Fekete, J.D.; Fellner, D. Visual Analysis of Large Graphs: State-of-the-Art and Future Research Challenges. *Comput. Gr. Forum* **2011**, *30*, 1719–1749.
4. Lex, A.; Streit, M.; Kruijff, E.; Schmalstieg, D. Caleydo: Design and Evaluation of a Visual Analysis Framework for Gene Expression Data in its Biological Context. In Proceedings of the 3rd IEEE Pacific Visualization Symposium, IEEE, PacificVis '10, Taipei, Taiwan, 2–5 March 2010; pp. 57–64.
5. Lex, A.; Partl, C.; Kalkofen, D.; Streit, M.; Gratzl, S.; Wassermann, A.M.; Schmalstieg, D.; Pfister, H. Entourage: Visualizing Relationships between Biological Pathways Using Contextual Subsets. *IEEE Trans. Vis. Comput. Gr.* **2013**, *19*, 2536–2545.
6. May, T.; Steiger, M.; Davey, J.; Kohlhammer, J. Using Signposts for Navigation in Large Graphs. *Comput. Gr. Forum* **2012**, *31*, 985–994.
7. Henry, N.; Fekete, J.D.; McGuffin, M.J. NodeTrix: A Hybrid Visualization of Social Networks. *IEEE Trans. Vis. Comput. Gr.* **2007**, *13*, 1302–1309.
8. Moscovich, T.; Chevalier, F.; Henry, N.; Pietriga, E.; Fekete, J.D. Topology-Aware Navigation in Large Networks. In Proceedings of the SIGCHI Conference on Human Factors in Computing Systems, Boston, MA, USA, 4–9 April 2009; pp. 2319–2328.
9. Heer, J.; Perer, A. Orion: A System for Modeling, Transformation and Visualization of Multidimensional Heterogeneous Networks. *Inf. Vis.* **2012**, *13*, 111–133.
10. Liu, Z.; Navathe, S.B.; Stasko, J.T. Ploceus: Modeling, Visualizing, and Analyzing Tabular Data as Networks. *Inf. Vis.* **2014**, *13*, 59–89.
11. Kucher, K.; Kerren, A. Text Visualization Techniques: Taxonomy, Visual Survey, and Community Insights. In Proceedings of the 8th IEEE Pacific Visualization Symposium, 2015, PacificVis '15, Hangzhou, China, 14–17 April 2015; pp. 117–121.
12. Federico, P.; Heimerl, F.; Koch, S.; Miksch, S. A Survey on Visual Approaches for Analyzing Scientific Literature and Patents. *IEEE Trans. Vis. Comput. Gr.* **2016**, doi:10.1109/TVCG.2016.2610422.
13. Görg, C.; Liu, Z.; Kihm, J.; Choo, J.; Park, H.; Stasko, J. Combining Computational Analyses and Interactive Visualization for Document Exploration and Sensemaking in Jigsaw. *IEEE Trans. Vis. Comput. Gr.* **2013**, *19*, 1646–1663.
14. Shen, Z.; Ogawa, M.; Teoh, S.T.; Ma, K.L. BiblioViz: A System for Visualizing Bibliography Information. In Proceedings of the 2006 Asia-Pacific Symposium on Information Visualisation, Tokyo, Japan, 1–3 February 2006; Volume 60, pp. 93–102.
15. Zhao, J.; Collins, C.; Chevalier, F.; Balakrishnan, R. Interactive Exploration of Implicit and Explicit Relations in Faceted Datasets. *IEEE Trans. Vis. Comput. Gr.* **2013**, *19*, 2080–2089.
16. Van Ham, F.; Wattenberg, M.; Viegas, F.B. Mapping Text with Phrase Nets. *IEEE Trans. Vis. Comput. Gr.* **2009**, *15*, 1169–1176.
17. Kairam, S.; Riche, N.H.; Drucker, S.; Fernandez, R.; Heer, J. Refinery: Visual Exploration of Large, Heterogeneous Networks through Associative Browsing. *Comput. Gr. Forum* **2015**, *34*, 301–310.
18. Chen, F.; Chiu, P.; Lim, S. Topic Modeling of Document Metadata for Visualizing Collaborations over Time. In Proceedings of the 21st International Conference on Intelligent User Interfaces, Sonoma, CA, USA, 7–10 March 2016; pp. 108–117.
19. Liu, S.; Chen, Y.; Wei, H.; Yang, J.; Zhou, K.; Drucker, S.M. Exploring Topical Lead-Lag across Corpora. *IEEE Trans. Knowl. Data Eng.* **2015**, *27*, 115–129.
20. Heimerl, F.; Han, Q.; Koch, S.; Ertl, T. CiteRivers: Visual Analytics of Citation Patterns. *IEEE Trans. Vis. Comput. Gr.* **2016**, *22*, 190–199.

21. Zimmer, B.; Kerren, A. Harnessing WebGL and WebSockets for a Web-Based Collaborative Graph Exploration Tool. In Proceedings of the 15th International Conference on Web Engineering, Rotterdam, The Netherlands, 23–26 June 2015; pp. 583–598.
22. Zimmer, B.; Kerren, A. OnGraX: A Web-Based System for the Collaborative Visual Analysis of Graphs. *J. Graph Algorithms Appl.* **2017**, *21*, 5–27.
23. Isenberg, P.; Heimerl, F.; Koch, S.; Isenberg, T.; Xu, P.; Stolper, C.; Sedlmair, M.; Chen, J.; Möller, T.; Stasko, J. Visualization Publication Dataset, 2015. Available online: http://www.vispubdata.org/ (accessed on 9 May 2017).
24. Chuang, J.; Manning, C.D.; Heer, J. Without the Clutter of Unimportant Words: Descriptive Keyphrases for Text Visualization. *ACM Trans. Comput.-Hum. Interact.* **2012**, *19*, 1–29.
25. Turney, P.D.; Pantel, P. From Frequency to Meaning: Vector Space Models of Semantics. *J. Artif. Intell. Res.* **2010**, *37*, 141–188.
26. Turian, J.; Ratinov, L.; Bengio, Y. Word Representations: A Simple and General Method for Semi-supervised Learning. In Proceedings of the 48th Annual Meeting of the Association for Computational Linguistics. Association for Computational Linguistics, Uppsala, Sweden, 11–16 July 2010; pp. 384–394.
27. Mikolov, T.; Sutskever, I.; Chen, K.; Corrado, G.S.; Dean, J. Distributed Representations of Words and Phrases and their Compositionality. In Proceedings of the Neural Information Processing Systems Conference, Lake Tahoe, Nevada, USA, 5–10 December 2013; pp. 3111–3119.
28. Pennington, J.; Socher, R.; Manning, C.D. GloVe: Global Vectors for Word Representation. In Proceedings of the 2014 Conference on Empirical Methods in Natural Language Processing, Dohar, Qatar, 25–29 October 2014; pp. 1532–1543.
29. Kanerva, P.; Kristofersson, J.; Holst, A. Random Indexing of Text Samples for Latent Semantic Analysis. In Proceedings of the 22nd Annual Conference of the Cognitive Science Society, Cognitive Science Society, Philadelphia, PA, USA, 13–15 August 2000; p. 1036.
30. Sahlgren, M.; Holst, A.; Kanerva, P. Permutations as a Means to Encode Order in Word Space. In Proceedings of the 30th Annual Conference of the Cognitive Science Society. Cognitive Science Society, Washington, DC, USA, 23–26 July 2008; pp. 1300–1305.
31. Sahlgren, M.; Gyllensten, A.C.; Espinoza, F.; Hamfors, O.; Holst, A.; Karlgren, J.; Olsson, F.; Persson, P.; Viswanathan, A. The Gavagai Living Lexicon. In Proceedings of the 10th International Conference on Language Resources and Evaluation, Portorož, Slovenia, 23–28 May 2016; pp. 344–350.
32. Harrower, M.; Brewer, C.A. ColorBrewer.org: An Online Tool for Selecting Colour Schemes for Maps. *Cartogr. J.* **2003**, *40*, 27–37.
33. Laboratories, K. KEGG: Kyoto Encyclopedia of Genes and Genomes. Available online: http://www.genome. jp/kegg/ (accessed on 9 May 2017).
34. Chau, D.H.; Kittur, A.; Hong, J.I.; Faloutsos, C. Apolo: Making Sense of Large Network Data by Combining Rich User Interaction and Machine Learning. In Proceedings of the SIGCHI Conference on Human Factors in Computing Systems, Vancouver, BC, Canada, 7–12 May 2011; pp. 167–176.
35. Gladisch, S.; Schumann, H.; Tominski, C. Navigation Recommendations for Exploring Hierarchical Graphs. In Proceedings of the 9th International Symposium on Advances in Visual Computing, Part II, Crete, Greece, 29–31 July 2013; pp. 36–47.
36. Roberts, J.C. State of the Art: Coordinated & Multiple Views in Exploratory Visualization. In Proceedings of the Fifth International Conference on Coordinated and Multiple Views in Exploratory Visualization, Zürich, Switzerland, 2 July 2007; pp. 61–71.
37. yWorks. yFiles for Java. Available online: https://www.yworks.com/products/yfiles-for-java. (accessed on 9 May 2017).
38. Khronos Group. WebGL Specification. Editor's Draft 24 February 2017. Available online: http://www. khronos.org/registry/webgl/specs/latest (accessed on 9 May 2017).

MDPI

Article

Sampling and Estimation of Pairwise Similarity in Spatio-Temporal Data Based on Neural Networks

Steffen Frey

Visualization Research Institute, University of Stuttgart; 70569 Stuttgart, Germany;
steffen.frey@visus.uni-stuttgart.de; Tel.: +49-(0)-711-6858-8629

Academic Editors: Achim Ebert and Gunther H. Weber
Received: 1 June 2017 ; Accepted: 18 August 2017; Published: 26 August 2017

Abstract: Increasingly fast computing systems for simulations and high-accuracy measurement techniques drive the generation of time-dependent volumetric data sets with high resolution in both time and space. To gain insights from this spatio-temporal data, the computation and direct visualization of pairwise distances between time steps not only supports interactive user exploration, but also drives automatic analysis techniques like the generation of a meaningful static overview visualization, the identification of rare events, or the visual analysis of recurrent processes. However, the computation of pairwise differences between all time steps is prohibitively expensive for large-scale data not only due to the significant cost of computing expressive distance between high-resolution spatial data, but in particular owing to the large number of distance computations ($O(|T|^2)$), with $|T|$ being the number of time steps). Addressing this issue, we present and evaluate different strategies for the progressive computation of similarity information in a time series, as well as an approach for estimating distance information that has not been determined so far. In particular, we investigate and analyze the utility of using neural networks for estimating pairwise distances. On this basis, our approach automatically determines the sampling strategy yielding the best result in combination with trained networks for estimation. We evaluate our approach with a variety of time-dependent 2D and 3D data from simulations and measurements as well as artificially generated data, and compare it against an alternative technique. Finally, we discuss prospects and limitations, and discuss different directions for improvement in future work.

Keywords: time-dependent data; neural networks; adaptive sampling; volume visualization

---

1. Introduction

Time-dependent data sets with increasing resolution in both time and space are generated at a fast rate, enabled by advances in parallel computing systems for simulations and high-accuracy measurement techniques. This data can feature millions of cells and thousands of time steps, and thus poses significant challenges for visual analysis. Even if the complete data—well-exceeding the available memory in most cases—could be presented to the user interactively, still numerous issues due to visual clutter and occlusion need to be prevented. A popular and natural choice to visualize the data without (temporal) occlusion and clutter issues is animation (i.e., sequentially rendering individual time steps). However , it has been shown to be ineffective as only a limited number of frames can be memorized by an observer (e.g., [1]). This motivates the development of visualization approaches that select and/or aggregate data in a data-driven way to enable efficient visual analysis and exploration.

For some of these type of data-driven visualization approaches, the computation of mutual distances between time steps is a fundamental operation for automatic analysis techniques. Recent examples of applications in visualization include the generation of a meaningful static overview visualization, the identification of rare events, as well as the visual analysis of recurrent processes.

Apart from that, visualizing the similarity information alone directly can also drive interactive visual exploration by indicating processes of interest to a user. In general, these application scenarios require the full computation of pairwise differences between all time steps. This is prohibitively expensive in particular in the context of large-scale data. This is not only due to (1) the significant cost of computing expressive distance between high-resolution time steps, but (2) especially owed to the large number of distance computations involved ($O(|T|^2)$), with $|T|$ being the number of time steps). This means that thousands of time steps already induce millions of (costly) distance computations.

In this work, we present and evaluate different strategies for the progressive computation of similarity information in a time series, as well as an approach for estimating missing distance information based on neural networks. Different strategies for sampling we consider range from purely random sampling over uniform (data-agnostic) to adaptive (data-driven) sampling strategies. With this sampled similarity information (i.e., a subset of a pairwise distances D of a time series T), we then aim to reconstruct the full set of pairwise distances D using a neural network. The goal of this approach is to let the neural network implicitly capture the special structure and properties of similarity information in spatio-temporal data. We then essentially combine both aspects by training neural networks for similarity estimation particularly on different sampling patterns of the different strategies. This eventually allows to automatically determines the sampling strategy yielding the best result in combination with trained networks for estimation.

The remainder of this paper is structured as follows. First, we review related work in Section 2. Then, we introduce our problem statement and give an overview of our approach in Section 3. We then discuss the two main parts of this work: different strategies for the sampling of similarity data (Section 4), as well as neural networks for the estimation of similarities (Section 5). Finally, we evaluate and discuss the properties of our approach in Section 6, before concluding our work in Section 7.

2. Related Work

2.1. Visualization of Spatio-Temporal Data

For the visualization of time-varying data, and extensions to many techniques could be applied to make them more general towards dealing with multi-field data. Lee and Shen [2] visualize trend relationships among variables in multivariate time-varying data. Joshi and Rheingans [3] evaluate illustration-inspired techniques for time-varying data, like speedlines or flow ribbons. One approach is to interpret the data as a space-time hypercube, and apply extended classic visualization operations like slicing and projection techniques [4] or temporal transfer functions [5] to it (cf. Bach et al. [6] for on overview). Another way to approach time-dependent data are feature-based techniques. Here, particularly Time Activity Curves (TAC) that contain each voxel's time series have been used as the basis for different techniques (e.g., [7]). Apart from such techniques working directly with scalar volume data, a large body of work in time-dependent volume visualization is based on feature extraction. Wang et al. [8] extract a feature histogram per volume block (typically hundreds to thousands of voxels). They then derive entropy-based importance curves that characterize the local temporal behavior of each block, and classify them via k-means clustering. Widanagamaachchi et al. [9] employ feature tracking graphs. Lee and Shen [10] visualize time-varying features and their motion on the basis of time activity curves (TAC) that contain each voxel's time series. Fang et al. [7] use TACs in combination with different similarity measures. Silver et al. [11] isolate and track representations of regions of interest. The robustness of this approach has been improved by Ji and Shen [12] with a global optimization correspondence algorithm based on the Earth Mover's Distance. Scale-space based methods and topological techniques have also been used here (e.g., [13,14]). Schneider et al. [15] compare scalar fields on the basis of the largest contours.

Another line of techniques is based on the direct comparison of time steps. The Earth Mover's Distance (EMD, also known as the Wasserstein metric) is a common metric to compute the difference between mass distributions (conceptually, it determines the minimum cost of turning one (mass) distribution into the other) [16]. For instance, Tong et al. [17] use different metrics to compute the distance between data sets, and employ dynamic programming to select the most interesting time steps accordingly. The field of video analysis also deals with related analysis problems, yet typically employing different methodologies. Specialized image and video metrics are used to compare frames (e.g., [18]), and distinct approaches were proposed to generate summaries of videos, e.g., based on the motion of actors over time [19]. In addition, illustrative techniques have been used to depict processes of interest. Lu and Shen [20] propose interactive storyboards composed of volume renderings and descriptive geometric primitives. While most techniques mentioned above deal with volume data, numerous approaches have been presented for flow visualization (cf. Post et al. [21] and McLoughlin et al. [22] for an overview). Note that different fields have developed different methodologies to quantify similarity for other application settings. For instance, to enable style-based search, Garces et al. [23] present a method for measuring the similarity in style between two pieces of vector art, based on the differences between four types of features: color, shading, texture, and stroke. Feature weightings are learned from crowdsourced experiments.

Frey and Ertl [24] presented a technique to generate transformations between arbitrary volumes, providing both expressive distances and smooth interpolates. On this basis, they presented a new approach for the streaming selection of time steps in temporal data that allows for the reconstruction of the full sequence with a user-specified error bound. An accelerated version with overall improved efficiency as well as an extension to manycore devices (i.e., GPUs) has been presented in a follow-up work [25]. We use this approach to quantify distances between time steps in this paper.

On the basis of similarity information between time steps, Frey and Ertl [26] adaptively select time steps from time-dependent volume data sets for an integrated and comprehensive visualization. This reduced set of time steps not only saves cost, but also allows to show both the spatial structure and temporal development in one combined rendering. The selection optimizes the coverage of the complete data on the basis of a minimum-cost flow-based technique to determine meaningful distances between time steps. An interactive volume raycaster produces an integrated rendering of the selected time steps, and their computed differences are visualized in a dedicated chart to provide additional temporal similarity information.

2.2. Similarity Matrices to Directly Visualize and Analyze Similarity Information

Their benefits and utility of recurrence plots and similarity matrices are discussed in detail by Marwan et al. [27]. There are also variants that extend those concepts from univariate to multivariate data. One possibility to apply these concepts to study the spatial structure of data is to separate the data into many one-dimensional data series, and to apply the recurrence analysis separately to each of these series [28]. Another possibility is the extension of the temporal approachof recurrence plots to a spatial one [29] at the cost of high-dimensional domains, e.g., a time-dependent 2D image is mapped to a 4D recurrence plot, which is, however, hard to visualize. Bautista et al. [30] analyze the difference between recurrence plots from different time series. For multi-field visualization, Frey et al. [31] presented an interactive approach on the basis of similarity matrices for extracting and exploring time-dependent phenomena, that allows to compare different locations, modalities, ensemble runs, or generally even data sets with no direct relation. It focuses on periodic and quasi-periodic behavior at single points, but was also used to analyze cross-correlations in ensemble and multi-variate data.

2.3. Machine Learning for Image Interpolation and Similarity Learning

Machine learning is popularly regarded as the only viable approach to building AI systems that can deal with (very) complicated environments [32]. In particular, in this work, we employ neural networks for estimating the similarity between different time steps. Image interpolation is a different task with different inherent characteristics, but there are also some related aspects. Hu et al. [33] propose an interpolation algorithm using a classification-based neural network approach with the goal to improve the image quality. Plaziac [34] compared two adaptive algorithms for image interpolation based on a multilayer perceptron. More recently, Chen et al. [35] used anisotropic probabilistic neural network on the basis of an anisotropic Gaussian kernel to provide high adaptivity of smoothness/sharpness during image/video interpolation.

Neural networks have also been applied to similarity learning, which belongs to the category of supervised machine learning in artificial intelligence. In general, this resembles our task of learning the similarities between time steps of spatio-temporal data, yet previous work has been done mostly for very different application scenarios. The goal is to learn from examples a similarity function that measures how similar or related two entities are. It has applications in ranking, in recommendation systems, visual identity tracking, face verification, and speaker verification. Guillaumin et al. [36] present two methods for learning robust distance measures for assessing the similarity of faces. Similarity learning is closely related to distance metric learning, in that metric learning is the task of learning a distance function over objects. Kulis [37] presents an overview of existing research in metric learning. Davis et al. [38] present an information-theoretic approach to learning a Mahalanobis distance function (the Mahalanobis distance is a measure of the distance between a point and a distribution).

3. Overview

The motivation behind this work is to gain insights from spatio-temporal data. This data can have different processes and structures, and be obtained via measurements and different types of CFD simulations (Figure 1). This means that this type of analysis is of interest to a wide variety of different fields. In this section, we first provide an introduction into similarity information from time series data in Section 3.1. We then cover the fundamentals of neural networks that are the basis for our similarity estimation approach discussed later in this work (Section 3.2). Finally, we give an outline of our approach and its different components in Section 3.3.

3.1. Similarity Information from Spatio-Temporal Data

We aim to analyze our time series data T on the basis of similarities between individual time steps $t \in T$. Here, the similarity between individual time steps is quantified by function $d : T \times T \to \mathbb{R}$, with the result being in the range $[0, \infty)$ (0 denotes identity). We further assume $d(\cdot, \cdot)$ to be symmetric, i.e., $d(t_0, t_1) = d(t_1, t_0)$ for $t_0, t_1 \in T$. In the following, we therefore only consider $d(t_0, t_1)$ for $t_0 < t_1$ (in the identity case of $t_0 = t_1$, $d(t_0, t_1) = 0$). As a basis for numerous analysis applications, in this work we are interested in obtaining all pair-wise similarities $d(t_0, t_1)$ between time steps $t_0, t_1 \in T$ with $t_0 < t_1$. This means that for $|T|$ time steps there are $|d(T)|$ (also denoted as $|D|$) time steps:

$$|d(T)| = \frac{(|T| - 1) \cdot |T|}{2} \qquad (1)$$

For the real-world data sets in Figure 1, the results are shown in Figure 2 in the form of similarity plots. We compute similarity information from the real-world data using the metric by Frey and Ertl [24,25] (cf. discussion later in Section 5.2). Essentially, it constitutes a fast computation method of the Earth Mover's Distance ([16], also known as the Wasserstein metric) that makes it computationally feasible to apply it directly to high-resolution data. While the computation of similarities between all pairs of time steps is very expensive, for training and testing purposes, we generated reference similarity information for the time series data introduced above over the course of several weeks on different machines (using both CPUs and GPUs). Most notably, due to the symmetric property of the distance quantification function d as discussed above (i.e., $d(t_0, t_1) = d(t_1, t_0)$ for $t_0, t_1 \in T$), not a full pair-wise matrix is shown but a only the cases for $t_0 < t_1$. In the end, this forms a triangle-shaped plot. Different similarity structures can be seen, with the Kármán and most notably the Supernova data set featuring processes at distant points in time that are very similar (in the Kármán, this can be observed in the bottom right where line structures with a small offset to the diagonal can be seen).

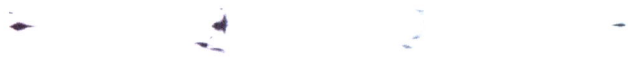

(a) **Bottle** (resolution 900×430, 160 time steps considered): laser pulse shooting through a bottle, captured via Femto Photography (Velten et al. [39]).

(b) **von Kármán** (resolution 301×101, 418 time steps): 2D time-dependent CFD simulation of a von Kármán vortex street.

(c) **Hot Room** (resolution 101×101, 265 time steps): air flow within a closed container, driven by buoyant forces imposed by a heated bottom plate and a cooled top plate. To provoke transient aperiodic flow, the container exhibits two barriers (one on the top, one on the bottom).

(d) **Droplet** (256^3, 155 time steps): two drops colliding asymmetrically (courtesy of C. Meister, Institute of Aerospace Thermodynamics, University of Stuttgart).

(e) **Supernova** (432^3, 60 time steps): result of a supernova simulation. The data set is made available by Dr. John Blondin at the North Carolina State University through US Department of Energy's SciDAC Institute for Ultrascale Visualization.

Figure 1. All data sets include scalar values, that are mapped to a representation that is shown here, and also used for the distance computation via a user-defined transfer function (respective distances of each time series are plotted in Figure 1).

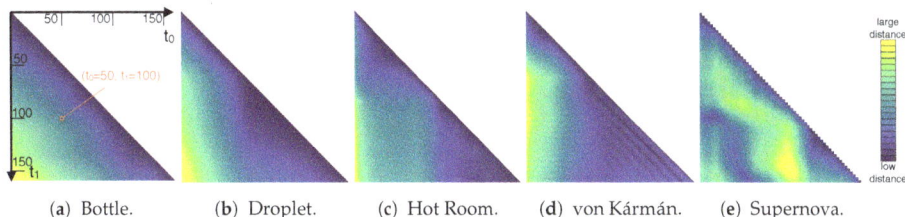

(a) Bottle.   (b) Droplet.   (c) Hot Room.   (d) von Kármán.   (e) Supernova.

Figure 2. Input similarity information from different data sets presented in the form of similarity matrices (with t_0 along the x-axis and t_1 along the y-axis, cf. (**a**)). Only one half of the matrix is visualized due the symmetry property of our distance metric (i.e., $d(t_0, t_1) = d(t_1, t_0)$). Values are mapped to colors using the viridis color map (low distances $\widehat{=}$ purple, medium $\widehat{=}$ green/blue, large $\widehat{=}$ yellow).

3.2. Neural Networks Basics (for Time Series Similarity Estimation)

In this work, we make use of so-called feedforward neural networks (aka multi- layer perceptrons (MLPs)) [32]. The goal of a feedforward network is to approximate some function f^*. In our case, we aim to estimate the result distance function $d(t_0, t_1)$ that determines the similarity between any two time steps in a time series $t_0, t_1 \in T$. In general, a feedforward network defines a mapping $y = f(x; \theta)$ and learns the value of the parameters θ that result in the best function approximation. In our application scenario, both input and out of the neural network is pair-wise similarity information in a time series.

These kind of models are called feedforward because information flows through the function being evaluated from x, through the intermediate operations used to define f, and eventually to the output f. There are no feedback connections in which outputs of the model are fed back into itself (as in recurrent neural networks, e.g., [32]).

Feedforward neural networks are typically represented by composing together different functions. The model is associated with a directed acyclic graph describing how the functions are composed together. For example, there might be three functions $f^{(1)}$, $f^{(2)}$, and $f^{(3)}$ connected in a chain, to form $f(x) = f^{(3)}(f^{(2)}(f^{(1)}))$. These chain structures are the most commonly used structures of neural networks. In this case, $f^{(1)}$ is called the first layer of the network, $f^{(2)}$ is called the second layer, and so on. While the first layer is denote as the input layer, the final layer of a feedforward network is called the output layer. During neural network training, we adjust $f(x)$ to match $f^*(x)$. Each example x is accompanied by reference result values $y = f^*(x)$. The training examples specify directly what the output layer must do at each point x; it must produce a value that is close to y. As the training data does not show the desired output for the layers between the input and the output layer, these layers are called hidden layers.

3.3. Approach Outline

A conceptual overview on our approach is given in Figure 3. Essentially, there are two distinct phases, that are indicated by the wide gray arrows: (1) select and (2) optimize. First, in selection, we evaluate different sampling strategies to sample similarity information. Sampling strategies define different techniques to approach the adaptive sampling of similarity information (cf. Section 4). We then use the full similarity information along with the sparse set generated by the adaptive sampling for training a neural network to reconstruct the full information. We then validate the generated model, which essentially results in an error value (subsequently denoted as *cost*). Using this information, we compare the obtained cost values of all sampling strategies, and choose the strategy yielding the lowest cost as our best strategy. With this, we enter our second phase in which we optimize the network belonging to the best strategy. While the selection just carries out one training and validation run for each strategy, the subsequent optimization step iteratively improves the network belonging to the best sampling strategy via continuous training.

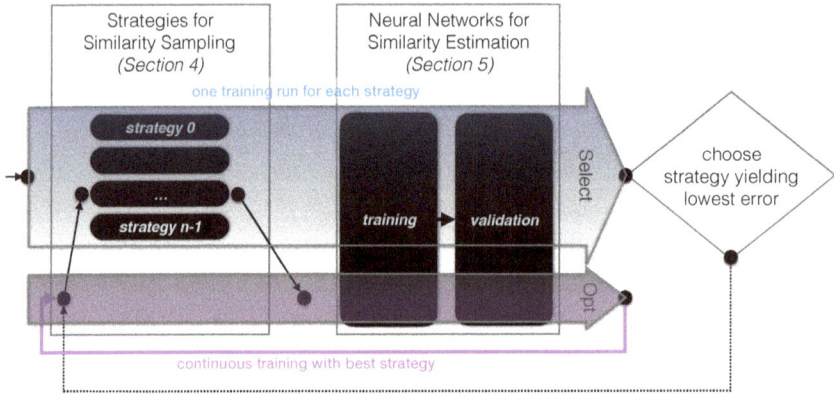

Figure 3. Overview of our approach for the adaptive sampling and estimation of similarities with neural networks (cf. Algorithm 2 for a more detailed description). The selection phase (**Select**) chooses the best strategy by carrying out one training and validation run for each strategy. Afterwards, the optimization phase (**Opt**) iteratively improves the network belonging to the best sampling strategy via continuous training with repeatedly updated training data.

4. Strategies for Similarity Sampling

As previously discussed in Section 3.1, we utilize and evaluate progressive approaches to compute similarity information between individual time steps $t \in T$ of a time series T. Here, progressive means that we iteratively add new similarity information (i.e., distances between pairs of time steps). The strategy basically for which time step pair its similarity is computed next. The strategies for determining the next time step pair to compute—subsequently denoted as similarity sampling—considered in this paper are now discussed in the following. Depending on their procedure, they belong to two time types of categories: **similarity pair-based** and **time step-based**.

Similarity pair-based denotes the concept that pairs of time steps can be selected in an arbitrary fashion, and therefore also completely independently from the samples taken so far (naturally, a pair of time steps only needs to be computed once). More formally, this means that the next time step pair (t_0, t_1) to compute the similarity for using metric $d(t_0, t_1)$ is chosen arbitrarily from the full set of time steps T (i.e., $t_0 \in T$ and $t_1 \in T$). Here, the only restriction is that we limit ourselves to $t_0 < t_1$ due the symmetry property of d (cf. discussion in Section 3.1).

In contrast, **time step-based** means that not individual pairs but time steps are progressively added into consideration, and the similarity between all pairwise combinations of considered time steps is computed before selecting new time steps. This means that only a subset $T^* \subset T$ is currently considered. Before adding a new time step $t \in T, t \notin T^*$, we first compute all combinations $t_0 \in T^* \times t_1 \in T^*$ (with $t_0 < t_1$). Only when all pairwise combinations are computed, we add a new time step to T^*.

The different sampling strategies (creating a set of similarity pairs P) we employ and evaluate in this paper on the basis of these different approaches outlined above are as follows ((**1**) and (**2**) follow similarity pair-based, (**3**)–(**7**) follow time step-based).

(**1**) **uniform pair.** In this strategy, the goal is to distribute samples in the temporal space $T \times T$ as evenly as possible (in a progressive fashion). Doing this, we start out with a random sample pair $P = \{(t_0, t_1)\}$ ($t_0 \in T$ and $t_1 \in T$). Subsequently, we then iteratively compute the new similarity pair (t_0, t_1) that has the maximum distance to any of the pairs $(t_0, t_1) \in P$ (i.e., to any of the pairs that have been computed so far).

(2) **random pair.** A time step pair—that has not been computed yet ($\notin P$)—is chosen randomly and processed next.

(3) **random time.** A random time step is selected that has not yet been considered ($\notin T^*$). As described above, before adding a new time step, first all pairwise combinations of time steps $\in T^*$ are computed before proceeding further.

(4) **uniform time.** Select the time step in between the largest interval range $t_{i+1} - t_i$ in T^* (with t_i and t_{i+1} denoting subsequent time steps $\in T$). In case there are multiple intervals with the same size, we choose one randomly.

(5) **distance-weighted time.** Choose a time step randomly (similar to (3)), but the probability of selecting an interval is weighted by $t_{i+1} - t_i - 1$ (akin to the selection criterion in (4)).

(6) **similarity time.** Consider the distance between two subsequent time steps in T: $d(t_i, t_{i+1})$. Add a time step in the interval with the largest distance.

(7) **similarity-weighted time.** Select an interval randomly to add a new time step to T^*, with the probability being weighted by $d(t_i, t_{i+1})$.

The different sampling patterns arising from these seven different strategies are exemplified in Figure 4 (distances which have not been computed yet are indicated in red). We achieve a large variety of strategies, following completely random, uniform, and similarity-adaptive approaches.

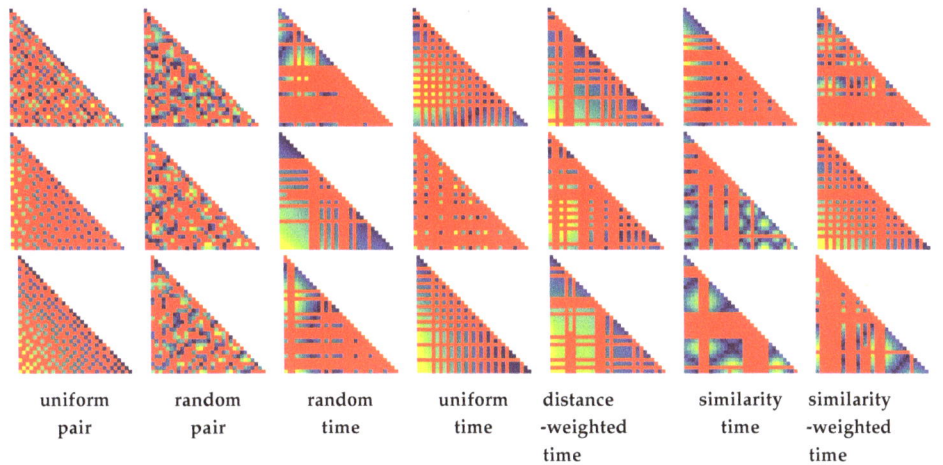

| uniform pair | random pair | random time | uniform time | distance -weighted time | similarity time | similarity -weighted time |

Figure 4. Different sampling strategies by example. Horizontally, we provide the results of a certain sampling strategy, while vertically these strategies are demonstrate by means of different input data sets. Similarity pairs that have not yet been computed by the sampling strategy are indicated in red.

5. Neural Networks for Similarity Estimation

In this section, we first discuss the basic model setup of our neural network (Section 5.1). We then outline how we obtain and generate the data used for its training and validation (Section 5.2). On this basis, we finally discuss our approach to train neural networks and to select the most appropriate respective sampling strategy (Section 5.3).

5.1. Model Setup

Our model is designed to estimate one missing similarity pair (t_0, t_1) on the basis of other available similarity information. The design choices described below are based on empirical tests, informally evaluating different model designs and neural network setups against each other. Note that while our resulting design is the best we found in our tests, we cannot consider it optimal due to the large search space (there is large variety of different ways to configure a neural network alone).

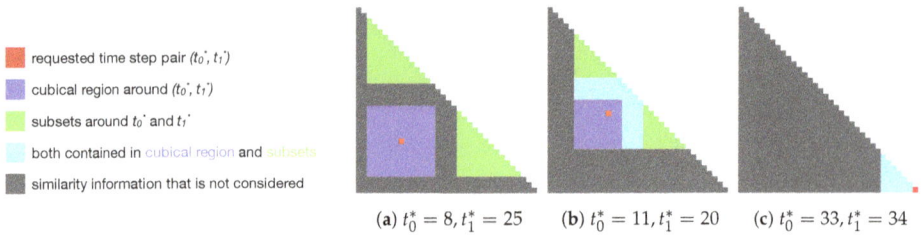

(a) $t_0^* = 8, t_1^* = 25$ (b) $t_0^* = 11, t_1^* = 20$ (c) $t_0^* = 33, t_1^* = 34$

Figure 5. Illustration of the different time step pairs that are considered for training at the example of different time step pairs (t_0^*, t_1^*) of interest (cf. Section 5.1, **Considered Input Data**).

Considered Input Data. For estimating the similarity information, we utilize a subset of the available similarity information from different regions. In more detail, we jointly consider two types of information in a (temporal) region of extent δ around the requested element (t_0^*, t_1^*) (individually, for each component of the pair, this results in a region of $T_0 = \{t_0 - \delta, \ldots, t_0 + \delta\}$ and $T_1 = \{t_1 - \delta, \ldots, t_1 + \delta\}$, respectively). An illustration of the considered time steps at different examples is shown in Figure 5.

1. The cubical region around (t_0^*, t_1^*), except for (t_0^*, t_1^*) itself (blue in Figure 5):

$$T_\delta(t_0^*, t_1^*) = T_0 \times T_1 \setminus (t_0^*, t_1^*). \tag{2}$$

   This results in $|T_0||T_1| - 1$ elements. This gives the similarity of close pairs in temporal space.
2. Two subsets of the similarity matrix, one around t_0 and one around t_1 (i.e., containing all pairwise combinations of time steps T_0 and T_1, respectively) (green in Figure 5). This results in a total of $|d(T_0)| + |d(T_1)|$ (according to Equation (1)), and gives the similarity to close time steps for each component of the pair.

In total, this accordingly results in I input elements (cf. Figure 5):

$$|I_\delta| = \underbrace{(2\delta + 1)^2 - 1}_{\text{cubical region (1.)}} + \underbrace{2|d(T_\delta)|}_{\text{similarity context per component (2., based on Equation (1))}} . \tag{3}$$

In cases where no similarity information is available (because it has not been sampled yet or it is out of the temporal range), the respective element gets a dedicated (missing) value of m (in our implementation we set $m = -1$, which clearly indicates a special value as similarity information is generally quantified by positive values). Similarity pair information from the two different types of information may also overlap (e.g., Figure 5b,c), in which case the respective similarity information is considered redundantly as input for the neural network.

Neural Network Structure. Neural networks have three types of layers: input, hidden, and output. There is exactly one input layer, with the number of neurons being determined by the size of the input data. Based on the previous discussion regarding considered input data, this means that we have a total number of $|I_\delta|$ input neurons (Equation (3)). The output layer also consists of exactly one layer. However, here, we just use single neuron as only one specific similarity pair is predicted at a time by our neural network. Regarding the hidden layers, there is much larger degree of freedom : how many hidden layers to actually have in the neural network and how many neurons (and which type) will be in each of these layers. Typically, these decisions have a significant impact on the results that can be achieved, but respective decisions come down to experience and trial-and-error to a certain extent. In this work, we use one hidden layer, with the number neurons equalling the number of neurons in the input layer. According to our experiments, this provides a good trade-off between underfitting and overfitting. For each neuron, we use the sigmoid as an activation function. Note that this is the design that worked best according to our experiments, but we do not consider it to be optimal or any kind of definite solution to the problem (but rather a step toward it).

5.2. Training and Validation Data Preparation

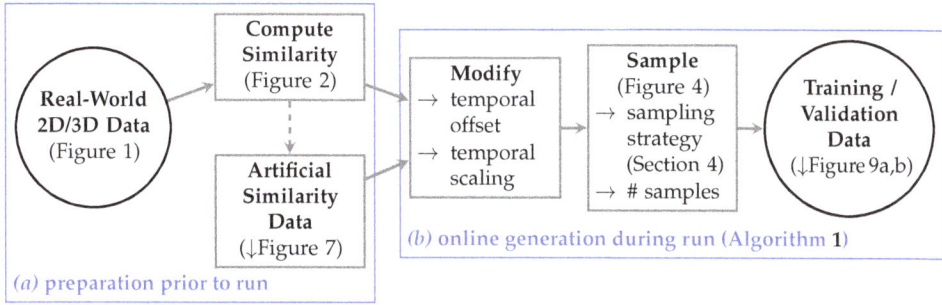

Figure 6. Pipeline for preparing training and validation data.

Algorithm 1 Generation of similarity data sets (for training, validation, and testing)

1: $\triangleright$ in: pool of similarity information P (composed of data sets P_i, featuring $P_i^{\text{\# time steps}}$ time steps)
2: $\triangleright$ in: number of sets n_X
3: $\triangleright$ in: number of time steps n_T
4: $\triangleright$ in: sampling strategy λ
5: $\triangleright$ out: sets of data X
6: **function** DATA_SET_GENERATION(P, λ, n_X, n_T)
7:     $X \leftarrow \varnothing$
8:     **while** $|X| < n_X$ **do**
9:         $\triangleright$ choose data set
10:         $i \leftarrow$ random_randint($0 \ldots |P| - 1$)
11:         $\triangleright$ choose random step size s (i.e., scaling of the time series)
12:         $s \leftarrow$ random_randint($1 \ldots P_i^{\text{\# time steps}}/n_T$)
13:         $\triangleright$ choose random offset o
14:         $o \leftarrow$ random_randint($0 \ldots s \cdot P_i^{\text{\# time steps}} - n_T - 1$)
15:         $\triangleright$ generate variant of P_i
16:         $P_i^* \leftarrow \varnothing$
17:         **for all** $e \in [0 \ldots n_T)$ **do**
18:             $P_i^* \leftarrow P_i^* \cup P_i[o + se]$
19:         **end for**
20:         $\triangleright$ determine number of samples to take, maximum number computed via Equation (1)
21:         $n_S \leftarrow$ random_randint($0 \ldots |d(n_T)|$)
22:         $X \leftarrow X \cup \{P_i^*, n_S\}$
23:     **end while**
24:     **return** X
25: **end function**

A large number of adequate training and validation data is crucial for the success when training neural networks. However, the computation of real-world distances is very expensive and can only be done for a few data sets. To overcome this issue, we generate additional artificial data and further modify the similarity information. In detail, we use the following multi-stage approach to generate a large variety of training and validation data (cf. Figure 6).

Real-World 2D/3D Data. As input, we use a set of typical real-world 2D/3D + time data sets (Figure 1).

Compute Similarity. To compute the pairwise distances between different time steps within each series, we use the approach proposed by Frey and Ertl [24,25]. It is used to make it computationally feasible to directly compute the similarity between high-resolution field data sets. Conceptually, it starts with an initial random assignment of so-called source elements from one data set to so-called target elements of the other data set (each element refers to a (scalar) mass unit given at a certain cell/position in the data). Then, this assignment is improved iteratively in the following. In each iteration, source elements exchange respectively assigned target elements under the condition that this improves the assignment. For this, the quality of an assignment is quantified by d, that essentially computes the sum of weighted distances of the assignments. Here, assignments are weighted by the scalar quantity that is transported. We use this value d directly (on the basis of Euclidean distances) to quantify the distance between the respective time steps. The respective results are shown in Figure 2. Please refer to Frey and Ertl [24,25] for a more detailed discussion.

Artificial Similarity Data. Only using a small number of data sets is not sufficient to cover the large variety of typical patterns of similarity information in general, and might also be dangerous in terms of training the network regarding the concrete data rather than generalizing for similarity estimation. Therefore, we added further, synthetically-generated time series data to supplement this. Here, the idea is to mimic the typical patterns that we have seen occurring in the similarity data, yet providing a larger variety to yield better generalization characteristics after learning. For this, we used the following equation ψ for $\bar{t} \in [0, 1)$, and three random values $\rho_0, \rho_1, \rho_2 \in [0, 1)$:

$$\psi(\bar{t}) = (\bar{t}\rho_0 + 0.5(1 - \bar{t}))\sin(\bar{t}\rho_1 + \rho_2(1 - \bar{t})) \tag{4}$$

We then compute similarity information from these, and use it during training and validation (Figure 7).

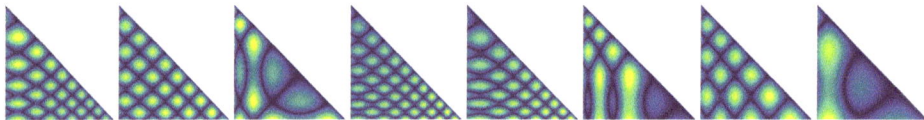

Figure 7. Input similarity information from synthetic data (Equation (4)).

Modify. We do not use the obtained similarity information directly, but randomly offset and scale the time series to get numerous variations on the basis of the available data. We outline our approach to prepare the training data X by means of Algorithm 1 (validation data is generated accordingly). We randomly pick data sets from our collection of real-world and artificial data (Line 10). To modify the data, we randomly choose a scaling factor s, that basically defines the step size with which time steps are considered (Line 12). Then, we use a random offset, which basically determines the first time step that is considered in a time series (Line 14). Finally, we employ this information to generate a new training element P_i^* (Lines 16–19), each one consisting of n_T time steps (we use $n_T = 35$ throughout this work).

Sample. We then take a random number of samples s from the modified similarity data using the respective sampling strategy (cf. discussion in Figure 4, Line 21).

Training / Validation Data. Finally, this yields the data that can be used for training and validation of the neural network. In more detail, each training / validation data element consists of a pair: (1) the original similarity data after **Modify**, and (2) the respective data after **Sample**. Each (1) and (2) contains pairwise similarity information between n_T time steps (a portion of this information has been removed from (2)).

5.3. Similarity Estimation

Algorithm 2 Our approach to estimate missing similarity information based on neural networks (see Figure 3 for a conceptual overview).

1: ▷ in: pool of similarity information P
2: ▷ in: number of sets n_X
3: ▷ in: number of time steps n_t
4: ▷ out: model for similarity estimation Θ
5: **function** TRAIN_SIMILARITY_ESTIMATION(P)
6:     $Y_{train} \leftarrow$ DATA_SET_GENERATION(P, n_{train}, n_T)
7:     $Y_{validate} \leftarrow$ DATA_SET_GENERATION($P, n_{validate}, n_T$)
8:     ▷ Select Sampling Model Λ
9:     $\check{c} \leftarrow \infty$
10:    ▷ Loop over all sampling strategies Λ (as discussed in Section 4)
11:    **for all** $\lambda \in \Lambda$ **do**
12:       ▷ obtain n_λ samples using sampling strategy λ
13:       $X_{train}^\lambda \leftarrow \lambda(Y_{train})$
14:       ▷ we use the Adam optimizer [40] for training the neural network
15:       $\Theta_\lambda \leftarrow$ TRAIN($X_{train}^\lambda, Y_{train}$)
16:       ▷ do validation
17:       $X_{validate}^\lambda \leftarrow \lambda(Y_{validate})$
18:       $c \leftarrow$ evaluate($X_{validate}, Y_{validate}$)
19:       **if** $c < \check{c}$ **then**
20:         $\check{c} \leftarrow c; \check{\lambda} \leftarrow \lambda$
21:       **end if**
22:    **end for**
23:    ▷ continue training with best selected model
24:    **loop**
25:       $Y_{train} \leftarrow$ DATA_SET_GENERATION(P, n_{train}, n_T)
26:       $X_{validate}^{\check{\lambda}} \leftarrow \check{\lambda}(Y_{validate})$
27:       $\Theta_{\check{\lambda}} \leftarrow$ TRAIN($X_{train}^{\check{\lambda}}, Y_{train}$)
28:    **end loop**
29:    **return** Θ
30: **end function**

Our overall approach to use neural networks to estimate missing similarity information and to select sampling strategies has been conceptually outlined already in Figure 3. In the following, we now aim to describe it in more detail by means of Algorithm 2.

First of all, we generate separate sets for training and validation using the procedure described above (Lines 6 and 7). On this basis, we then aim to determine the sampling strategy with the lowest cost $\check{c}$ (Lines 9–22). For each sampling strategy $\lambda \in \Lambda$ (Line 11), we then obtain a sampled (i.e., sparse) variant of the similarity information (Line 13). We then use this as input for training (Line 15). During validation (Lines 17–18), we determine a cost c that we then compare against the cost obtained by other strategies. If it is smaller than the smallest cost $\check{c}$ determined so far (Line 19), we save the respective model λ of the respective strategy as the best one so far (Line 20). After testing all models, we continue refining the model that corresponds to the sampling strategy that led to smallest validation cost $\check{c}$ (Line 27).

6. Results

In this section, we first discuss our evaluation setup (Section 6.1). We then evaluate the results with different sampling strategies (Section 6.2) as well as similarity estimation with neural networks for the selected strategy (Section 6.3). Finally, we discuss properties and limitations of our approach (Section 6.4).

6.1. Evaluation Setup

Parameters, Software and Hardware Setup. For our implementation, we use Python and TensorFlow [41] (r0.11). For training, we used the GPU implementation on the basis of CUDA using a GTX1070 on an Ubuntu 16.04 system with 32GB of RAM and an Intel Core i7-4770 CPU. Furthermore, we employ a batch size of 4096 and 1024 training iterations. In total, this means that there are four million training cases in each epoch. For validation, we use 16384 cases. As discussed above, each individual case consists of a reference (i.e., a modified version of a real-world or a synthetic data set) as well as a sampled version of it. In our evaluation, we randomly vary the number of samples such that they cover between 10% and 50% of all pairwise similarity information. We use squared distances to assess the difference of the estimated similarity to the reference. In this work, as mentioned above, we evaluate our approach by considering time windows of size $n_T = 35$. In total, the generation of training sets, training, and validation takes around six hours for testing each sampling strategy using our setup described above. Note that overall the goal here is to train a network that is able to predict similarity for a wide range of temporal data, which is why we train and validate considering a large variety of generated training data. This means that the training process only needs to be done once, and the resulting neural network can then be applied to estimating similarity data as-is. Evaluating the trained network can be done very quickly and yields comparable performance to other types of estimation considering a similar amount of data for estimation (e.g., via inverse distance weighting as discussed below).

Comparison against Inverse Distance Weighting for Similarity Estimation. We compare our approach for estimating missing similarities with neural networks against a standard approach for scattered data interpolation, namely inverse distance weighting. Here, the similarity information corresponding to time step pairs $d'(t_0, t_1)$ is calculated with a weighted average of the values available for the known pairs. The neighborhood considered here is the same as is used by the neural network (i.e., as specified in Equation (2)). With this, the value estimation $d'(t_0, t_1)$ for a missing (i.e., not yet sampled) value is computed as follows (D denotes a map of previously computed similarity information, with m being returned for unknown pairs):

$$d'_\delta(t_0, t_1) = \frac{\sum_{(\bar{t}_0, \bar{t}_1) \in T_\delta(t_0, t_1)} \omega(\bar{t}_0, \bar{t}_1) D(\bar{t}_0, \bar{t}_1)}{\sum_{(\bar{t}_0, \bar{t}_1) \in T_\delta(t_0, t_1)} \omega(\bar{t}_0, \bar{t}_1)}, \tag{5}$$

with

$$\omega(t_0, t_1) = \begin{cases} \left(\frac{1}{\sqrt{(t_0 - t_0^*)^2 + (t_1 - t_1^*)^2}} \right)^p & \text{, if } D(t_0, t_1) \neq m \\ 0 & \text{, else.} \end{cases} \tag{6}$$

Here, p is a positive real number that specifies the power parameter. Weight decreases as distance increases from the interpolated points. Larger values for p assign greater influence to values closest to the interpolated point, with the result converging toward nearest neighbor interpolation for large values of p.

6.2. Sampling Strategies

For evaluating the different sampling strategies (cf. Section 4) regarding their performance in the context of a neural networks for similarity estimation, we train a neural network over 2048 epochs with respectively sampled data.

For each sampling strategy, we further compare the validation results of the neural network to the results achieved with inverse distance weighting for different power parameters. The respective results are shown in Figure 8. Most prominently, it can be seen that the sampling strategies relatively perform similarly across all estimation approaches: **uniform pair** and **uniform time** yield the best results (lowest validation cost), while **random time** and **similarity time** yield the worst results here. On the basis of the different sampling patterns in Figure 4, we assume that the main reason behind this are the resulting larger temporal regions in which no similarity information is available. For these, it is much more difficult across all similarity estimation approaches to yield reasonably good results. Note that we consider a number of samples that is randomly chosen to be between 10% and 50% of the full sampling. In preliminary tests with a larger number of samples, adaptive approaches performed much better relatively. This indicates that a more complex combination of strategies (or more advanced sampling strategies overall) could be worthwhile to consider in this context. However, a closer investigation and evaluation of this remains for future work.

The best result overall in our evaluation setting is achieved by our neural network-based similarity estimation with the **uniform time** sampling strategy. Not only in this case but within each sampling strategy, it can be seen that our neural network-based approach consistently yields better results (i.e., lower validation cost) than any inverse distance weighting variant. Please refer to the upcoming section for a closer discussion of the different reasons behind this at the example of the **uniform time** sampling strategy.

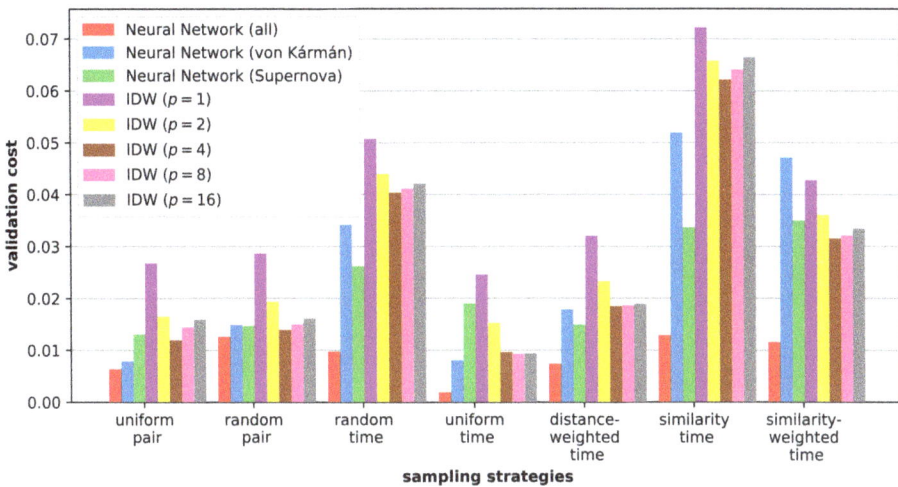

Figure 8. Results of different adaptive sampling strategies and interpolation schemes after the initial phase of sampling strategy selection (**Select** in Figure 3). Respective costs are given for the estimation with the trained neural network as well as inverse distance interpolation (*IDW*) for different power parameters p. For the neural network, not only the results for training with all data sets (*Neural Network (all)*) are presented, but also the validation costs are provided for neural networks that have only be trained with either the von Kármán data set (*Neural Network (von Kármán)* on the basis of Figure 2d) or the Supernova data set (*Neural Network (Supernova)* on the basis of Figure 2e).

For analyzing the utility of using a variety of different data sets for training, we also include the results of networks that have just been trained on the basis of one data set. For this evaluation, we use the von Kármán data set (*Neural Network (von Kármán)* in Figure 8, employing similarity information from Figure 2d) as well as the Supernova data set (*Neural Network (Supernova)*, on the basis of Figure 2e). In both cases, still the same total number of training data sets is generated via modification and sampling. This means that the only difference is that just a single real-world data set (and no artificial similarity data) is employed for training, but the same validation process is used as for *Neural Network (all)* (i.e., all data sets are always considered for validation). It can be seen in Figure 8 that the respective neural networks trained with a single data set deliver worse results than the neural network that has been trained more diversely with all data sets (*Neural Network (all)*). However, they still produce reasonable results that are comparable to the quality generated by inverse distance weighting. Comparing *Neural Network (von Kármán)* and *Neural Network (Supernova)* against each other, it can be seen that their relative performance depends significantly on the sampling strategy as well. Essentially, this indicates that how successful different sampling strategies are also depends on the type of properties and characteristics of the similarity information that is employed for training. Among others, this supports the approach—as discussed in Section 3.3—of taking a variety of strategies into account and using an automatic approach to select the best one for a provided collection of data sets of interest. A more exhaustive evaluation of respective properties remains for future work.

6.3. Similarity Estimation

Next, we discuss the results of similarity estimation with the uniform time sampling strategy that has been determined to deliver the best results in our setup. Reference similarity information, samples, and the reconstructed information with the estimated similarities for our neural network as well as for inverse distance interpolation are shown in Figure 9 (at the example of a subset of the validation set). Overall, as reflected by the small cost/error value, it can be seen that even in cases where a large portion of the data is missing, the trained network performs well in filling in the missing information.

Good results can be achieved with our neural network-based approach over a large variety of cases. Despite potentially only a fraction of the similarity information being available and/or temporal changes occurring at a high rate, we are still able to yield a good approximation of the actual values.

In comparison, inverse distance weighting struggles particularly in the case of a higher rate of changes (e.g., case 5 and case 10). As we consider a fairly large neighborhood ($\delta = 6$), the lower power variants for inverse distance interpolation that also give further away samples a significant weight yield insufficient results, in particular for the cases with a large variation (i.e., $p = 1$ and $p = 2$). In turn, a large power parameter ($p = 16$) effectively only considers the closest points, which yields blocky (non-smooth) results which is particularly noticeable in some smoother cases (e.g., cases 7 and 8). The best performance of inverse distance weighting overall is achieved in-between with $p = 4$ and $p = 8$, that shares both issues of a high and a low power parameter, yet to a lesser extent. However, generally inferior results are achieved in comparison to the similarity estimation by the neural network.

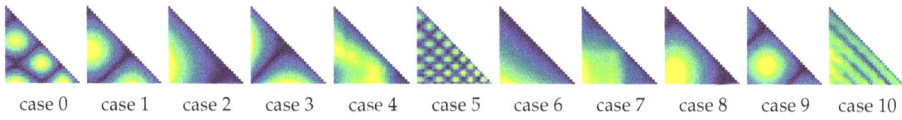

case 0   case 1   case 2   case 3   case 4   case 5   case 6   case 7   case 8   case 9   case 10

(a) Reference.

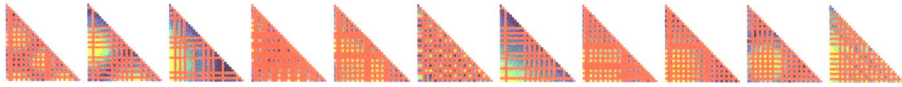

(b) Sampling (missing colors indicated in red).

0.00064   0.00042   0.00050   0.00092   0.00064   0.00142   0.00034   0.00072   0.00021   0.00014   0.00325

(c) Neural Network (all).

0.03727   0.01514   0.00362   0.01534   0.01293   0.06094   0.00347   0.00546   0.00457   0.01209   0.01596

(d) Inverse Distance Weighting ($p = 1$).

0.01575   0.00573   0.00138   0.00842   0.00760   0.05185   0.00124   0.00328   0.00227   0.00448   0.01319

(e) Inverse Distance Weighting ($p = 2$).

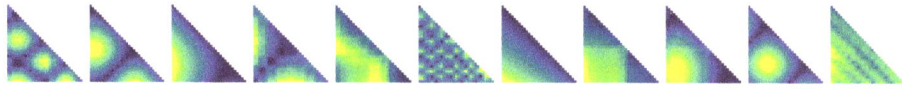

0.00512   0.00102   0.00025   0.00562   0.00493   0.04255   0.00019   0.00269   0.00123   0.00086   0.00987

(f) Inverse Distance Weighting ($p = 4$).

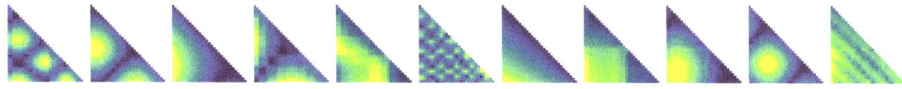

0.00434   0.00066   0.00014   0.00570   0.00440   0.04080   0.00011   0.00280   0.00114   0.00055   0.00910

(g) Inverse Distance Weighting ($p = 8$).

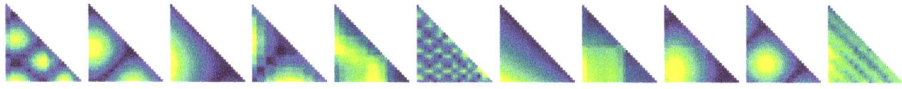

0.00435   0.00066   0.00014   0.00588   0.00433   0.04075   0.00011   0.00284   0.00115   0.00054   0.00909

(h) Inverse Distance Weighting ($p = 16$).

Figure 9. Reference, sampling, and estimation of similarities for different interpolation strategies. The numbers below the plots with estimated similarity give the respective validation cost ((c)–(h)).

6.4. Discussion

While the approach presented in this paper delivers promising results, there are also a variety of limitations and shortcomings that we aim to briefly discuss in the following. On this basis, we also indicate plans and directions for future work to overcome some of these limitations.

First of all, there is a large parameter space in setting up, configuring, training, etc. of neural networks. Typically, these decisions have a significant impact on the results that can be achieved, but respective decisions come down to experience and trial-and-error to a certain extent. The paper at hand presents the best design of a network for spatio-temporal similarity estimation we determined so far according to our experiments. However, we do not consider it to be optimal or any kind of definite solution to the problem. To find a better solution, we aim to systematically try different types of neurons, different numbers of layers and numbers of neurons in each layer, different batch sizes and learning rates, etc. As discussed above, we also found that more advanced sampling similarity strategies beyond the rather basic ones considered in this paper could be worthwhile to pursue and potentially yield better results with the same number of samples. Likewise, although the formulation of the synthetic data is comparably simple (Equation (4)), it can create a variety of different structures we see in practice (in combination with the additional modifications applied to the data). However, a more elaborate method could be able to more comprehensively represent a wider variety of time series characteristics. To develop such a more advanced artificial representation of similarity data in future work, we aim to systematically consider a large and diverse set of similarity information in different kinds of time series data.

Conceptually, the task that we solve with our neural network can be seen to be the generation of the full set of pairwise similarity information from a given set of samples. This problem can be described as a general reconstruction or interpolation problem from sparse samples, and essentially a wide variety of methods can be applied to that problem. Naturally, this includes inverse distance weighting that we use for the sake of comparison above, but also more complex techniques from signal processing like compressed sensing. However, note that there are special properties that differentiate time series similarity data from standard signals or images. These properties depend on the specific metric that is used for assessing the difference between two time steps. For instance, for the distance operator used in this paper, there is the symmetry property (cf. Section 3.1), and $d(t_a, t_b) + d(t_b, t_c) \approx d(t_a, t_c)$ holds for linear behavior in between t_a and t_c. We consider our basic approach described in this paper to be generally applicable for the training and estimation of different types of data and metrics, as we do not explicitly encode these properties, but rather aim to let the network adapt to this through training. The only exception to this is the symmetry property that we explicitly exploit for the sake of efficiency (resulting in the triangle-shaped similarity plots), but the extension to non-symmetric metrics is straight-forward. In general, we assume that the more cross-relations there are in the data, the more do we benefit from using a learning-based approach that is able to capture these respective properties. However, a closer evaluation of these aspects is subject to future work. Also, while in this paper we limit ourselves to comparing our neural network-based approach to inverse distance weighting methods (that are the most popular approach in the field of computer graphics and visualization for scattered data interpolation), advanced approaches from different domains (particularly signal processing) could also yield good results in our considered scenario. A thorough comparison against (different implementations of) these methods also remains for future work.

Furthermore, we aim to evaluate the utility of our approach in real-world visualization scenarios. As discussed above, there is a variety of visual analysis applications that can directly use the estimated similarity information for automatic selection and aggregation of data, and we would like to evaluate the impact of the quality of the distance estimation on the final visualization result. While we believe that our approach is generally applicable in terms of methodology, we particularly aim to more thoroughly evaluate the generality of a trained network. In our evaluation (cf. Figure 8 and respective discussion in Section 6.2), we already consider the extreme example of a network that has just been trained on the basis of one data set (that has however been modified as discussed in Section 5.2). Here, it can be seen that, while delivering significantly worse results than our comprehensively trained network, it still yields decent results that are comparable in quality to the reference inverse distance weighting techniques. While this can be interpreted as a small indicator that we are able to achieve good results for general similarity information from a small set of data sets, a more extensive evaluation is required to thoroughly analyze respective properties.

7. Conclusions

In this work, we presented different strategies for the progressive computation of similarity information in spatio-temporal data, along with an approach for estimating missing distance information. For similarity estimation, we proposed to use a neural network design that directly takes the already available similarity information of a time series into account. We then automatically determined the sampling strategy that yields the best result in combination with respectively trained networks for estimation. For training and validation, we used a variety of time-dependent 2D and 3D data from simulations and measurements as well as artificially generated data.

We could demonstrate that we achieve good results already with our proposed approach, with further improvements being subject to future work. In particular, we further aim to further explore the huge parameter space inherent to the setup and training of neural networks by systematically testing different types of neurons, different numbers of layers and numbers of neurons in each layer, different batch sizes and learning rates, etc. to further improve our results. We also plan to to evaluate the impact of the quality of estimated distance estimation on the final result of different types of visualization applications. Finally, we aim to compare our approach to a larger variety of alternative approaches for similarity estimation, develop more advanced techniques for artificial data generation, and conduct a more comprehensive evaluation regarding generalization properties.

Acknowledgments: The author would like to thank the German Research Foundation (DFG) for supporting the project within project A02 of SFB/Transregio 161 and the Cluster of Excellence in Simulation Technology (EXC 310/1) at the University of Stuttgart.

Conflicts of Interest: The author declares no conflict of interest.

References

1. Joshi, A.; Rheingans, P. Illustration-inspired techniques for visualizing time-varying data. In Proceedings of the VIS 05, IEEE Visualization, Minneapolis, MN, USA, 23–28 October 2005; pp. 679–686.
2. Lee, T.Y.; Shen, H.W. Visualization and Exploration of Temporal Trend Relationships in Multivariate Time-Varying Data. *IEEE Vis. Comput. Graph.* **2009**, *15*, 1359–1366.
3. Joshi, A.; Rheingans, P. Evaluation of illustration-inspired techniques for time-varying data visualization. *Comput. Graph. Forum* **2008**, *27*, 999–1006.
4. Woodring, J.; Wang, C.; Shen, H.W. High dimensional direct rendering of time-varying volumetric data. In Proceedings of the IEEE Visualization, Seattle, WA, USA, 19–24 October 2003; pp. 417–424.
5. Balabanian, J.P.; Viola, I.; Möller, T.; Gröller, E. Temporal Styles for Time-Varying Volume Data. In Proceedings of the 3DPVT'08—The Fourth International Symposium on 3D Data Processing, Visualization and Transmission, Atlanta, GA, USA, 18–20 June 2008; Gumhold, S., Kosecka, J., Staadt, O., Eds.; 2008; pp. 81–89.
6. Bach, B.; Dragicevic, P.; Archambault, D.; Hurter, C.; Carpendale, S. A Descriptive Framework for Temporal Data Visualizations Based on Generalized Space-Time Cubes. *Comput. Graph. Forum* **2016**, doi:10.1111/cgf.12804.
7. Fang, Z.; Möller, T.; Hamarneh, G.; Celler, A. Visualization and Exploration of Time-varying Medical Image Data Sets. In Proceedings of the Graphics Interface 2007, Montreal, QC, Canada, 28–30 May 2007; ACM: New York, NY, USA, 2007; pp. 281–288.
8. Wang, C.; Yu, H.; Ma, K.L. Importance-Driven Time-Varying Data Visualization. *IEEE Vis. Comput. Graph.* **2008**, *14*, 1547–1554.

9. Widanagamaachchi, W.; Christensen, C.; Bremer, P.T.; Pascucci, V. Interactive exploration of large-scale time-varying data using dynamic tracking graphs. In Proceedings of the IEEE Symposium on Large Data Analysis and Visualization (LDAV), Seattle, WA, USA, 14–15 October 2012; pp. 9–17.

10. Lee, T.Y.; Shen, H.W. Visualizing time-varying features with TAC-based distance fields. In Proceedings of the 2009 IEEE Pacific Visualization Symposium, Beijing, China, 20–23 April 2009; pp. 1–8.

11. Silver, D.; Wang, X. Tracking and Visualizing Turbulent 3D Features. *IEEE Vis. Comput. Graph.* **1997**, *3*, 129–141.

12. Ji, G.; Shen, H.W. Feature Tracking Using Earth Mover's Distance and Global Optimization. *Pac. Graph.* **2006**.

13. Weber, G.; Dillard, S.; Carr, H.; Pascucci, V.; Hamann, B. Topology-Controlled Volume Rendering. *IEEE Vis. Comput. Graph.* **2007**, *13*, 330–341.

14. Narayanan, V.; Thomas, D.M.; Natarajan, V. Distance between extremum graphs. In Proceedings of the IEEE Pacific Visualization Symposium, Hangzhou, China, 14–17 April 2015; pp. 263–270.

15. Schneider, D.; Wiebel, A.; Carr, H.; Hlawitschka, M.; Scheuermann, G. Interactive Comparison of Scalar Fields Based on Largest Contours with Applications to Flow Visualization. *IEEE Vis. Comput. Graph.* **2008**, *14*, 1475–1482.

16. Rubner, Y.; Tomasi, C.; Guibas, L. The Earth Mover's Distance as a Metric for Image Retrieval. *Int. J. Comput. Vis.* **2000**, *40*, 99–121.

17. Tong, X.; Lee, T.Y.; Shen, H.W. Salient time steps selection from large scale time-varying data sets with dynamic time warping. In Proceedings of the 2012 IEEE Symposium on Large Data Analysis and Visualization (LDAV), Seattle, WA, USA, 14–15 October 2012; pp. 49–56.

18. Seshadrinathan, K.; Bovik, A.C. Motion Tuned Spatio-temporal Quality Assessment of Natural Videos. *IEEE Trans. Image Process.* **2010**, *19*, 335–350.

19. Correa, C.D.; Ma, K.L. Dynamic Video Narratives. *ACM Trans. Graph.* **2010**, *29*, 88.

20. Lu, A.; Shen, H.W. Interactive Storyboard for Overall Time-Varying Data Visualization. In Proceedings of the 2008 IEEE Pacific Visualization Symposium, Kyoto, Japan, 5–7 March 2008; pp. 143–150.

21. Post, F.H.; Vrolijk, B.; Hauser, H.; Laramee, R.S.; Doleisch, H. The State of the Art in Flow Visualisation: Feature Extraction and Tracking. *Comput. Graph. Forum* **2003**, *22*, 775–792.

22. McLoughlin, T.; Laramee, R.S.; Peikert, R.; Post, F.H.; Chen, M. Over Two Decades of Integration-Based, Geometric Flow Visualization. *Comput. Graph. Forum* **2010**, *29*, 1807–1829.

23. Garces, E.; Agarwala, A.; Gutierrez, D.; Hertzmann, A. A Similarity Measure for Illustration Style. *ACM Trans. Graph.* **2014**, *33*, 93.

24. Frey, S.; Ertl, T. Progressive Direct Volume-to-Volume Transformation. *IEEE Trans. Vis. Comput. Graph.* **2017**, *23*, 921–930.

25. Frey, S.; Ertl, T. Fast Flow-based Distance Quantification and Interpolation for High-Resolution Density Distributions. In Proceedings of the EG 2017 (Short Papers), Lyon, France, 24–28 April 2017.

26. Frey, S.; Ertl, T. Flow-Based Temporal Selection for Interactive Volume Visualization. *Comput. Graph. Forum* **2016**, doi:10.1111/cgf.13070.

27. Marwan, N.; Carmenromano, M.; Thiel, M.; Kurths, J. Recurrence plots for the analysis of complex systems. *Phys. Rep.* **2007**, *438*, 237–329.

28. Vasconcelos, D.; Lopes, S.; Viana, R.; Kurths, J. Spatial recurrence plots. *Phys. Rev. E* **2006**, *73*, doi:10.1103/PhysRevE.73.056207.

29. Marwan, N.; Kurths, J.; Saparin, P. Generalised recurrence plot analysis for spatial data. *Phys. Lett. A* **2007**, *360*, 545–551.

30. Bautista-Thompson, E.; Brito-Guevara, R.; Garza-Dominguez, R. RecurrenceVs: A Software Tool for Analysis of Similarity in Recurrence Plots. In Proceedings of the Electronics, Robotics and Automotive Mechanics Conference 2008, Morelos, Mexico, 30 September–3 October 2008; pp. 183–188.

31. Frey, S.; Sadlo, F.; Ertl, T. Visualization of temporal similarity in field data. *IEEE Trans. Vis. Comput. Graph.* **2012**, *18*, 2023–2032.

32. Goodfellow, I.; Bengio, Y.; Courville, A. *Deep Learning*; The MIT Press: Cambridge, MA, USA, **2016**.

33. Hu, H.; Holman, P.M.; de Haan, G. Image interpolation using classification-based neural networks. In Proceedings of the IEEE International Symposium on Consumer Electronics, Reading, UK, 1–3 September 2004; pp. 133–137.

34. Plaziac, N. Image Interpolation Using Neural Networks. *IEEE Trans. Image Proccess.* **1999**, *8*, 1647–1651.

35. Chen, C.H.; Kuo, C.M.; Yao, T.K.; Hsieh, S.H. Anisotropic Probabilistic Neural Network for Image Interpolation. *J. Math. Imag. Vis.* **2014**, *48*, 488–498.
36. Guillaumin, M.; Verbeek, J.J.; Schmid, C. Is that you? Metric learning approaches for face identification. In Proceedings of the 2009 IEEE 12th International Conference on Computer Vision, Kyoto, Japan, 29 September–2 October 2009; pp. 498–505.
37. Kulis, B. Metric Learning: A Survey. *Found. Trends Mach. Learn.* **2013**, *5*, 287–364.
38. Davis, J.V.; Kulis, B.; Jain, P.; Sra, S.; Dhillon, I.S. Information-theoretic metric learning. In Proceedings of the 24th International Conference on Machine Learning, Corvalis, OR, USA, 20–24 June 2007; pp. 209–216.
39. Velten, A.; Wu, D.; Jarabo, A.; Masia, B.; Barsi, C.; Joshi, C.; Lawson, E.; Bawendi, M.; Gutierrez, D.; Raskar, R. Femto-photography: Capturing and Visualizing the Propagation of Light. *ACM Trans. Graph.* **2013**, *32*, 44.
40. Kingma, D.P.; Ba, J. Adam: A Method for Stochastic Optimization. *axiv* **2014**, arXiv:1412.6980.
41. Abadi, M.; Agarwal, A.; Barham, P.; Brevdo, E.; Chen, Z.; Citro, C.; Corrado, G.S.; Davis, A.; Dean, J.; Devin, M.; et al. TensorFlow: Large-Scale Machine Learning on Heterogeneous Systems. Available online: tensorflow.org (accessed on 22 August 2017).

informatics

MDPI

Article

Multidimensional Data Exploration by Explicitly Controlled Animation

Johannes F. Kruiger [1], Almoctar Hassoumi [2], Hans-Jörg Schulz [3], AlexandruC Telea [4] and
Christophe Hurter [5,*]

1   Institute Johann Bernoulli, University of Groningen, 9727 Groningen, The Netherlands; j.f.kruiger@rug.nl
2   ENAC/DEVI, University of Toulouse, 31055 Toulouse, France; almoctar.haiz@gmail.com
3   Institute for Computer Science, University of Rostock, 18051 Rostock, Germany;
     hans-joerg.schulz@uni-rostock.de
4   Institute Johann Bernoulli, University of Groningen, 9727 Groningen, The Netherlands; a.c.telea@rug.nl
5   ENAC/DEVI, University of Toulouse, 31055 Toulouse, France; christophe.hurter@enac.fr
*   Correspondence: christophe.hurter@enac.fr

Academic Editors: Achim Ebert and Gunther H. Weber
Received: 30 June 2017; Accepted: 14 August 2017; Published: 20 August 2017

Abstract: Understanding large multidimensional datasets is one of the most challenging problems
in visual data exploration. One key challenge that increases the size of the exploration space is the
number of views that one can generate from a single dataset, based on the use of multiple parameter
values and exploration paths. Often, no such single view contains all needed insights. The question
thus arises of how we can efficiently combine insights from multiple views of a dataset. We propose
a set of techniques that considerably reduce the exploration effort for such situations, based on the
explicit depiction of the view space, using a small multiple metaphor. We leverage this view space
by offering interactive techniques that enable users to explicitly create, visualize, and follow their
exploration path. This way, partial insights obtained from each view can be efficiently and effectively
combined. We demonstrate our approach by applications using real-world datasets from air traffic
control, software maintenance, and machine learning.

Keywords: information visualization; small multiple; big data; animation

1. Introduction

Information visualization (infovis) aims to leverage user ability to retrieve insight from data
representation thanks to the usage of computer-supported, interactive, and visual representations [1].
One of the main challenges of infovis is providing insight into high-dimensional datasets. Such datasets
consist of many elements (observations), each having multiple dimensions.

Many visualization techniques for high-dimensional data can be explained as *element-based* plots.
In such a plot, every element of the dataset is depicted separately (and in the same way as the
other elements). Examples of such plots are the classical 2D or 3D scatterplots (every element is a
point), parallel coordinate plots (every element is a polyline), multidimensional projections, and graph
drawings (every element is a graph edge).

When the number of input dimensions is high, no single such plot can be created that shows the
entire information present in the input data. This problem is typically solved by generating multiple
plots, for various parameter combinations, each of them showing a *partial* insight. For example, a 2D
scatterplot shows the correlation of only two of all the input dimensions, and a graph drawing can
generally show an uncluttered view of the input data for only a limited portion of a large graph.

One way to address this problem is to resort to *interactive* exploration, by allowing the user
to create multiple views, each which shows a different part, or aspect, of the data. The problem

then becomes how to *navigate* the (very large) space of all potential views and combine the partial insights provided by them in an efficient and effective way. To address this data exploration issue, we propose a set of innovative techniques where controlled animation plays a central role. Our techniques considerably reduce the exploration effort by allowing the user to directly sketch the exploration path over a visual depiction of the view space created by a small multiple metaphor. This way, a potentially infinite set of intermediate views can be created easily and intuitively. Real-time linking of the view-space navigation and the display of the intermediate views allows one to go forward, backtrack, or change the exploration path. Finally, data patterns found in different views can be selected and interactively combined to generate new views on the fly.

In summary, our contributions are

- a novel representation of the view space based on a small multiple metaphor
- a set of interaction techniques to continuously navigate the view space and combine partial insights obtained from different views

We demonstrate our proposed techniques with the visual exploration of real-world multidimensional datasets represented by multidimensional projections and bundled graph drawings taken from the domains of air traffic control, software maintenance, and machine learning.

The remainder of this paper is structured as follows. Section 2 overviews related techniques for visualizing multidimensional data using element-based plots, as well as techniques for interactive view-space exploration. Section 2.2 introduces the design and implementation of our proposed visualization and interaction. Section 4 shows applications of our techniques to different types of element-based plots. Section 5 discusses our proposal and results. Finally, Section 6 concludes the paper.

2. Related Work

Next, we overview related work to our proposal, structured along three directions: the types of methods used for visualizing element-based plots Section 2.1) and the usage of animation for multidimensional data exploration (Section 2.2).

2.1. Element-Based Plots

Let $D = \{d_i\}$, $1 \leq i \leq N$, be a dataset containing n-dimensional elements $d_i \subset \mathbb{R}^n$. Further, let $P = \{p_i\} \in \mathcal{P}$ be the set of parameters that controls a visualization and $V(P, D) \in \mathcal{V}$ the resulting view for the given dataset D. In the following, we use the term *visualization* to denote the function that creates a depiction (image) of a dataset, and the term *view* to denote such a depiction. That is, a visualization is a function that inputs datasets, is controlled by parameters, and outputs views. Moreover, for notation simplicity, we will shorten $V(P, D)$ to $V(P)$ when the input dataset D is constant and only the visualization parameters P vary, and respectively to $V(D)$ when the visualization parameters P are constant but only the input datasets D change.

We define an element-based plot as a 2D drawing $V(D) = \{V(d_i)\} \subset \mathbb{R}^2$ consisting of N visual shapes $V(d_i)$, each mapping a dataset element d_i. Introduced as a concept by Hurter et al. [2], element-based plots generalize many classical infovis techniques: for scatterplots, d_i are nD points and $V(d_i)$ are 2D points; for table lenses, d_i are nD points and $V(d_i)$ are bar-chart-like displays of n numerical values [3,4]; for parallel coordinate plots, d_i are nD points and $V(d_i)$ are 2D polylines; for 2D and 3D scalar fields like color images or 3D data cubes (volumes), d_i are points in $\mathbb{R}^2$, respectively $\mathbb{R}^3$, and $V(d_i)$ are color-mapped pixels, respectively voxels [2,5,6]; and for graphs or trail-sets, d_i are weighted relations or trajectories in some Euclidean space, and $V(d_i)$ are 2D curves [7]. Key to grouping all these visualization techniques under the denomination of element-based plots is the fact that every data item d_i is mapped one-to-one to an *independent* visual element $V(d_i)$. As we shall see in Section 2.2, this allows us to manipulate the visualization $V(D)$ in very flexible ways.

Below we discuss two types of element-based plots which are particularly important in our context, as they are those on which we validate our animation-based exploratory technique in the remainder of this paper: multidimensional projections and graph drawings.

Multidimensional projections (MPs) are a particularly important type of element-based plots. Here, d_i are nD points (as for table lenses, parallel coordinate plots (PCPs) [8,9], or scatterplots [10]), and $V(d_i)$ are 2D points (as for scatterplots). Hence, MPs are as visually scalable and clutter-free as scatterplots, and more visually scalable and clutter-free than table lenses and PCPs. Moreover, MPs improve upon scatteplots since scatterplots are constructed by using only two of the n dimensions of D, whereas MPs are constructed by considering all n dimensions. As such, scatterplots visually encode the similarity of points $d_i \in D$ according to only two of their n dimensions, whereas MPs encode the similarity according to all n dimensions.

Many MP techniques exist, having various trade-offs between the simplicity, speed, and accuracy of encoding n-D point similarities. They can be grouped in two main classes. *Similarity-based* projections require only a $N \times N$ real-valued similarity (distance) matrix encoding the pairwise similarities of points $d_i \in D$ to construct $V(D)$. In this class, Multidimensional Scaling (MDS) and its variants compute $V(D)$ using an optimization process similar to force-based schemes akin to those used in graph layouts [11]. A different approach is proposed by t-SNE [12], which defines the probabilities of picking point pairs in D and minimizes the Kullback-Leibler divergence between those probabilities and those that have the same point pairs as neighbors in the 2D projection. This results in projections that successfully show which n-D points are neighbors to each other, which then helps in visually finding clusters of similar points. In contrast, *Attribute-based* projections require access to the nD dimensions, or attributes, of the observations $d_i \in D$ to compute the projection $V(D)$. The simplest, and arguably most known, projection in this class is the simple 2D scatterplot created by selecting two dimensions of D. Another well known member of this class is principal component analysis (PCA), which projects p_i on the plane defined by the axes that describe most of the variance in D [13]. However, PCA is notoriously inaccurate in encoding similarity when D resides on highly *curved* manifolds in n-D. ISOMAP [14] first determines which n-D points are neighbors to build a neighborhood graph, then computes the distances over the neighborhood graph, and finally performs MDS with those distances. Finally, LAMP [15] uses a small number of so-called landmark points that are projected to 2D with a classical MP technique (such as MDS), while the remaining points are arranged around the landmarks using locally affine transformations.

We also note that other classification of MP techniques exist. For instance, these can be grouped into global (scatterplots, PCA, star coordinates [16,17], orthographic star coordinates [18], biplots [19,20], radial visualizations [21,22]), and local (LAMP, LLE [23], t-SNE [12]). Global techniques use the same transformation to project all points to the target (2D) space. They are simpler and faster to compute but may generate large projection errors. Local techniques may use different transformations for different point neighborhoods. They are in general more complex and slower than local techniques, but they preserve the so-called 'data structure' better after the projection than local techniques.

Among all the aforementioned multidimensional visualization techniques (scatterplots, table lenses, PCP's, and projections), the latter are the most visually scalable and clutter free, as they map an n-D point to a 2D point. However, projections are fundamentally weak in accurately showing similarities between n-D points for the *entire* input dataset D [24]. As such, given such a dataset D, there is typically no *single* projection $V(D)$ that can faithfully show all patterns D. We shall show in Section 4 how we effectively address this problem by interactively combining insights obtained from different projections of a given dataset D.

Graph drawings (GDs) are a second important type of element-based plots. In detail, a graph is a dataset $D = (N, E)$ with nodes $n \in N$ and edges $e \in E$. Both nodes and edges can have data attributes, thereby making D a multidimensional dataset. A graph drawing is a visualization $V(D) = (V(N), V(E))$, where $V(D)$ is typically a 2D scatterplot of points $V(n)|n \in N$. $V(D)$ is typically created by embedding methods such as force-directed layouts [25] but also, as shown recently, multidimensional projections [26]. $V(E)$ can be a set of straight line segments $V(e)|e \in E$. However, $V(e)$ can also be 2D curves, as follows. Drawing large graphs (thousands of edges or more) with straight lines easily creates massive clutter which renders such drawings close to useless. One prominent method to simplify, or tidy up, such drawings is to *bundle* their edges, thereby trading clutter for overdraw—that is, many bundled edges will overlap in $V(E)$. This creates empty space between the edge bundles which allows one to easily follow them visually, thereby assessing the coarse-scale connections between groups of related nodes in $V(N)$ more easily than in an unbundled drawing. The drawback of bundling is that individual edges in $V(E)$ are harder to distinguish due to the overlap. Numerous edge bundling (EB) methods exist [27–32]. However, as a recent survey points out [7], no such method is optimal from all perspectives. For instance, some methods offer a very precise control of the shape and positions of the bundled edges, but only handle particular types of graphs [27,32]. Other methods can handle graphs of any kind and are very scalable but offer far less control on the resulting look-and-feel of the bundled drawing [28,29]. Yet other methods fall in-between the above two extremes but generate a very wide set of drawing styles [33].

2.2. Navigating Multidimensional Data Visualizations

Exploring the space of visualizations that one can create for a given multidimensional dataset is a wide topic. Several technique classes exist to this end, as follows.

Animation is a prominent technique that supports exploring the parameter space $\mathcal{P}$ and the related view space $\mathcal{V}$. Animation has a long history in data exploration [34,35]. In this context, animation corresponds to a (smooth) change of the visual variables used to encode the input data $V(D)$ [1,36].

Animation has proven to help users transition between visual configurations [37] while maintaining the mental map of the data exploration [38]. The simplest form of animation, used by virtually all visualization tools, lets an user vary the value of p_i interactively by a classical user interface (GUI) while watching how the $V(P)$ changes. While simple, this type of animation does not 'guide' the user in the exploration of the parameter space $\mathcal{P}$ and the related view space $\mathcal{V}$ in any way. Within the scope of multidimensional visualization, parameter-space animations include rolling-the-dice [39], where the user controls the plane on which the multidimensional data is projected; the grand tour [40], where a large sequence of 2D projections are displayed from a multidimensional dataset in a flip-book manner (the parameters P controlling projection-plane orientation in nD being varied randomly); the class tour [41], which refines the grand tour so as to generate projections which preserve class separation of the data points; combinations of the grand tour with projection pursuit [42] (the parameters P controlling the projection-plane orientation being varied along the derivatives of the so-called projection pursuit index, so as to drive the tour through interesting projections); and drawing faded trails that connect two consecutive views in a tour to give a feeling of how the projection plane changed in between [43]. A limitation of most grand tour techniques is that they only handle attribute-based projections [44].

Small multiples: Explicitly showing a sampling of $\mathcal{V}$ partially addresses this issue. One can show a history of views $V(P_i)$ obtained for various parameter settings P_i used in the exploration so far. The history can be shown as a linear, grid-like, or hierarchical set of thumbnails depicting $V(P_i)$, a metaphor also called 'projection board' [44]. By clicking on the desired thumbnail, the user can go back to the corresponding state P_i and associated view $V(P_i)$, and continue exploration from there. Approaches that utilize such a methodology are Ma's Image Graphs [45] and Elmqvist et al.'s Data Meadow [37]. Most projection pursuit variants also fall into this category [42,46], every view V_i being one deemed 'interesting' from the perspective of the projection pursuit index, a metric which typically measures the distance from a given projection to an uninteresting generic Gaussian-like scatterplot. However, a limitation of most projection pursuit variants is that, just like grand tour variants, they require attribute-based projections [44]. Projection pursuit methods have been recently enhanced by limiting the number of interesting views to $n/2$ [47].

Many scagnostics techniques also use the small-multiple metaphor to show interesting projections, where interest is defined in a wide variety of ways [48–51]. Besides using the exploration history, the views $V(P_i)$ can be picked by optimizing for diversity of this set, by subsampling an initial large random sampling of $\mathcal{V}$, or by iteratively refining a given set of views to improve their diversity [52]. Other methods add aesthetic constraints, such as a symmetry score, to exclude unsuitable views right from the view set [53]. The thumbnails $V(P_i)$ can also be placed in 2D by projecting the set of presets P_i by using Multi-Dimensional Scaling (MDS) methods [52,53], similar to those discussed in Section 2.1, or simply arranging them in increasing distance to the currently shown one [54]. Other arrangements use a regular grid layout that allows performing additional view transformations by a spreadsheet-like metaphor [55]. All above metaphors assume a small set of preset views $V(P_i)$, so that they can all be displayed simultaneously as thumbnails without taking too much screen space. If users are not satisfied with these proposed views, they can usually only adjust the parameters P_i that generated them in the hope to get better ones. Another way to refine the set $V(P_i)$ is to mark the views that are closest to what the user has in mind and supersample $\mathcal{V}$ around such 'interesting' views to get better ones [56].

Preset controller: In contrast to small multiples, Van Wijk et al. [57] show several values of specific parameter-sets P_i by a 2D scatterplot $S = \{x(P_i)\}$, where $x(P_i) \in \mathbb{R}^2$ is the projection of the point P_i. Next, one can manipulate a point of interest $x \in \mathbb{R}^2$, or the scatterplot points $x(P_i)$, and generate a corresponding parameter-set value $p \in \mathcal{P}$ by using Shepard interpolation of the values P_i based on the distances $\|x - x(P_i)\|$. While the preset controller is very simple to use and is scalable in the number of parameters $|P|$, it does not explicitly depict the view space $\mathcal{V}$ but, rather, only an abstract view of the parameter space $\mathcal{P}$.

Direct manipulation: Animation can be also controlled directly in the view space $\mathcal{V}$ rather than in the parameter space $\mathcal{P}$. For this, the user directly manipulates the depicted visual elements in $V(P)$ to modify the parameters P. Examples of such manipulations are deformation, focus-and-context, and semantic lens techniques, all of which typically linearly interpolate between two parameter-set values P_1 and P_2 and show the corresponding animation of $V(P_1)$ to $V(P_2)$. Such animations have been applied to large element-based plots such as bundled graphs and scatterplots [6,58]. Smooth real-time animations of large datasets have been made possible by using GPU-based techniques [59]. For multidimensional projections, we have the following direct manipulation techniques. Control-point-based projections, such as LSP [60], PLMP [61], generalized Sammon mapping [62], hybrid MDS [63], and LAMP [15], allow users to interactively (dis)place a small subset of $V(D)$, called control points, on the 2D view plane, after which they arrange the remaining points around these controls so as to best preserve the nD data structure. This effectively allows users to customize their projections, at the risk of creating visual structures that do not relate well to the data. Targeted projection pursuit (TPP) [64] allows users to drag elements in a multidimensional projection plot $V(D)$ to, for example, better separate classes. From the resulting scatterplot $V^{user}(D)$, it seeks the parameters P for an actual projection $V(P, D)$ that is close to $V^{user}(D)$. While powerful as an interaction mechanism, TPP limits itself to only linear projections. Recently, ProjInspector use a preset controller, where the k presets correspond to user-chosen DR projections $V_i(D), 1 \leq i \leq k$. When one drags the point of interest in the controller, the tool generates a view $V(D)$ that blends all $V_k(D)$ by means of mean values coordinates interpolation [44]. ProjInspector is arguably the closest technique to the one we present here, as such, we will discuss the similarities and differences in detail in Section 5.

Interaction techniques: All the above-mentioned visualization techniques use a mix of interaction techniques to enable exploration. While these are not specific to multidimensional data exploration, it is worth mentioning them here, as they next allow us better placing our contribution. First, as already explained, the parameters P can be changed by means of classical GUIs: the elements of the view $V(D)$, the control points of a preset controller [44,57], and the control points of a projection

[15,60–63] are changed by simple mouse-based click-and-drag. Techniques can be next classified as single-view or multiple-view. Single-view techniques are either interaction-less, e.g., some grand tour and projection pursuit variants, or require direct manipulation in the single view, as explained earlier. Multiple-view techniques display several views $V_i(D)$ of the data. If direct manipulation is implemented in one view $V_i(D)$, then the elements undergoing change should be updated in all other views $V_j(D)$ – a well-known technique under the name of *linked views* [65] or *coordinated views* [66]. Depending on the exact semantics of the views, either unidirectional or bidirectional linking can be used [67]. For instance, a preset controller is unidirectionally linked to the data view(s) it controls. Finally, brushing and selection are ubiquitous techniques for exploring the view space, by showing details of the data element under the mouse and click-and-drag (typically) to select a subset of $V(D)$ for special treatment, respectively [68]. We will use all these techniques in the design of our exploratory visualization in Section 2.2.

Other approaches: Visualization presets have also been investigated for graph datasets [69]. Separately, exploratory visualization approaches have used the view space in a foresighted manner to sketch possible next steps along the visual exploration path. Several such approaches exist, which can be subsumed by the term 'visualization by example' [70]. All such approaches allow one to select a desired view $V(P_i)$ from a range of candidates, the main distinction being how these candidates are picked from the view space $calV$ and how V is presented to the user.

Our proposal: We combine several of the advantages, and reduce some of the limitations, of the above-mentioned techniques for navigating a view space V constructed from a high-dimensional dataset D by means of a parameter space P, as follows:

- *Genericity:* We handle all types of element-based plots (Section 2.1), e.g., scatterplots, graph/trail drawings, and DR projections, in an uniform way and by a single implementation.
- *View by example:* We provide an explicit small-multiple-like depiction $V(P_i)$ of the view space V.
- *Continuity:* We allow a continuous change of the current view based on smooth interpolation between the small-multiple views $V(P_i)$ without having to bother about understanding the explicit abstract parameter space P. This allows generating an infinite set of intermediary views in V.
- *Free navigation:* The view generation is in the same time controlled by the user (one sees along which existing views one navigates) and unconstrained (one can freely and fully control the shape of the navigation path).
- *Ease of use and scalability:* We generate our intermediary views by simple click-and-drag of a point in the view space; these views are generated in real-time for large datasets D (millions of elements).
- *Control:* Most importantly, and novel with respect to all approaches discussed so far, we propose a simple mechanism for changing only *parts* of the current view, while keeping other parts fixed. This enables us to combine insights from different views $V(P_i)$ on-the-fly, to accumulate insights on the input dataset D.

3. Proposed Method

We next present our animation-based exploration of visualization spaces for element-based plots, following the discussion in Section 2. To better outline the added-value of our proposal, we first classify the types of such animation-based explorations along two axes that explain how the transition between views in V can be created: explicit vs. implicit, and guided vs. free, as follows (see also Figure 1).

The first axis (vertical in Figure 1) describes the type and number of *preset views* $V(P_i)$ between which the user can navigate, and has two values (options):

- **Guided:** The set of preset views between which the user can choose $V(P_i)$ is limited by construction, and depends on the dimensionality of the input dataset D.

- **Free:** The set of preset views is fully configurable by the user, who can choose any number and type of views in $\mathcal{V}$ to animate between.

The second axis (horizontal in Figure 1 characterizes how the user can *control* the animation during the transition between two views $V_1 \in \mathcal{V}$ and $V_2 \in \mathcal{V}$, and has tho values:

- **Implicit:** Once the transition (animation) between V_1 and V_2 is triggered by the user, the generation of intermediate views between V_1 and V_2 happens automatically (usually via some type of linear interpolation). The user can specify V_1 and V_2, but not the *path* in the view-space $\mathcal{V}$ along which the animation evolves nor can he slow/accelerate/pause the animation.
- **Explicit:** The user can choose the path along which the animation evolves, and also the speed thereof.

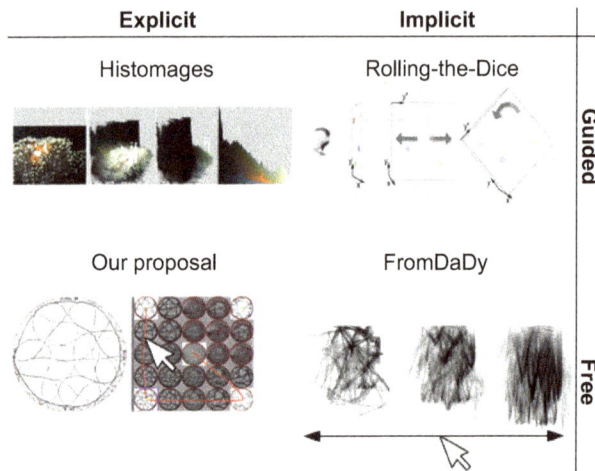

Figure 1. Animation design space between two data views V_1 and V_2. The user can control or not the transition (controlled vs. automatic), and the transition is defined by the user or predefined by the tool (explicit vs. implicit). The presented tools are Histomages [71], Rolling-the-Dice [39], FromDaDy [58], and our proposal.

To clarify the above, some examples of existing animations follow.

Implicit and guided transition: Rolling-the-dice [39] is a metaphor of a 3D rolling die which allows smooth transitions between 2D scatterplots obtained from an nD dataset. The preset views V_i are implicitly defined by the pair-wise combinations of data dimensions in the input dataset D ($n^2/2$ in total). The transitions are automatic linear interpolations between two such scatterplots.

Implicit and free transition: In FromDady [58], the user can interactively define views to animate between, based on desired combinations of the dataset's attribute-pairs. In contrast to ScatterDice, the user can control the transition speed and direction (V_1 to V_2 or conversely) with a mouse drag gesture. However, the set of preset views V_i is given by the pair-wise combinations of dimensions in the input data. Moreover,the transition is not entirely free, although one can control its direction and speed, its path is still a linear interpolation between V_1 and V_2.

Explicit and guided transition: The user can define an open set of preset views V_i that can be interpolated. For example, in the preset controller, these views are arbitrary parameter-sets P_i that generate corresponding views $V(P_i)$ [57]. However, the transitions between views are still automatically determined, typically by linear interpolation. Further examples hereof are [2,6,71].

Explicit and free transition: To our knowledge, no existing technique allows *freely* defining both the endpoints $V_i \subset \mathcal{V}$ and the controllable interpolation *path* for the animation. Our proposal, described next, fills this gap.

3.1. Our Proposal

Our key idea is to combine the main strengths of the explicit and free animation methods existing in prior work, in an efficient and effective way. To this end, we choose to

- Depict the view presets $V(P_i)$ by a simple grid-like small-multiple metaphor (as in [55]). This way, users see directly which visualizations $V(P_i)$ they can interpolate between, rather than seeing the more abstract parameters P_i that would generate these visualizations (as in [57]).
- Allow one to freely sketch the interpolation path between two such view presets $V(P_i)$, in an interactive and visual way (as in [57]), rather than automatically controlling the interpolation via a linear formula.

We proceed as follows. Figure 2 illustrates the process for a multidimensional dataset D that we visualize using multidimensional projections, which are introduced in Section 2.1. We proceed by allowing users to specify any (small) set of views V_i for the given dataset D. These can be created either by the same visualization method, but using different parameter values P_i, or by completely different methods. The only restriction is that these should be *element*-based plots, i.e., consist each of N visual elements $V_i(d_j)$, $1 \le j \le N$, one for each data element $d_j \in D$. We organize these view presets in a simple square grid, much like [55], to minimize the used screen space to display them (see Figure 2b).

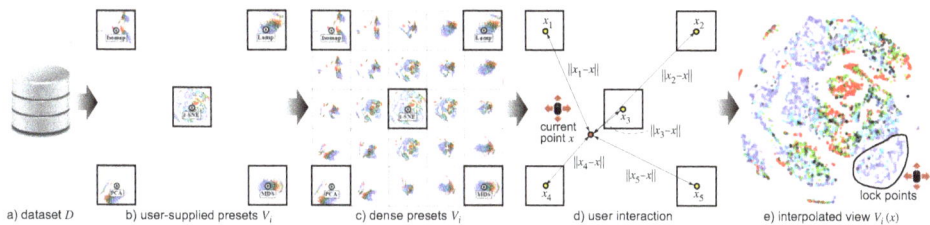

a) dataset D b) user-supplied presets V_i c) dense presets V_i d) user interaction   e) interpolated view $V_i(x)$

Figure 2. Proposed animation-based exploration pipeline. See Section 3.1.

As explained in Section 2.2, manually generating a rich set of view presets V_i can be delicate. Typically, one starts with a small set of just a few salient view presets, for instance, one for each type of visualization method considered for a given dataset. In our running example in Figure 2, we have five such presets, corresponding to five multidimensional projection methods (PCA [13], Isomap [14], LAMP [15], MDS [11], and t-SNE [12]). In general, these views can be very different. In other words, they are a very sparse sampling of the rich view space $\mathcal{V}$. We do not know what lies in between, so, navigating between them can be unintuitive.

To alleviate this, and also to realize our animation-based exploration mechanism presented next, we propose a view-space interpolation mechanism, as follows. Let $x \in \mathbb{R}^2$ be a point in the 2D grid space where we place the view thumbnails. Placement of thumbnails can be freely specified by the user, based on perceived similarities between the views and following the design of the original preset controller [57]. Let now x_i be the centers of the thumbnails V_i in this grid. We use an Inverse Distance Weighting (IDW) method, such as the Shepard method [72], to compute the elements $V(x, d_j)$ of the interpolated view at position x as:

$$V(x, d_j) = \begin{cases} \dfrac{\sum_{i=1}^{N} w_i(\mathbf{x}) V_i(d_j)}{\sum_{i=1}^{N} w_i(\mathbf{x})}, & \text{if } \|\mathbf{x} - \mathbf{x}_i\| \neq 0 \text{ for all } i \\ V_i(d_j), & \text{if } \|\mathbf{x} - \mathbf{x}_i\| = 0 \text{ for some } i \end{cases} \qquad (1)$$

Here, $\| \cdot \|$ denotes 2D Euclidean distance, and $w_i(\mathbf{x}) = \frac{1}{\|\mathbf{x}-\mathbf{x}_i\|^p}$ is the interpolating basis function controlled by the power parameter $p > 1$. Typically we use $p = 2$, which leads to classical inverse quadratic Shepard interpolation. Key to our idea is the fact that all element-based plots consist of sets of simple graphical elements $V_i(d_j)$ such as dots or line segments. These can be thus easily interpolated using Equation (1). If desired, different types of (smooth) scattered-point interpolation can be used, such as radial basis functions or mean values coordinates [44].

Having now a way to interpolate between views, we supersample the view space $\mathcal{V}$ to generate additional views between the presets provided by the user and display these additional interpolated views along with the provided ones. Figure 2c shows this: here, from the original five presets, we create an additional number of 20 presets, yielding to a total grid of $5 \times 5 = 25$ views. The interpolated views now give a better idea of what the animation-based exploration will produce when we will navigate the view space between the originally provided views.

Having now the densely-sampled view space, we provide an interactive way for users to generate arbitrary views. For this, the user drags a so-called *focus* point x over the thumbnail grid. Note that x can assume any pixel position over the extent of the grid, i.e., is not constrained to the centers of the grid cells only. We then generate the view $V(x)$ corresponding to this point, using the same interpolation (Equation (1)) as for the view-space supersampling, based on the distance of x to the centers of the originally-provided views (Figure 2d), and display this view at full resolution in a separate window (Figure 2e).

As noted, the original views can be freely arranged over the thumbnails grid. This, and changing the power p in Equation (1), offers a flexible way of specifying how the original views are mixed to yield the view $V(x)$ at some position x. To better understand this, Figure 3 shows the effects of these two changes. Here, we have five views $V_1 \ldots V_5$, arranged on a five by five thumbnails grid. Each entire view (and thus thumbnail) is simply a colored pixel in RGB space, for illustration purposes: V_1 = red, V_2 = green, V_3 = yellow, V_4 = purple, V_5 = blue, and V_6 = purple. As such, we can render the interpolated views $V(x)$ at all the positions x over the thumbnails grid. In other words, the color images in Figure 3 show the entire view space $\mathcal{V}$ that can be generated from the five presets, something we cannot do, of course, for actual views of complex datasets. Figure 3a shows how the interpolated views (colors) smoothly vary between the five presets. Figure 3b shows how the view space changes if we drag the five preset views to be in the same horizontal grid row, as indicated by the arrows in Figure 3a. Finally, Figure 3c,d shows the effect of changing the value p from the default $p = 2$ used in the first two images (a,b) to $p = 1$ and $p = 0.5$, respectively. This allows us to control the shape of the zone of influence of each view over the thumbnails grid.

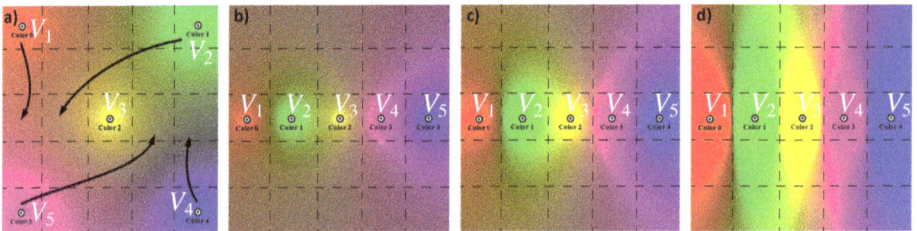

Figure 3. Thumbnail grid and view interpolation. Each image shows a set of five preset views $V_1 \ldots V_5$ arranged on a five by five grid. The views are simple color pixels for illustration purposes. Each grid pixel is colored to reflect what the interpolated view would be at that position. (**a**) Initial placement of the preset views; (**b**) Effect of dragging the views along the arrows shown in (**a**); (**c,d**) Effect of setting $p = 1$ and $p = 0.5$, respectively. The grids (**a,b**) use the default $p = 2$.

Several types of exploration of the view space are now possible using these mechanisms (see also Figure 4). The simplest is to click on one of the cells of the thumbnails grid. This sets the current view to the preset view V_i corresponding to that cell (Figure 4a). Separately, one can drag the mouse to follow a path in the view space. Making this path pass through a number of preset views V_i essentially constructs a 'visual story' that leads the user through the insights provided by these views, in visiting order (Figure 4b). In this sense, our technique is a specific instance of the design space model proposed in [73]. We offer the possibility of saving such a path and explicitly drawing it atop the thumbnails grid. This allows one to revisit an earlier-inspected view in the view space, thus, to backtrack the exploration, following well-known principles in information visualization [74] (Figure 4c).

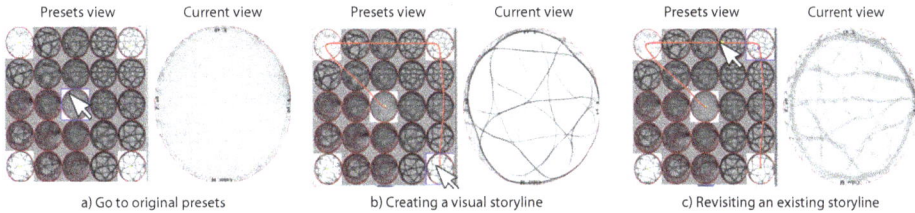

a) Go to original presets   b) Creating a visual storyline   c) Revisiting an existing storyline

Figure 4. (a–c) Three different types of navigation of the view space.

Besides arranging the preset views in the thumbnails grid and interactively dragging the focus point over the grid to generate a current view, we provide a third (and last) interaction mechanism called the *lock tool*. This allows the user to select points in the current view by lasso selection (see Figure 2e). The underlying data points $d_i \in D$ corresponding to these selected points are then excluded from the interpolation in Equation (1) when the focus point is dragged. This offers the possibility to the user of locking interesting visual patterns that have been discovered, during the animation, in one or several current views. This way, the user can effectively interactively create a custom view that blends insights obtained from different parts of the exploration process.

3.2. Implementation Details

We next detail the implementation of our proposal. Since we heavily rely on animation, we need to be able to generate the current view $V(x)$, by using the interpolation in Equation (1), in real-time for large element-based plots containing hundreds of thousands of points or more. We developed our exploration tool in C# using .NET 4.5 and using OpenGL for rendering. To support scalable animation, we investigated different solutions. We first found that the fixed OpenGL pipeline (used earlier for similar animation-based explorations [2]) and the render-to-texture OpenGL extension (also used earlier for similar goals [58]) do not provide enough scalability. We next investigated the use of the OpenGL transform feedback (also used earlier for similar goals [6]). This technique addresses scalability issues and is faster than the render-to-texture solution. However, the implementation becomes quite complex, as one has to code all the interaction and interpolation in a fragment shader whose code becomes hard to manage. We therefore explored another solution based on coding Equation (1) in NVIDIA's CUDA parallel programming language. This solution proved to be even faster than the transform feedback one, while allowing for a simple implementation too. As a result, our tool can generate fluent animations of datasets having up to one million visual elements (points) on a modern GPU.

4. Applications

We now demonstrate the working of our proposed animation-based exploration techniques by applying them to two types of element-based plots—multidimensional projections (Section 4.1) and bundled graph drawings (Section 4.2). Additional video material illustrating our animation-based exploration is available online [75].

4.1. Multidimensional Projections

Our datasets D are here tables where each row corresponds to an observation and each column to a different (numerical) property measured for the respective observations. Hence, D can be seen as point sets in a high-dimensional space. The preset views V_i we start with are various projections of such nD datasets to two dimensions. We consider three such datasets, and associated projections, as follows.

4.1.1. Software Dataset

In this dataset, observations correspond to $N = 6733$ software repositories from SourceForge.net. The $n = 12$ dimensions are different quality metrics of the software repositories, such as the number of lines per method, coupling between objects, and the cyclomatic complexity of methods. For each repository (observation), metric values are averaged over all its units of computation (classes, methods, files). Details on these metrics and the data are available [76]. Additionally, every observation is labeled by its type (e.g., small library, large library, monolithic application).

One interesting question for this dataset is whether projections can help us separate repositories of one kind (say, small class libraries) from those of other kinds (say, large monolithic systems). Note that none of the metric values present in the data contains *absolute* size information.

Figure 5 shows a five by five thumbnail grid based on the Isomap, LAMP, t-SNE, PCA, and MDS projections. Here and next, we chose these projections as they are well-known in the literature and often used in practice when analyzing multidimensional data. The projected points are colored based on their class (repository kind) attribute, which is not used by the projection. In the thumbnails, but even more so when interactively navigating the view space by clicking anywhere in the thumbnail grid, we see that no single view can separate well points of one color (class) from the others. The view that yields the best separation is t-SNE, also shown in detail in the right image in Figure 5. However, even in this view, we see no clear segregation of points of one class from the others. We conclude that the 11 recorded metrics are not a good predictor of the type of software system in a repository—in other words, systems of different kinds can have similar qualities, and conversely. Interestingly, the same result was noted by the original paper that analyzed this dataset [76], which did not use multidimensional visualization. Separately, a different work also showed that using 2D projections cannot achieve the desired separation but that such a separation is better possible using a 3D projection [77].

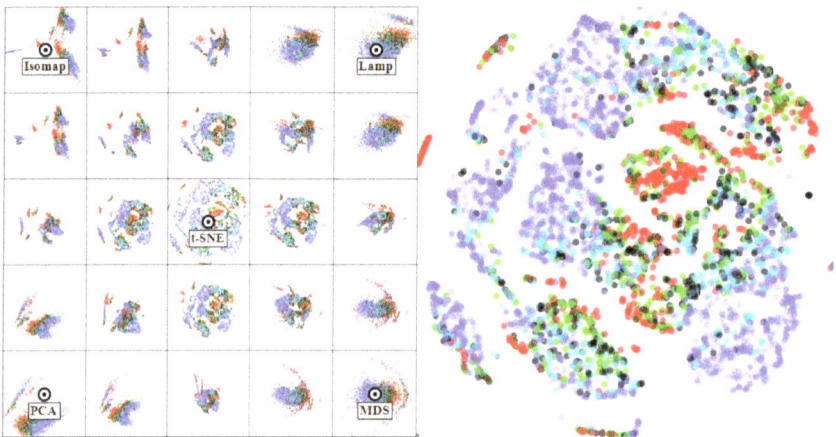

Figure 5. Software quality dataset (Section 4.1.1). No projection in the view space is able to separate well repositories of one kind from those of other kinds.

4.1.2. Segmentation Dataset

In this use-case, the dataset D is a set of seven images. Every dataset element $d_i \in D$ corresponds to a three by three block of pixels from one of the images, that is manually labeled into six classes (e.g., brickface, sky, grass, etc). Each of the $N = 2300$ elements has 17 dimensions which describe image features computed on the block of pixels. This dataset is well known in the machine learning literature , in the context of designing classifiers able to predict the six classes from the 17 measured dimensions [78]. To do this, one way is to engineer discriminating features using the raw 17 ones present in the data. Projections can help us in determining how good the engineered features are: if we find a projection where same-class points are well separated into clusters then the features that the projection has used as input are a good start for building a good classifier [79].

Figure 6 shows our method applied on this dataset, with the right image showing the t-SNE projection. As visible, several of the intermediate views (between the five presets) are able to separate well one or more classes from the others but no single view can do this for all classes. More interesting, we see that different views can separate well different classes. This is a key insight that we interactively refine further (see Figure 7). We start from the central t-SNE view (a). Next, we see in the tumbnail grid that the LAMP view V_1 separates cluster 1 (red) very well from the rest (f). We navigate to V_1, where we can easily select the red points and lock them for transitions. (b) From here, we notice another view (V_2, image (g)) where clusters 2 (gray-blue) and 3 (purple) are well separated. We navigate from V_1 to V_2. Since cluster 1 is locked, it will stay well separated during this process. We now can easily select clusters 2 and 3 and lock them. Next, we repeat the process by finding V_3 where we can separate cluster 4, and finally V_4, which separates the two remaining clusters 5 and 6. The entire process takes around two minutes and requires only basic click, drag, and brush mouse interaction. The iterative lock-and-view-change process is equivalent with an iterative classifier design where one finds specific features (given by the projections corresponding to the views V_i), which are good to separate one or a few classes from the rest. While we have not created and tested such a 'cascading' classifier, doing so should be relatively straightforward now that we know which configurations are good in separating one class from the rest, and the order of these configurations.

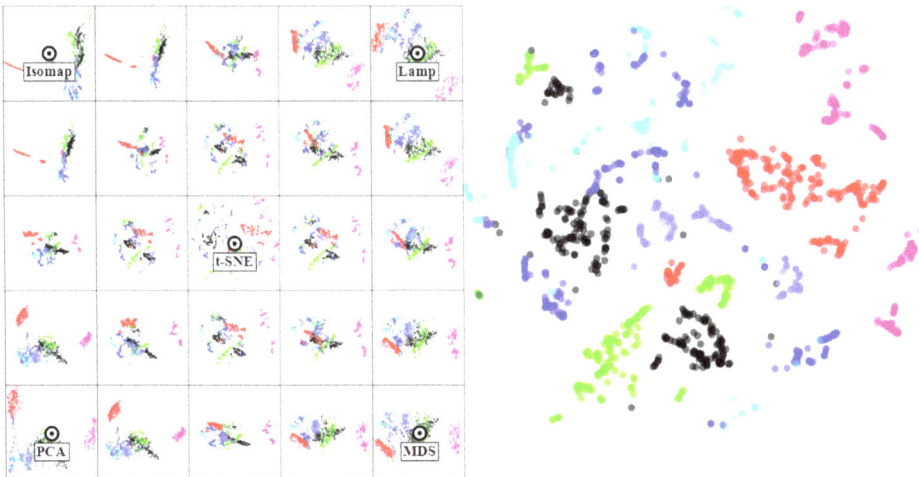

Figure 6. Segmentation dataset (Section 4.1.2). Different views can separate well one or more of the classes, but no view succeeds this for all classes.

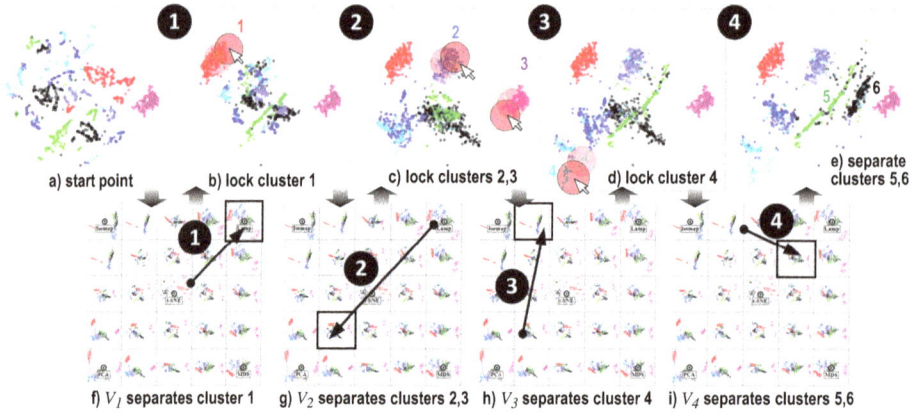

Figure 7. Creating a mixed view which separates well the six classes present in the segmentation dataset.

4.2. Bundled Graph Drawings

Graphs are used to encode large relational datasets, such as the dependencies between components in a software system. As outlined in Section 2, a graph drawing can also be seen as an element-based plot, where the drawn edges are the elements. As also explained there, edge bundling (EB) is an established effective instrument for reducing clutter in large graph drawings, allowing one to follow easier the coarse structure of a graph. However, we have outlined that no single bundling method is ideal. The situation here is very similar to the visual exploration of multidimensional projections: we can generate a right set of bundled drawings; each drawing may be good for conveying partial insights in the input graph; but no drawing is optimal in this respect. Hence, having a way to merge the insights obtained from multiple EB drawings is useful.

We can use our method to mix several drawings created by different EB methods, as follows. For this, we consider a dataset having 1024 nodes, which are functions of a software system, and $N = 4021$ edges, which model function calls. The nodes are laid out in a radial way (for details, see the *radial* dataset in [28,29]). Figure 8 (b) shows the unbundled graph, where one clearly cannot see any structure. The thumbnail grid is based on five presets, containing the bundling of the graph by the KDEEB method [29] (images (c) and (d); we used here two parameter sets for KDEEB, the first by pre-clustering graph edges based on their spatial similarity, and the second by using no clustering), the SBEB method [28] (image (e)), and the ADEB method [30] (image (f)). We start our navigation from the unbundled view to view 2, where we see three bundles which are well separated from the others. As such, we want to keep these, so we lock them. We continue the same process by going through views 3 to 5, each time locating bundles that are well readable in the current view, and locking these. The final image (g) shows the locked bundles in the context of a view that is very close to the original unbundled graph. This represents thus a 'patchwork', or mix, between four types of bundling, applied selectively on different parts of the data and the original data. Obtaining such a view is not possible by using any of the EB algorithms in existence.

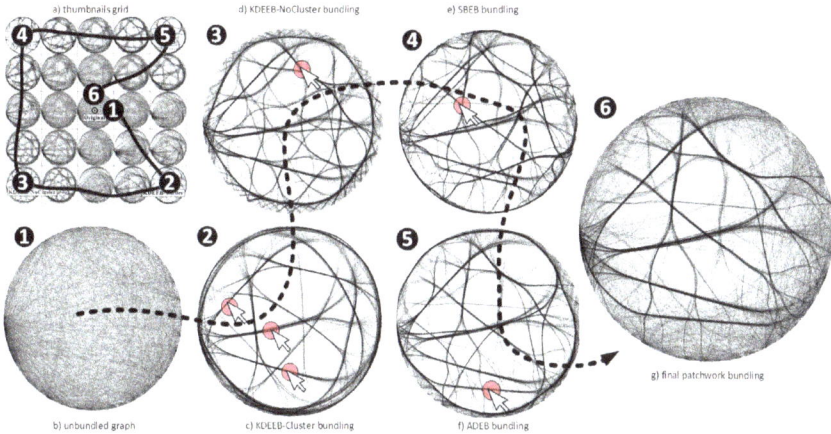

Figure 8. Extracting relevant structures from four types of bundling algorithms to create a new bundling view for a graph dataset.

A second use-case regards the visual comparison of EB techniques. Given several such techniques (or results of the same technique but obtained for different parameter values), an important question is which are the main differences between them. Knowing this helps users in, for example, assessing which techniques behave similarly and thus can be substituted for each other in an application. Figure 9 shows how we can answer this question. We use here the same thumbnails grid and dataset as in Figure 8. However, we only perform four simple navigations, by going several times to-and-fro between the four preset views in the grid corners. The right image in Figure 9 shows intermediate frames during these animations. We see here that going from SBEB to ADEB or from SBEB to KDEEB-NoCluster creates relatively structured in-between frames. Hence, the methods SBEB, ADEB, and KDEEB-NoCluster provide similar layouts in terms of their visual signatures. In contrast, going from KDEEB-Cluster to KDEEB-NoCluster or to ADEB creates intermediate views that look completely messy. This tells us that the KDEEB-Cluster method has a very different visual signature (style of result) than the other methods. A second use-case for this scenario is to compare the quality of two EB techniques *A* and *B*: if we know that, for example, *A* is of high quality, and our animation shows only small differences between *A* and *B*, then we can infer that *B* is also of good quality (the converse being also true).

Informatics **2017**, *4*, 26

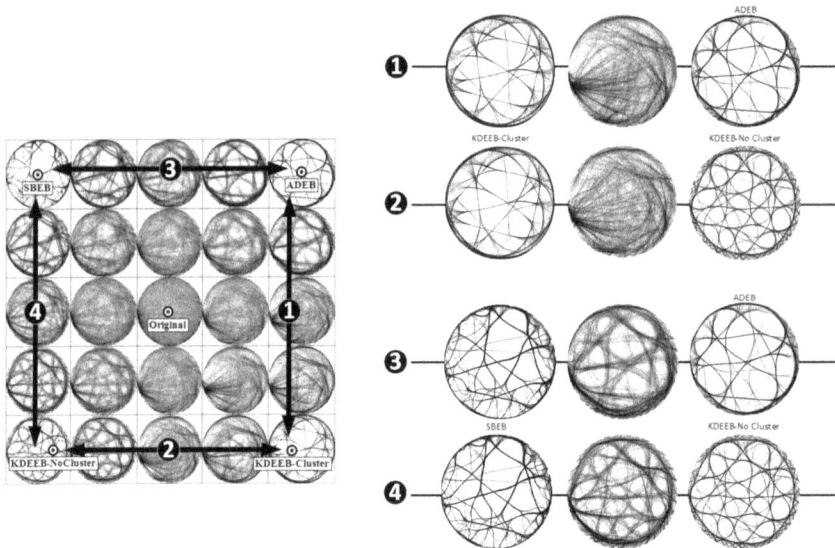

Figure 9. Understanding differences between several EB techniques. The transition between KDEEB-cluster and any other technique produces fuzzy intermediate states (1 and 2), while the transition between SBEB and the othe techniques produces sharper images (3 and 4). This shows that KDEEB-Cluster is visually very different from the other clustering techniques considered.

In the examples above, we had as input data *general* graphs and, as such, we used general-graph bundling methods such as KDEEB, SBEB, and ADEB. Hence, even if the nodes are laid out on a circle, the bundling methods we use should not be confused with, for example, HEB [27]; the only resemblance is that both bundling methods read an input graph whose nodes were laid out on a circle. However, this does not mean that we cannot use HEB or any other method for hierarchical graphs as a preset view, as long as the input graph is hierarchical, of course.

As a final note, we remark that our animation technique provides, for free, the standard relaxation introduced by Holten [27] and since then implemented by virtually all other EB techniques. Briefly put, this technique generates intermediate views between the fully unbundled and fully bundled one by linear interpolation. Our technique does this by default if we add the unbundled view as a preset and then simply animate our focus point between this view and any preset corresponding to a fully bundled view (see Figure 10)).

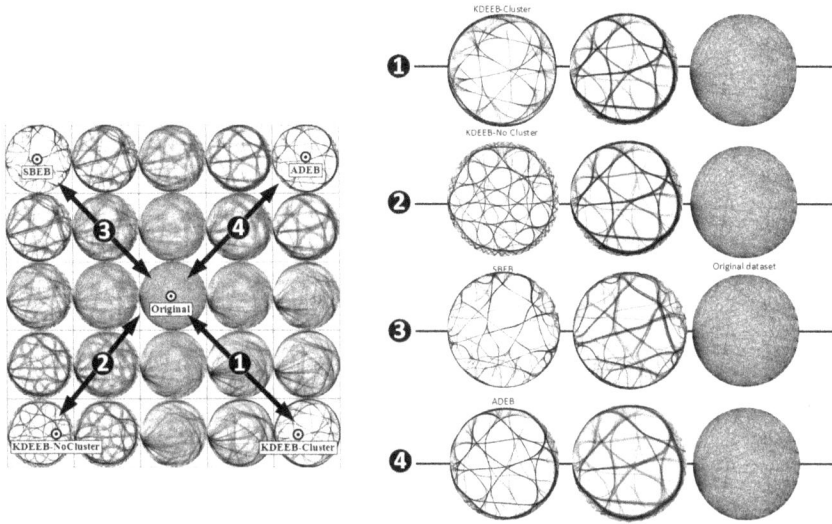

Figure 10. Bundling relaxation provided for four bundling techniques (presets in the corners of the grid) by our animation technique.

5. Discussion

We next discuss several aspects of our proposed exploration method.

Generality: Our technique can be applied to any so-called element-based plots, i.e., visualizations that consist of a (large) number of simple geometric objects such as points or lines. Scatterplots, projections, graph drawings, Cartesian uniformly-sampled fields, and parallel coordinate plots (the latter two not illustrated in this paper) fall into this class. All that one needs is a number of such plots expressed as a set of 2D primitives.

Another generic aspect is the generation of the preset views used for the thumbnails grid. For this, we used here a different visualization method (projection technique or bundling method) for each preset view. However, many more options are possible. For example, one can use any suitable scagnostics or projection pursuit-like technique to generate a (small) set of interesting views (given any application-specific definition of interestingness) and directly use them as presets. In particular, the projections of high-dimensional data of Lehmann and Theisel [47] is interesting, as it captures (much of) the structure of an nD dataset with only $n/2$ projections. As above, the only constraint here is that all these presets are element-based plots of the same dataset.

Scalability and simplicity: Our technique is simple to implement, requiring only the application of Equtaion (1) on the sets of 2D primitives corresponding to the preset views. Our current CUDA-based implementation can handle interpolation of datasets of roughly 10 preset views, each having about one million points, in real time on a modern GPU.

Ease of use: The technique is fully automatic, requiring no user parameter setting, apart from organizing the available preset views in the thumbnails matrix—which can be done by a simple click-to-place process. Apart from that, all interactions are done via mouse dragging (to control the animation) and brushing (to lock or unlock elements in the current view).

Related techniques: We have discussed several related techniques in Section 2. It is insightful to visit these after the presentation of our own method to pinpoint similarities and differences.

Grand tours: As this family of techniques, we aim to explore a nD space by means of projections, and use animation to transition from a projection to another. However, our techniques are applicable to *any* element-based plot, beyond projections, e.g., bundled graph drawings; for projections, we can handle any projection type, whereas tours are (to our knowledge) limited to attribute-based projections; we *explicitly* show the view space $\mathcal{V}$, and exploration path therein, in the thumbnail grid, wheras tours do not do this; we propose the lock mechanism to combine insights from *different* views in a synthesized view.

Projection pursuit techniques: As this family of techniques, we pre-select a small set of preset views to start the exploration from. We do not propose any specific ways to choose these presets, whereas the key goal of pursuit techniques is precisely how to find interesting projections. However, animation for exploration of the view space between these presets is our key contribution, whereas pursuit techniques typically present these interesting views in a static (table like) fashion. Separately, we mention a similarity between targeted projection pursuit (TPP) [64] and our work in the sense that both techniques use direct manipulation to modify the view. However, TPP does this by moving *elements* (points) in the view and then *automatically* searches for a projection that best fits the user-modified view. In contrast, we *lock* elements in the current view and we go to the next view(s) under *user control*. Finally, we note that TPP only handles linear projections, whereas we handle any projection type, and actually any element-based plot.

Projection inspector: As already mentioned, ProjInspector [44] is arguably the closest technique to ours. Both techniques use a weighted interpolation of a small number of precomputed projections to generate, on-the-fly, a new projection. The interpolation is directly controlled by the user using a preset controller metaphor. Yet, key differences exist: (a) ProjInspector's preset controller must be a regular k-sided convex polygon (where k is the number of presets), since they use mean values coordinates to compute the preset weights. In contrast, we use Shepard-type interpolation, which does not have such constraints. The difference is very important. It allows us to arrange our presets in any way so as to flexibly define regions over the thumbnail grid where only certain presets have a large influence. The importance of this flexibility was demonstrated earlier by Van Wijk et al. [57]. In contrast, ProjInspector only allows permuting presets along what is essentially a sampled circle. (b) ProjInspector does not use animation for data exploration. In contrast, we allow users to define multiple exploration paths, which are explicitly drawn and visualized in the thumbnail grid, and then (re)run the exploration along them. (c) ProjInspector only handles control-point-based projections [15,60–63]. As outlined several times, we handle any projection, and actually any element-based plot. (d) ProjInspector is designed to work interactively for datasets of moderate size; the largest dataset at which the authors report interactive exploration has 19029 points. Our GPU-based implementation allows handling up to one million points at interactive rates (Section 3.2).

Limitations and open points: Our technique has, however, several limitations, as follows.

First, different placements of the preset views in specific places in the thumbnails grid can result in significantly different view-space explorations, due to the fact that the interpolation in Equation (1) considers the distances form the current focus point to these preset views. As such, certain intermediate views can or cannot be possible to realize, depending on where the presets are placed. Finding the optimal placement of presets for a given exploration of the view space is a challenging, and not yet solved, problem. However, we note that the same issue exists for all comparable techniques, most notably the original preset controller [57] and ProjInspector [44].

Secondly, we note that some intermediate views created by interpolation typically may not correspond to an actual view of the input dataset realized by any given visualization method. On the one side, this is a positive thing, in the sense that they let us create views on our data which would not have been possible by using any of the considered existing visualization methods. However, this can also be dangerous, in the sense that the created views may reflect the actual data in misleading ways. Whether the latter is the case has to be determined on a case-by-case basis, using the semantics of the actual visualizations at hand and what is important that any view preserves from the data. We should also note that the same problem is present in basically all control-point-based projection methods [15,60–63]: an important use-case thereof allows users to *freely* place the controls, with no relation to the underlying data similarity. As such, the resulting projection(s) may be misleading. However, this freedom is important when the underlying data dimensions do not fully reflect the true similarity of the observations, e.g., when the dimensions are features extracted automatically from complex data such as images, video, or music [15]. Manually arranging the controls allows users to craft a projection that best matches their own understanding of similarity. Precisely the same freedom is offered by our interactive navigation combined with the lock mechanism.

6. Conclusions

In this paper, we presented a new technique to support the interactive visual exploration of multidimensional datasets. For this, we leverage the user's cognitive ability in terms of discovering stable and/or changing patterns in an animated view that is created under direct control of the user, by interpolation from a number of so-called preset views. Our technique combines and generalizes earlier animation and interaction techniques for exploring multidimensional data from a variety of viewpoints. Our proposal is generic (it works for any element-based plot consisting of a set of 2D point or line primitives), simple to implement and use, and scalable up to roughly one million data elements. We demonstrate our technique by applying it on multidimensional projections and bundled graph layouts and on several real-world datasets.

Future work can target a number of directions. First, it is interesting to explore semi-automatic ways for detecting patterns of interest in the views created by animation, so that the user can select and lock these more easily when creating mixed views. Secondly, a more flexible layout than the grid one can be considered for the presets. Finally, and most challengingly, extending our technique to handle more types of visualizations than element-based plot would open its applicability to even wider areas of information visualization.

Acknowledgments: The authors acknowledge the support of the French National Agency for Research (Agence Nationale de la Recherche —ANR) under the grant ANR-14-CE24-0006-01 project "TERANOVA", the SESAR Research and Innovation Action Horizon 2020 under project "MOTO" (The embodied reMOte TOwer) and the "Remote Tower" project from the "Universite Federale Toulouse Midi-Pyrenees" under the grant "Initiatives d'Excellence IDEX UNITI Actions Thèmatiques Strategiques (ATS) 2014".

Author Contributions: H.-J.S. and C.H. had the original idea for this visualization technique which was further extended thanks to J.F.K. and the multidimensional data projection. A.H. developed the interpolation algorithm. C.H. is mainly responsible for the implementation of the visualization system. A.T. and C.H. wrote the paper.

Conflicts of Interest: The authors declare no conflict of interest.

References

1. *Readings in Information Visualization: Using Vision to Think*; Card, S.K.; Mackinlay, J.D.; Shneiderman, B. (Eds.) Morgan Kaufmann Publishers Inc.: San Francisco, CA, USA, 1999.
2. Hurter, C.; Telea, A.; Ersoy, O. MoleView: An attribute and structure-based semantic lens for large element-based plots. *IEEE TVCG* **2011**, *17*, 2600–2609.
3. Rao, R.; Card, S.K. The Table Lens: Merging Graphical and Symbolic Representations in an Interactive Focus+Context Visualization for Tabular Information. In Proceedings of the SIGCHI Conference on Human Factors in Computing Systems, Boston, MA, USA, 24–28 April 1994; pp. 318–322.
4. Telea, A. Combining Extended Table Lens and Treemap Techniques for Visualizing Tabular Data. In Proceedings of the Eighth Joint Eurographics/IEEE VGTC conference on Visualization, Lisbon, Portugal, 8–10 May 2006; pp. 51–58.
5. Brosz, J.; Nacenta, M.A.; Pusch, R.; Carpendale, S.; Hurter, C. Transmogrification: Causal manipulation of visualizations. In Proceedings of the 26th Annual ACM Symposium on User Interface Software and Technology, St. Andrews, UK, 8–11 October 2013; pp. 97–106.
6. Hurter, C.; Taylor, R.; Carpendale, S.; Telea, A. Color tunneling: Interactive exploration and selection in volumetric datasets. In Proceedings of the 2014 IEEE Pacific Visualization Symposium, Yokohama, Japan, 4–7 March 2014; pp. 225–232.
7. Lhuillier, A.; Hurter, C.; Telea, A. State of the Art in Edge and Trail Bundling Techniques. *Comput. Graph. Forum* **2017**, doi:10.1111/cgf.13213.

8. Inselberg, A. The plane with parallel coordinates. *Vis. Comput.* **1985**, *1*, 69–91.
9. Inselberg, A. *Parallel Coordinates*; Springer: Berlin, Germany, 2009.
10. Cleveland, W.S. *Visualizing Data*; Hobart Press: Summit, NJ, USA, 1993.
11. Borg, I.; Groenen, P. *Modern Multidimensional Scaling: Theory and Applications*, 2nd ed.; Springer: Berlin, Germany, 2005.
12. Van der Maaten, L.J.P.; Hinton, G.E. Visualizing High-dimensional Data using t-SNE. *JMLR* **2008**, *9*, 2579–2605.
13. Jolliffe, I.T. *Principal Component Analysis*, 2nd ed.; Springer: Berlin, Germany, 2002.
14. Tenenbaum, J.B.; De Silva, V.; Langford, J.C. A global geometric framework for nonlinear dimensionality reduction. *Science* **2000**, *290*, 2319–2323.
15. Joia, P.; Coimbra, D.; Cuminato, J.; Paulovich, F.; Nonato, L.G. Local Affine Multidimensional Projection. *IEEE TVCG* **2011**, *17*, 2563–2571.
16. Kandogan, E. Visualizing Multi-Dimensional Clusters, Trends, and Outliers Using Star Coordinates. In Proceedings of the Seventh ACM SIGKDD International Conference on Knowledge Discovery and Data Mining, San Francisco, CA, USA, 26–29 August 2001.
17. Kandogan, E. Star Coordinates: A Multi-Dimensional Visualization Technique with Uniform Treatment of Dimensions. In Proceedings of the IEEE InfoVis 2000, Salt Lake City, UT, USA, 9–10 October 2000.
18. Lehmann, D.J.; Theisel, H. Orthographic Star Coordinates. *IEEE TVCG* **2013**, *19*, 2615–2624.
19. Rubio-Sánchez, M.; Sanchez, A.; Lehmann, D.J. Adaptable Radial Axes Plots for Improved Multivariate Data Visualization. *CGF* **2017**, *36*, 389–399.
20. Gabriel, K.R. The biplot graphic display of matrices with application to principal component analysis. *Biometrika* **1971**, *58*, 453–467.
21. Rubio-Sanchez, M.; Raya, L.; Diaz, F.; Sanchez, A. A comparative study between RadViz and Star Coordinates. *IEEE TVCG* **2016**, *22*, 619–628.
22. Daniels, K.M.; Grinstein, G.; Russell, A.; Glidden, M. Properties of normalized radial visualizations. *Inf. Visual.* **2012**, *11*, 273–300.
23. Roweis, S.T.; Saul, L.K. Nonlinear Dimensionality Reduction by Locally Linear Embedding. *Science* **2000**, *290*, 2323–2326.
24. Martins, R.; Coimbra, D.; Minghim, R.; Telea, A. Visual analysis of dimensionality reduction quality for parameterized projections. *Comput. Graph.* **2014**, *41*, 26–42.
25. Tollis, I.; Battista, G.D.; Eades, P.; Tamassia, R. *Graph Drawing: Algorithms for the Visualization of Graphs*; Prentice Hall: Upper Saddle River, NJ, USA, 1999.
26. Kruiger, J.F.; Rauber, P.; Martins, R.; Kerren, A.; Kobourov, S.; Telea, A. Graph Layouts by t-SNE. *CGF* **2017**, *36*, 1745–1756.
27. Holten, D. Hierarchical edge bundles: Visualization of adjacency relations in hierarchical data. *IEEE TVCG* **2006**, *12*, 741–748.
28. Ersoy, O.; Hurter, C.; Paulovich, F.; Cantareiro, G.; Telea, A. Skeleton-based edge bundling for graph visualization. *IEEE Trans. Vis. Comput. Graph.* **2011**, *17*, 2364–2373.
29. Hurter, C.; Ersoy, O.; Telea, A. Graph bundling by kernel density estimation. *Comput. Graph. Forum* **2012**, *31*, 865–874.
30. Peysakhovich, V.; Hurter, C.; Telea, A. Attribute-Driven Edge Bundling for General Graphs with Applications in Trail Analysis. In Proceedings of the 2015 IEEE Pacific Visualization Symposium (PacificVis), Hangzhou, China, 14–17 April 2015; pp. 39–46.
31. Lhuillier, A.; Hurter, C.; Telea, A. FFTEB: Edge Bundling of Huge Graphs by the Fast Fourier Transform. In Proceedings of the 10th IEEE Pacific Visualization Symposium, PacificVis 2017, Seoul, Korea, 18–21 April 2017.
32. Hofmann, J.; Groessler, M.; Pichler, P.P.; Lehmann, D.J. Visual Exploration of Global Trade Networks with Time-Dependent and Weighted Hierarchical Edge Bundles on GPU. *Comput. Graph. Forum* **2017**, *36*, 1545–1556.
33. van der Zwan, M.; Codreanu, V.; Telea, A. CUBu: Universal real-time bundling for large graphs. *IEEE TVCG* **2016**, *22*, 2550–2563.
34. Tversky, B.; Morrison, J.B.; Betrancourt, M. Animation: Can It Facilitate? *Int. J. Hum.-Comput. Stud.* **2002**, *57*, 247–262.

35. Chevalier, F.; Riche, N.H.; Plaisant, C.; Chalbi, A.; Hurter, C. Animations 25 Years Later: New Roles and Opportunities. In Proceedings of the International Working Conference on Advanced Visual Interfaces (AVI '16), Bari, Italy, 7–10 June 2016; pp. 280–287.

36. Bertin, J. *Semiology of Graphics*; University of Wisconsin Press: Madison, WI, USA, 1983.

37. Elmqvist, N.; Stasko, J.; Tsigas, P. DataMeadow: A visual canvas for analysis of large-scale multivariate data. *Inf. Vis.* **2008**, *7*, 18–33.

38. Archambault, D.; Purchase, H.; Pinaud, B. Animation, Small Multiples, and the Effect of Mental Map Preservation in Dynamic Graphs. *IEEE TVCG* **2011**, *17*, 539–552.

39. Elmqvist, N.; Dragicevic, P.; Fekete, J.D. Rolling the dice: Multidimensional visual exploration using scatterplot matrix navigation. *IEEE TVCG* **2008**, *14*, 1141–1148.

40. Asimov, D. The grand tour: A tool for viewing multidimensional data. *SIAM J. Sci. Stat. Comput.* **1985**, *6*, 128–143.

41. Dhillon, I.S.; Modha, D.S.; Spangler, W. Class visualization of high-dimensional data with applications. *Comput. Stat. Data Anal.* **2002**, *41*, 59–90.

42. Huber, P.J. Projection pursuit. *Ann. Stat.* **1985**, *13*, 435–475.

43. Huh, M.Y.; Kim, K. Visualization of Multidimensional Data Using Modifications of the Grand Tour. *J. Appl. Stat.* **2002**, *29*, 721–728.

44. Pagliosa, P.; Paulovich, F.; Minghim, R.; Levkowitz, H.; Nonato, G. Projection inspector: Assessment and synthesis of multidimensional projections. *Neurocomputing* **2015**, *150*, 599–610.

45. Ma, K.L. Image graphs—A novel approach to visual data exploration. In *Visualization'99: Proceedings of the IEEE Conference on Visualization*; Ebert, D., Gross, M., Hamann, B., Eds.; IEEE Computer Society: Washington, DC, USA, 1999; pp. 81–88.

46. Friedman, J.H.; Tukey, J.W. A projection pursuit algorithm for exploratory data analysis. *IEE Trans. Comput.* **1974**, *100*, 881–890.

47. Lehmann, D.J.; Theisel, H. Optimal Sets of Projections of High-Dimensional Data. *IEEE TVCG* **2015**, *22*, 609–618.

48. Seo, J.; Shneiderman, B. A rank-by-feature framework for unsupervised multidimensional dataexploration using low dimensional projections. In Proceedings of the IEEE Symposium on Information Visualization, Austin, TX, USA, 10–12 October 2004; pp. 65–72.

49. Sips, M.; Neubert, B.; Lewis, J.P.; Hanrahan, P. Selecting good views of high-dimensional data using class consistency. *CGF* **2009**, *28*, 831–838.

50. Schreck, T.; von Landesberger, T.; Bremm, S. Techniques for precision-based visual analysis of projected data. *Inf. Vis.* **2010**, *9*, 181–193.

51. Lehmann, D.J.; Albuquerque, G.; Eisemann, M.; Magnor, M.; Theisel, H. Selecting coherent and relevant plots in large scatterplot matrices. *CGF* **2012**, *31*, 1895–1908.

52. Marks, J.; Andalman, B.; Beardsley, P.; Freeman, W.; Gibson, S.; Hodgins, J.; Kang, T.; Mirtich, B.; Pfister, H.; Ruml, W.; et al. Design Galleries: A General Approach to Setting Parameters for Computer Graphics and Animation. In *ACM SIGGRAPH'97: Proceedings of the International Conference on Computer Graphics and Interactive Techniques*; Whitted, T., Ed.; ACM Press/Addison-Wesley Publishing Co.: New York, NY, USA, 1997; pp. 389–400.

53. Biedl, T.; Marks, J.; Ryall, K.; Whitesides, S. Graph Multidrawing: Finding Nice Drawings Without Defining Nice. In *GD'98: Proceedings of the International Symposium on Graph Drawing*; Whitesides, S.H., Ed.; Springer: Berlin, Germany, 1998; Volume 1547, pp. 347–355.

54. Wu, Y.; Xu, A.; Chan, M.Y.; Qu, H.; Guo, P. Palette-Style Volume Visualization. In Proceedings of the EG/IEEE Conference on Volume Graphics, Prague, Czech Republic, 3–4 September 2007; pp. 33–40.

55. Jankun-Kelly, T.; Ma, K.L. Visualization exploration and encapsulation via a spreadsheet-like interface. *IEEE Trans. Vis. Comput. Graph.* **2001**, *7*, 275–287.

56. Shapira, L.; Shamir, A.; Cohen-Or, D. Image Appearance Exploration by Model-Based Navigation. *Comput. Graph. Forum* **2009**, *28*, 629–638.

57. Van Wijk, J.J.; van Overveld, C.W. Preset based interaction with high dimensional parameter spaces. In *Data Visualization*; Springer: Berlin, Germany, 2003; pp. 391–406.

58. Hurter, C.; Tissoires, B.; Conversy, S. FromDaDy: Spreading data across views to support iterative exploration of aircraft trajectories. *IEEE TVCG* **2009**, *15*, 1017–1024.

59. Hurter, C. Image-Based Visualization: Interactive Multidimensional Data Exploration. *Synth. Lect. Vis.* **2015**, *3*, 1–127.

60. Paulovich, F.V.; Nonato, L.G.; Minghim, R.; Levkowitz, H. Least square projection: A fast high-precision multidimensional projection technique and its application to document mapping. *IEEE TVCG* **2008**, *14*, 564–575.

61. Paulovich, F.V.; Silva, C.T.; Nonato, L.G. Two-phase mapping for projecting massive data sets. *IEEE TVCG* **2010**, *16*, 12811290.

62. Pekalska, E.; de Ridder, D.; Duin, R.; Kraaijveld, M. A new method of generalizing Sammon mapping with application to algorithm speed-up. In Proceedings of the Annual Conference Advanced School for Computer Image (ASCI), Heijen, The Netherlands, 15–17 June 1999; pp. 221–228.

63. Morrison, A.; Chalmers, M. A pivot-based routine for improved parent-finding in hybrid MDS. *Inf. Vis.* **2004**, *3*, 109–122.

64. Faith, J. Targeted Projection Pursuit for Interactive Exploration of High-Dimensional Data Sets. In Proceedings of the 11th International Conference on Information Visualization, Zurich, Switzerland, 4–6 July 2007.

65. Roberts, J.C. Exploratory Visualization with Multiple Linked Views. In *Exploring Geovisualization*; Elsevier: Amsterdam, The Netherlands, 2004; pp. 149–170.

66. North, C.; Shneiderman, B. *A Taxonomy of Multiple Window Coordination*; Technical Report T.R. 97-90; School of Computing, Univercity of Maryland: College Park, MD, USA, 1997.

67. Baldonaldo, M.; Woodruff, A.; Kuchinsky, A. Guidelines for using multiple views in information visualization. In Proceedings of the Working Conference on Advanced Visual Interfaces (AVI), Palermo, Italy, 24–26 May 2000; pp. 110–119.

68. Becker, R.; Cleveland, W. Brushing scatterplots. *Technometrics* **1987**, *29*, 127–142.

69. Schulz, H.J.; Hadlak, S. Preset-based Generation and Exploration of Visualization Designs. *J. Vis. Lang. Comput.* **2015**, *31*, 9–29.

70. Liu, B.; Wünsche, B.; Ropinski, T. Visualization by Example—A Constructive Visual Component-based Interface for Direct Volume Rendering. In Proceedings of the International Conference on Computer Graphics Theory and Applications, Angers, France, 17–21 May 2010; pp. 254–259.

71. Chevalier, F.; Dragicevic, P.; Hurter, C. Histomages: Fully Synchronized Views for Image Editing. In Proceedings of the 25th Annual ACM Symposium on User Interface Software and Technology, Cambridge, MA, USA, 7–10 October 2012; pp. 281–286.

72. Shepard, D. A Two-dimensional Interpolation Function for Irregularly-Spaced Data. In Proceedings of the 23rd ACM National Conference, Las Vegas, NV, USA, 27–29 August 1968; pp. 517–524.

73. Schulz, H.J.; Nocke, T.; Heitzler, M.; Schumann, H. A Design Space of Visualization Tasks. *IEEE TVCG* **2013**, *19*, 2366–2375.

74. Shneiderman, B. The Eyes Have It: A Task by Data Type Taxonomy for Information Visualizations. In Proceedings of the 1996 IEEE Symposium on Visual Languages, Boulder, CO, USA, 3–6 September 1996; pp. 336–343.

75. Kruiger, H.; Hassoumi, A.; Schulz, H.-J.; Telea, A.; Hurter, C. Video Material for the Animation-Based Exploration. 2017. Available online: http://recherche.enac.fr/~hurter/SmallMultiple/ (accessed on 17 August 2017).

76. Meirelles, P.; Santos, C., Jr.; Miranda, J.; Kon, F.; Terceiro, A.; Chavez, C. A study of the relationships between source code metrics and attractiveness in free software projects. In Proceedings of the 2010 Brazilian Symposium on Software Engineering, Bahia, Brazil, 27 September–1 October 2010; pp. 11–20.

77. Coimbra, D.; Martins, R.; Neves, T.; Telea, A.; Paulovich, F. Explaining three-dimensional dimensionality reduction plots. *Inf. Vis.* **2016**, *15*, 154–172.

78. Lichman, M. UCI Machine Learning Repository. University of California, Irvine, School of Information and Computer Sciences. 2013. Available online: http://archive.ics.uci.edu/ml (accessed on 17 August 2017).

79. Rauber, P.E.; Falcao, A.X.; Telea, A.C. Projections as Visual Aids for Classification System Design. *Inf. Vis.* **2017**, doi:10.1177/1473871617713337.

![informatics logo] *informatics*

MDPI

Article

Big Data Management with Incremental K-Means Trees–GPU-Accelerated Construction and Visualization

Jun Wang [1], Alla Zelenyuk [2], Dan Imre [3] and Klaus Mueller [1],*

[1] Visual Analytics and Imaging Lab, Computer Science Department, Stony Brook University, Stony Brook, NY 11794, USA; junwang2@cs.stonybrook.edu
[2] Chemical and Material Sciences Division, Pacific Northwest National Laboratory, Richland, WA 99352, USA; alla.zelenyuk-imre@pnnl.gov
[3] Imre Consulting, Richland, WA 99352, USA; dimre2b@gmail.com
* Correspondence: mueller@cs.stonybrook.edu

Academic Editors: Achim Ebert and Gunther H. Weber
Received: 1 June 2017; Accepted: 26 July 2017; Published: 28 July 2017

Abstract: While big data is revolutionizing scientific research, the tasks of data management and analytics are becoming more challenging than ever. One way to remit the difficulty is to obtain the multilevel hierarchy embedded in the data. Knowing the hierarchy enables not only the revelation of the nature of the data, it is also often the first step in big data analytics. However, current algorithms for learning the hierarchy are typically not scalable to large volumes of data with high dimensionality. To tackle this challenge, in this paper, we propose a new scalable approach for constructing the tree structure from data. Our method builds the tree in a bottom-up manner, with adapted incremental k-means. By referencing the distribution of point distances, one can flexibly control the height of the tree and the branching of each node. Dimension reduction is also conducted as a pre-process, to further boost the computing efficiency. The algorithm takes a parallel design and is implemented with CUDA (Compute Unified Device Architecture), so that it can be efficiently applied to big data. We test the algorithm with two real-world datasets, and the results are visualized with extended circular dendrograms and other visualization techniques.

Keywords: data management; hierarchy construction; parallel computing; visualization

1. Introduction

Big data is everywhere we turn today, and its volume and variety are still growing at an unprecedented speed. Every minute, massive volumes of data and information are produced from various resources and services, recording and affecting everyone and everything—the internet of things, social networks, the economy, politics, astronomy, health science, military surveillance—just to name a few. The vast development of modern technology has meant that data has never been so easy for humankind to acquire. This is especially meaningful for scientific research, which has been revolutionized by big data in the past decade. Both Nature and Science have published special issues on big data dedicated to discussing the opportunities and challenges [1,2].

Nevertheless, with the growing volume, managing and extracting useful knowledge from big data is now more complex than ever. Big data needs big storage, and the immense volume makes operations such as data retrieval, pre-processing, and analysis very difficult and hugely time-consuming. One way to meet such difficulties is to obtain the multilevel hierarchy (the *tree* structure, or *dendrogram*) embedded in the data. Knowing the data hierarchy not only helps users gain a deeper insight of the data under investigation, it can also often serve as the first step to making many further analyses scalable, e.g., nearest neighbor searching [3], hierarchical sampling [4], clustering [5], and others.

However, traditional algorithms for constructing such tree structures, whether bottom-up (*agglomerative*) or top-down (*divisive*), typically are not able to process large volumes of data of with high dimensionality [5]. Although the steadily and rapidly developing Graphics Processing Units (GPU) technologies have been offering effective acceleration to accomplish high-speed data processing, the task of adapting each specific application with parallel computing is still a heavy burden for most users.

To tackle these unsolved challenges, we propose in this paper a new scalable algorithm that runs cooperatively on the Central Processing Units (CPU) and GPU. The algorithm takes a *bottom-up* approach. Each level of the tree is built taking advantage of an algorithm called *parallel incremental k-means*, which iteratively reads unclustered points, and in parallel clusters them into small batches of clusters. Centers in a cluster batch are initially found with the CPU, which is also in charge of managing the tree hierarchy and updating the parameters for building the next level. The distribution of pairwise distances between sample points is calculated on the GPU, and can be visualized in a panel for users to select the proper thresholds. GPU-computed standard deviations are employed to remove irrelevant dimensions that contribute very little to the clustering, to further boost the computing efficiency. By adjusting these algorithm parameters, users can control the complexity of the resulted hierarchy, i.e., the tree's height and branching factors.

The controlled hierarchy complexity is especially meaningful for the visualization of big data, considering that the scalability and multilevel hierarchy is regarded by some as one of the top challenges in extreme-scale visual analytics [6]. To communicate the results, we visualize the trees as circular dendrograms enhanced with visual hints of tree node statistics. Other explorative visualization techniques are also employed to provide a deeper investigation of the hierarchical relationship of the data. We use two real-world datasets to demonstrate the visualization, as well as the effect of the new parallel construction algorithm.

The remainder of the paper is structured as follows. Section 2 briefly reviews the related work. Section 3 presents our new parallel algorithm. The detailed GPU implementation is given in Section 4. We test the algorithm with real-world datasets in Section 5, and present the visualization accordingly at the same time. Section 6 discusses the results and closes with conclusions.

2. Related Work

Clustering is usually the first step in big data analysis. A broad survey of clustering algorithms for big data has recently been given by Fahad et al. [5]. In general, these algorithms can be categorized into five classes: partition-based (some well-known ones are k-means [7], PAM [8], FCM [9]), hierarchical-based (e.g., BIRCH [10], CURE [11], Chameleon [12]), density-based (DBSCAN [13], OPTICS [14]), grid-based (CLIQUE [15], STING [16]), and model-based (MCLUST [17], EM [18]). Among all, hierarchical-based approaches, including agglomerative and divisive, can learn the multilevel structure embedded in data, which is often required in big data analysis. However, as these algorithms are designed for clustering, they typically contain mechanisms for ensuring cluster quality, and metrics for cutting the tree. These designs might not be necessary if we only require the hierarchical relationship of the data, but maintaining them can add extra algorithms and complexity.

Another approach targeted at learning the tree structure is the *k-d tree* [19,20]. The classic version of the k-d tree iteratively splits data along the dimension with the highest variance, resulting in a binary tree structure. An improved version is the *randomized k-d tree* [21], which randomly selects the split dimension from the top N_D dimensions with the highest variance, and builds multiple trees instead of just one to accelerate searching along the tree. One shortcoming of the k-d tree is that, due to the binary branching, the depth of the hierarchy grows very fast, with an increase in data volume and cluster complexity. Managing such hierarchies, especially with visualization tools, is typically very difficult.

In contrast, the k-means tree algorithm [22] (also called hierarchical k-means) allows users to control the branching factor, and hence the depth, of the tree by adjusting the parameter k. Variations of this type of algorithm have been widely used in solving computer vision problems [3,23,24]. However, determining a good value for k is usually difficult without knowing the details of the data, and forcing all tree nodes to have the same fixed branching factor may also sometimes be problematic. On the other hand, the incremental clustering algorithm proposed by Imrich et al. [25] controls the cluster radius such that a flexible number of clusters can form accordingly. Our work adapts this algorithm to construct the tree hierarchy, which can overcome the mentioned deficiency of previous methods.

Since NVIDIA released CUDA (https://developer.nvidia.com/cuda-toolkit) in 2007, GPU-based computing has become much friendlier for developers, and thus widely applied in various high-performance platforms. General parallel algorithms have been devised for sorting [26], clustering [8,27], classification [28,29], and neural networks training [30]. Our work also takes a parallel design and is implemented cooperatively on the CPU and GPU, referencing the implementation of the previous work.

The tree structure is often visualized as a dendrogram laid out vertically or horizontally [31]. As low-level nodes of a complex tree may easily become cluttered, a circular layout can be advantageous. Such visualization has been adopted by several visual analytic systems [25,32], as well as in our own research. Other big data visualization techniques, for instance, t-SNE [33], are also related to our work, as we will utilize them to assist in analyzing data structural relationships.

3. Construction of the Tree Hierarchy

As mentioned, our algorithm for constructing the tree hierarchy takes a bottom-up style. Each level of the tree is built upon the lower level, with the parallel incremental k-means algorithm such that the value of k does not need to be predetermined. The distance threshold used for each level is decided adaptively according to the distribution of pairwise distances of lower level points. We also applied basic dimension reduction techniques to make the implementation more efficient in both time and memory.

3.1. GPU-Accelerated Incremental K-Means

The pseudo code for the CPU incremental k-means algorithm proposed by Imrich et al. [25] is given in Algorithm 1. The algorithm starts with making the first point of a dataset the initial cluster center, and then scans each unclustered data point p and looks for its nearest cluster center c with a certain distance metric calculated via the function $distance(c, p)$. p will be clustered into c if they are close enough under a certain threshold t, otherwise, p will become a new cluster center for later points. The process stops when all points are clustered. The distance threshold t acts as the regulator for the clustering result, such that a larger t leads to a smaller number of clusters each with more points, while a small t could result in many small clusters.

Algorithm 1: Incremental K-Means

Input: data points P, distance threshold t
Output: clusters C
C = empty set
for each *un-clustered point p in* P
    if *C is empty* **then**
        Make p a new cluster center and add it into C
    else
        p = next un-clustered point
        Find the cluster center c in C *that is closest to p*
        d = distance(c, p)
        if *d < t* **then** *Cluster p into c*
        else *Make p a new cluster center and add it into* C
        end if
    end if
end for
return C

One important advantage of incremental k-means over other common clustering algorithms is that it can handle streaming data. Each new data point in a stream can be simply added to the nearest cluster or made into a new cluster center depending on the distance threshold t. However, Algorithm 1 is not very scalable, and can gradually become slower with as the data size and number of clusters grow, as the points coming at a later time will have to be compared against all the cluster centers that came before it. This can be extremely compute-intensive in the big data context, especially when the points are of high dimensionality and the cost of calculating the distance metric takes a non-negligible amount of time.

To solve the scalability issue of Algorithm 1, a parallelized version of the incremental k-means that can run on the GPU was devised by Papenhausen et al. [27]. The pseudo code of an adapted version used in our work is given in Algorithm 2, which clusters points of a dataset iteratively. In each iteration step, the algorithm first runs Algorithm 1 on the CPU to detect a batch of b cluster centers, and then in parallel computes the distances between each unclustered point to each center on the GPU. The nearest center for each point is found at the same time, so that a point can be assigned with a label of its nearest cluster if their distance is within the threshold. However, a point can be officially assigned to a cluster only on the CPU after the labels are passed back from the GPU. After each iteration step, a batch of at most b clusters is generated and added to the output set. Clustered points will not be scanned again in later iterations. At last, the process stops when all points are clustered.

The batch size b controls the workload balance between CPU and GPU. A larger b means fewer iterations but leans towards being CPU bound, which means the GPU may have more idle time waiting for the CPU to complete, while a small b could result in GPU underutilization. The value of b is also suggested to be a multiple of 32 to avoid divergent warps under a CUDA implementation.

It is worth noting that the original algorithms in [25,27] also track small clusters that have not been updated for a while and consider them as outliers. Then, a second pass is performed to re-cluster points in these outlier clusters. However, in our work, we decided to preserve these small clusters for users to judge whether they are actually outliers or not, thus, all recognized clusters will be returned directly without a second pass.

Algorithm 2: Parallel Incremental K-Means

Input: data points P, distance threshold t, batch size s
Output: clusters C
C = *empty set*
while *number of un-clustered points in $P > 0$*
    Perform Alg. 1 until a number of s clusters B emerge
    in parallel:
        for each *un-clustered point p_i*
            Find the cluster center b_i in B that is closest to p_i
            $d_i = distance(b_i, p_i)$
            $c_i = b_i$ *if $d_i < t_i$, otherwise $c_i = null$*
        end for
    end parallel
    Assign each point p_i to c_i if c_i exists
    Add B to C
end while
return C

3.2. Construction of the Tree Hierarchy

By running Algorithm 2 with a proper value of t, we can typically generate a redundant number of clusters, each containing a small number of closely positioned data points. If we consider these clusters as the first level of the tree hierarchy, the higher levels of the tree can be built in a bottom-up manner. To build the next level, we see each cluster center as a point in the current level, and then simply run incremental k-means on them with an increased distance threshold. The pseudo code of such an algorithm is given in Algorithm 3.

The height of the tree, as well as its branching factors, are controlled by the initial distance threshold t and the function *update_threshold(t)*, which computes the distance threshold for raising the next level. The initial t can be chosen according to the distribution of the point distances, e.g., the value indicating the average intra-cluster distance of clusters. Then, a practical strategy for updating it could simply be setting up a *growth rate* so that *update_threshold(t)* = t * *growth rate*, i.e., the threshold will grow linearly. However, this can be problematic sometimes, and result in a t' that is either too large (which merges all points into a single node at an early phase) or too small (which turns each point into a single a cluster in the worst case). Thus, to make it more flexible, we also specify a range of *shrinking rate*, e.g., from 0.05–0.8, which is the ratio of the number of nodes between the new level and the current level. A threshold t' is considered ineffective whenever the shrink rate of the new level falls out of the range. In such a case, we predefine the function to compute t' again in the same manner as the initial t, but based on nodes of the current level. The construction process stops when the current level contains only the root node, or has fewer nodes than a predetermined number. In the latter case, we make the root node with the last level as its children.

Algorithm 3: Incremental K-Means Tree

Input: data points **P**, initial distance threshold t, batch size s
Output: root node of the tree
*Set each data point in **P** as a leaf node*
L = leaves
t' = t
while *L does not meet the stop condition*
    Perform Alg. 2 with t' and s to generate C clusters (nodes) from L
    *Make nodes in **L** as children of the corresponding nodes in **C***
    t' = update_threshold(t)
    L = C
end while
if *L contains only one node* **then**
    Return L[0] as the root
else
    *Make and return the root with **L** as its children*

3.3. Determine the Distance Threshold

Since Algorithm 2 was merely designed for sampling points in each cluster to downscale the data size [27], the distance threshold t could be specified by the user accordingly to control cluster numbers and sizes. However, as t can directly affect the structure of the result hierarchy in Algorithm 3, it must be assigned carefully.

When looking for a proper value of t for constructing the next level of the tree, we first compute a histogram of the pairwise distances between points in the current level. However, computing the distance matrix could be very compute-intensive, especially given a dataset of high dimensionality. We took two approaches to ease this issue. First, we reduce the number of dimensions by removing the irrelevant ones that have little contribution to the clustering process. As we typically use the Euclidean distance metric, the dimension reduction can be done simply by ignoring dimensions with small standard deviations. Second, when there are too many points, we only use a random sampling of, say, 20,000–50,000 points, instead of all of them, to compute the histogram. The process of computing the distance distribution as well as computing the dimension standard deviations are all GPU-accelerated to further boost the speed. Details of the implementation are given in Section 4.

In our experience, the distribution of point distances in a dataset typically has several peaks and gaps, which indicate internal and external distances between point clusters. A good value of t can then be selected through referencing these values. Although the histogram's bin width may affect the judgment, we find that the clustering result is not very sensitive to small variations of t. We typically chose 200 bins in our experimentation, while the exact number can be varied for other datasets.

4. Low Level Implementation

The algorithms described in the previous section were implemented on a single NVIDIA GPU with CUDA. We now introduce the GPU kernel implementation in detail.

4.1. Kernel for Parallel Clustering

For our parallel incremental k-means, each GPU thread block has a thread dimension of 32 by 32, so that we lunch $N/32$ thread blocks if we have a total of N data points. Figure 1 briefly illustrates the GPU thread block access pattern. Each thread block compares 32 points to a batch of s cluster centers (s must be a multiple of 32). The x coordinate of a thread tells which point it will be operating on, and the y coordinate is mapped to a small group of $s/32$ cluster centers. For example, if we set $s = 128$, each thread will process four cluster centers ($128/32 = 4$). The cluster centers and points are stored in memory as two matrices with the same x dimension. Then, each thread will compute the distances between the corresponding point and the small group of cluster centers, and store the nearest cluster index and its distance value in the shared memory for further processing.

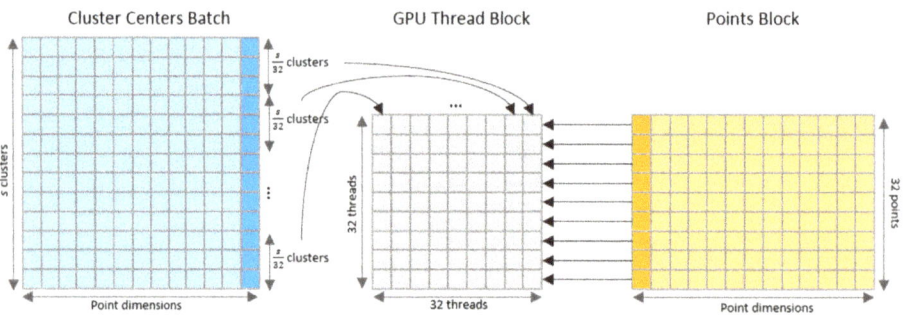

Figure 1. GPU access pattern for the parallel clustering algorithm.

Figure 2 gives the pseudo code for the GPU accelerated algorithm, where the two 32 by 32 shared memories are denoted *distance[][]* and *cluster[][]*. Each row of the *distance[][]* stores the distances between a point and its nearest cluster center in the small group of centers. That is to say, if the batch size is 128, each element of *distance[][]* stores the distance of the nearest center among the four that are compared against each other in a thread. Meanwhile, the indexes of the corresponding clusters are saved in *cluster[][]*. Then, after synchronizing all the threads in the block so that all shared memories are filled with stable results, each thread with x id of 0 will scan through one row of the shared memory, looking for the minimum distance and the nearest cluster. A point will be labeled with the nearest cluster's id if the distance is within the threshold, otherwise the point will be labeled -1, indicating that the point is not clustered in the current iteration of the algorithm (see Algorithm 2). After all the thread blocks finish their job, the labels of all the points will be returned and used for CPU to officially assign points to clusters.

As mentioned, the cluster center batch size s can directly influence the per-thread workload. A larger s means each thread will have to process more cluster centers. The workload of the GPU kernel is also affected by the dimensionality of the data and the computing complexity of the distance metric. Our typical cases are to run the algorithm with datasets of about 400–800 dimensions, and with the Euclidean distance metric. As a result, we found that setting $s = 128$ could reach the best workload balance between the GPU and CPU, although the choice for other datasets may vary.

```
pid = blockDim.y * blockIdx.x + tid.x                    // point id
C = centers to compare

distance[tid.y][tid.x] = minimum distance between point[pid] and centers in C
cluster[tid.y][tid.x] = id of the nearest cluster center in C
syncthreads()
if tid.x == 0 then
    min_dis = minimum of the row distance[tid.y]
    if min_dis < distance threshold then
        label[pid] = the corresponding cluster id stored in the row cluster[tid.y]
    else label[pid] = -1
    end if
end if
```

Figure 2. The pseudo code of GPU kernel for parallel clustering.

4.2. Parallel Computing of Standard Deviations and Pairwise distances

The computation of the dimension standard deviations is done on the GPU with an optimization technique called *parallel reduction* [34], which takes a tree-based approach within each GPU thread block. As the data of one dimension forms a very long vector, the calculation of the mean, as well as the standard deviation of the vector, can be transferred into a vector reduction operation (The standard deviation formula is $\sigma = \sqrt{\sum(x_i - \mu)^2/N}$, which requires vector mean μ. The part inside the squared root can be transferred into vector summing, which can be done with vector reduction.). For our CUDA implementation, dimensions are mapped to the y-coordinates of blocks. We launch 512 threads in a block, each mapped to the value of one data point in one dimension. Thus, the block dimension is D by $N/512$, where N is the number of points and D is the data dimensionality. The iterative process of the parallel reduction is illustrated in Figure 3a. Each value mapped to a thread is initialized in the beginning. The initialization depends on the goal of the function, i.e., for calculating the vector mean μ, each value is divided by N, and for calculating the vector variance, each value x_i is mapped to $(x_i - \mu)^2/N$. And then in each iteration step, the number of active threads in a block is halved, and the values of the second half of the shared memory are added to the first half, until there is only one active thread getting the final result of the block and storing it into the output vector. The pseudo code of the GPU kernel is given in Figure 3b.

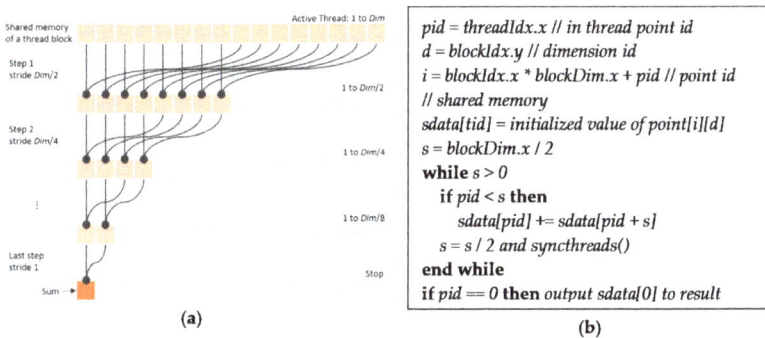

```
pid = threadIdx.x // in thread point id
d = blockIdx.y // dimension id
i = blockIdx.x * blockDim.x + pid // point id
// shared memory
sdata[tid] = initialized value of point[i][d]
s = blockDim.x / 2
while s > 0
    if pid < s then
        sdata[pid] += sdata[pid + s]
    s = s / 2 and syncthreads()
end while
if pid == 0 then output sdata[0] to result
```

(a)   (b)

Figure 3. The parallel reduction. (**a**) Illustration of GPU thread block iterations. (**b**) The pseudo code of the GPU kernel implementation.

The result of the reduction operation is a vector downscaled by 512 times of the input. If the resultant vector is still very large, we can use the parallel reduction again until we reach the final output of a single summed value. However, as the cost of memory transfers may be higher than the benefit from the GPU parallelization when processing a short vector, a single CPU scan would be more than sufficient in such a case. Although it depends on the data volume, at most, two parallel reductions would usually suffice for computing dimension means and standard deviations in our applications.

As we implemented all the algorithms on a single multi-core GPU, a practical difficulty is that there may not be enough GPU memory to hold all the data, especially those of high dimensionality. Even if it is possible to set up a super-large GPU memory, the length of the data array may go beyond the maximum indexable value so that they cannot be accessed. Our solution is to divide data into blocks of the size that can be held in memory of a single GPU, and operate parallel reduction on each of them. Then, an extra CPU scan is operated on results from data blocks to summarize the final output.

The computation of the pairwise distance histogram faces a similar problem. Although we can sometimes fit the sampled data points in GPU memory, the length of the resultant vector can easily go beyond the indexable range (e.g., the length of the vector from pairwise distances of 50,000 points is 1,249,975,000). Then again, we divide sampled points into blocks of fixed size. As we only need the histogram, we update the statistics on the CPU whenever the distances of points from two blocks are returned by the GPU, and then drop the result to save memory. The access pattern of GPU thread blocks for computing pairwise distances is straightforward; each thread calculates one pair of distances. As we use 32×32 thread blocks, there will simply be $N/32 \times N/32$ blocks launched.

5. Experiments and Visualization

We experimented with two datasets, and in the following, we present our visualizations for communicating the results. For both datasets, we use the Euclidean distance to compare points and cluster centers, based on which data hierarchies are built.

5.1. The MNIST Dataset

The first experiment used the MNIST dataset (http://yann.lecun.com/exdb/mnist/). It contains 55,000 handwritten digit images, each with a resolution of 28×28 pixels, i.e., each data point is of 784 dimensions. The greyscale value of a pixel ranges from 0 to 1. As pixels near the edge of an image are usually blank, we filter out these dimensions via the standard deviation scheme, leaving 683 dimensions for use. The histogram of pairwise distances from 20,000 samples is shown in Figure 4. Here, we can see that the histogram generally follows a Gaussian distribution. Considering the multiclass nature of the dataset, the single peak actually implies the range of the interclass distance, and suggests that data points are widespread on the hypersphere.

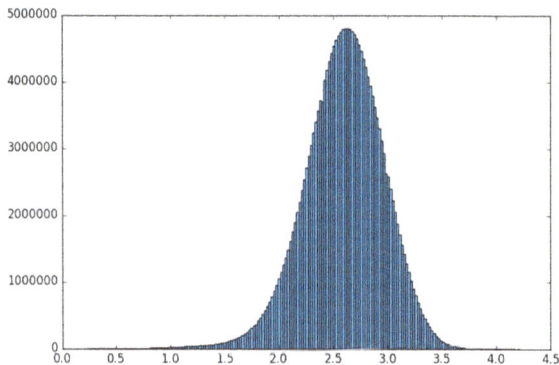

Figure 4. Distribution of sampled points' pairwise distances from the MNIST dataset.

Referencing Figure 4, we decide to build the tree with an initial distance threshold of 1.4 and a distance growth rate of 1.1. Such a small initial threshold means that the first level of the tree mainly serves for data downscaling. We can then construct a tree hierarchy with a total of seven levels by running our parallel algorithm. The distance thresholds used, and the number of nodes that built each level of the tree are both listed in Table 1.

As upper levels of the tree usually contain more meaningful knowledge, we cut the tree at level three to generate the final output. We then visualize the result with the extended circular dendrogram, as illustrated in Figure 5. Here, only significant nodes with more than five data points are rendered. In the dendrogram, the root node is located in the center, and nodes on levels farther from the root are located in outer circles. A node is named after its level and its ID on the level. The size and the color of a node signify the number of data points it contains, decided by the formula

$$s = \theta(n/l), \tag{1}$$

where n is the number of member points, l is the level of the node, and $\theta(\cdot)$ scales and regulates the range of the output. We use a color map from white to blue regarding the RGB value. This means that a larger node with a bluer color owns more member points than a node on the same level of the dendrogram, but with a smaller size and a lighter color. Nodes between two neighboring levels are connected with edges. The width of an edge indicates the amount of point merged from a child node to its parent node, where a wide edge from the parent node connects a child of large number of members. By such, dominant branches of the tree can be easily recognized.

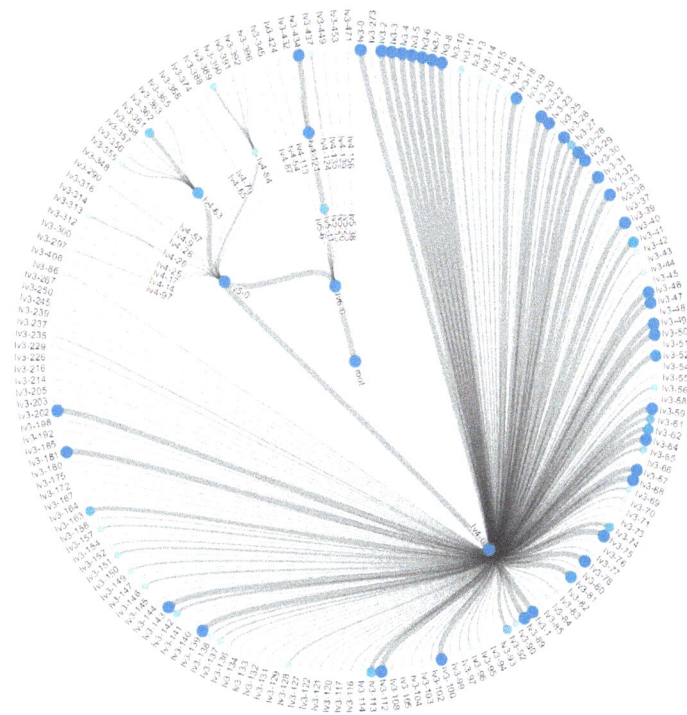

Figure 5. The extended circular dendrogram visualizing the hierarchy built from the MNIST dataset.

Table 1. Distance thresholds and the number of nodes for each tree level of the MNIST dataset.

Level	1	2	3	4	5	7	8 (*root*)
Distance threshold	1.4	1.54	1.70	1.86	2.05	2.25	2.48
Number of nodes	20,610	2928	521	205	83	19	1

There are some interesting structures we can observe in Figure 5. First, node *lv4-0* is a very big node, which absorbs most level-three nodes, including a few large ones. This could mean that many clusters generated on level three have already been very similar to each other. However, there are other nodes, e.g., *lv3-434*, going up to the root via other branches. To have a deeper investigation, we look at the images included in the different nodes. Then, we find that even though *lv3-434* and *lv3-3* are all nodes containing images of digit 0, the images in *lv3-3* (Figure 6b) are much more skewed than those in *lv3-434* (Figure 6a), so that they are nearer to images of other nodes merged into *lv4-0*, e.g., node *lv3-7* (Figure 6c), regarding the Euclidean distance.

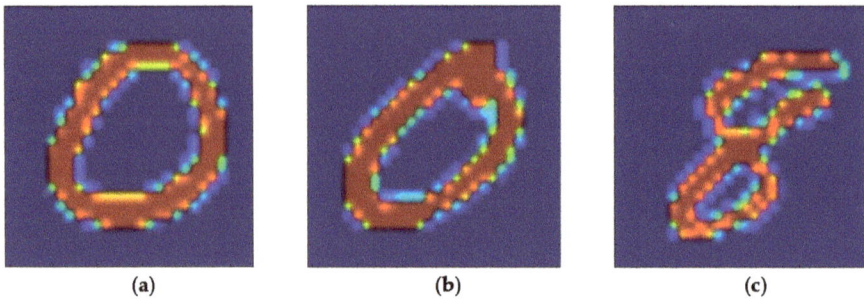

(a)	(b)	(c)

Figure 6. Images contained in different level three nodes. (a) An image of digit 0 from node *lv3-434*. (b) An image of digit 0 from node *lv3-3*. (c) An image of digit 8 from *lv3-7*. Considering the Euclidean distance, (b) and (c) could be closer than (a) and (b).

5.2. The Aerosol Dataset

The Aerosol dataset was collected for atmospheric chemistry to understand the processes that control the atmospheric aerosol life cycle. The dataset was acquired by a state-of-the-art single particle mass spectrometer, termed SPLAT II (see [35] for more detail). Each data point in the dataset is a 450-dimensional mass spectra of an individual aerosol particle. The dataset we use contains about 2.7 million data points. The standard deviation of each dimension is shown in Figure 7, where we can observe that a few dimensions are apparently more variant than others. By setting a threshold of 0.01 of the max standard deviation, 36 dimensions are selected and marked with red bars in Figure 7, while the remaining bars are colored blue (most of which are too small to be observed clearly). We sampled 20,000 points to compute the distance histogram. Here, we demonstrate two histograms respectively calculated with all 450 dimensions (Figure 8a), and only the 36 dimensions (Figure 8b). The two histograms are almost identical, implying that the reduction of dimension will not affect the result.

The histograms in Figure 8 clearly shows several peaks, where the leftmost peak indicates the intraclass distance of most clusters, and the rest imply interclass distances. Hence, we set the initial distance threshold $t = 5000$, which is the leftmost peak's distance value, so that these small dense clusters can be recognized. Since the volume of the dataset is quite large, we set the growth rate of the distance threshold to be 2.0, i.e., the distance threshold will be doubled whenever building the next level of the hierarchy. In this way, the distance threshold can quickly grow such that there are fewer nodes with a large branching factor in each level, and hence, the result tree structure may have a controllable height.

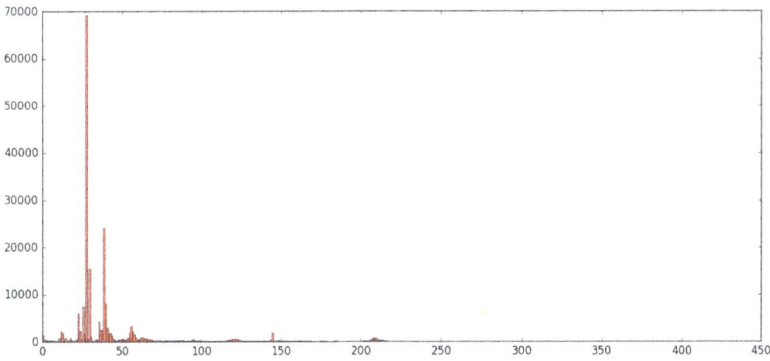

Figure 7. Standard deviations of each dimension of the Aerosol dataset. By applying a threshold of 0.01 of the max standard deviation, only 36 dimensions are selected for use.

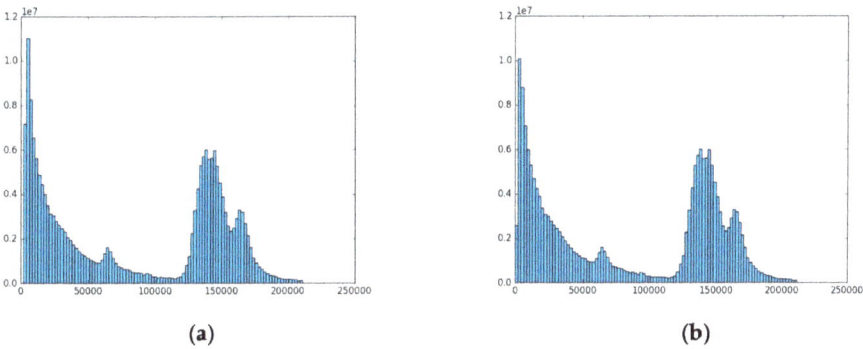

Figure 8. The distribution of point distances in the Aerosol dataset. (**a**) The histogram calculated with all 450 dimensions. (**b**) The histogram calculated with the selected 36 dimensions. The two are almost identical, implying that dimension reduction will not affect the result.

The distance thresholds for constructing each level of the tree, as well as the resultant number of nodes in each level, are given in Table 2. Although the dataset is large, the resulted hierarchy has only seven levels in total. Due to the fast growth of the distance threshold, the number of nodes shrinks rapidly between levels. Given the background knowledge that there are actually about 20 types of particles in the dataset, we cut the tree at level four. This also makes the visualization more manageable. We visualize the final result again with the extended circular dendrogram in Figure 9.

Table 2. Distance thresholds and the resulted number of nodes for each level of the Aerosol dataset.

Level	1	2	3	4	5	6	7 (*root*)
Distance threshold	5 k	10 k	20 k	40 k	80 k	160 k	320 k
Number of nodes	196,591	4549	709	176	45	9	1

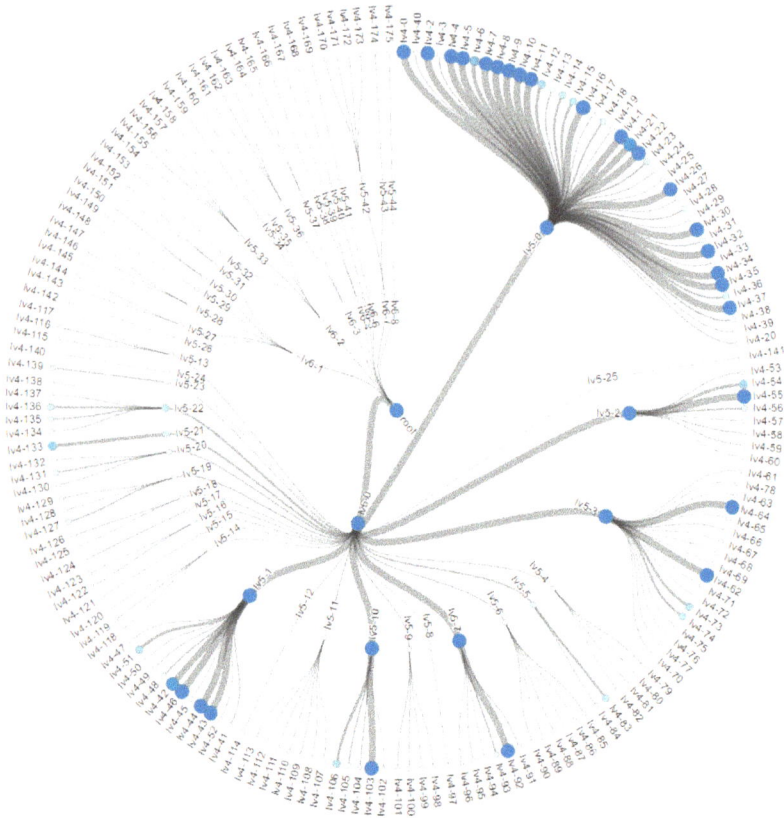

Figure 9. The extended circular dendrogram visualizing the hierarchy constructed from the Aerosol dataset.

From Figure 9, we can see that while node *lv6-0* is the dominating child for the root, it has a few large sub-branches (referencing the thick edge to *lv6-0* from *lv5-0*, *lv5-1*, *lv5-10*, *lv5-3*, etc.). Also by observing that all other nodes on level six are comparably very small (with small node sizes and thin edges), some good clustering of different particles may have been recognized on level five. However, it seems that some level-five nodes (e.g., *lv5-0* and *lv5-1*) are still too inclusive and variant, as they own some thick branches and many of their children are also big nodes. This implies that some clusters are mixed due to their closely located centers.

Figure 10 demonstrates the t-SNE layout [33] of all level-four nodes. The point sizes and colors (from white to blue) in the scatter plot are scaled regarding the nodes' number of members. Some of the largest nodes are labeled. In Figure 10, we see that children of the node *lv5-0* are all closely located, while other nodes (e.g., *lv4-43*, child of *lv5-1*, and *lv4-55*, child of *lv5-2*) are farther away. This confirms the correctness of the tree structure. Among children of *lv5-0*, there are seemingly two sub-clusters: one around *lv4-0*, and one around *lv4-10*. This could imply different types of particles.

A further analysis may reveal more knowledge of the dataset, but doing so would go beyond the focus and scope of this paper. However, the presented examples have sufficiently demonstrated the validity of our algorithm, as well as the effectiveness of the visualization.

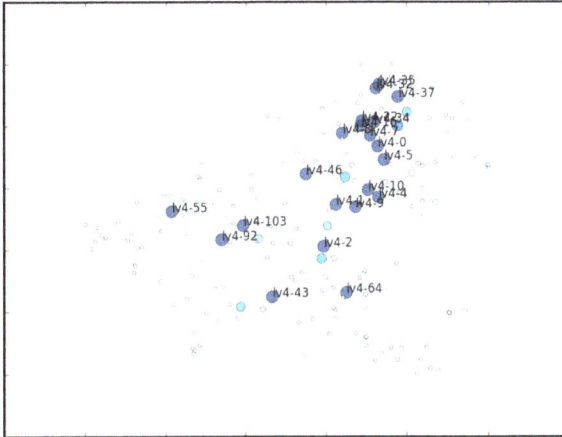

Figure 10. The t-SNE layout of nodes on level four.

6. Discussion and Conclusions

We have presented the algorithm of incremental k-means tree for big data management, as well as its GPU-accelerated construction and visualization. Its effectiveness has been demonstrated by experimenting on two real-world datasets. Nevertheless, there are a few issues worth discussion.

First, we have not conducted experiments testing the exact timings of our algorithm under different scales of data volume. Although this could be a potential future work, the actual time cost is also related to the settings of the initial distance threshold and its growth. For constructing each level, a large distance threshold not only means a smaller number of clusters to be learned, but also more points will be clustered in one scan of data, and hence, fewer iterations and shorter running time. In addition, the number of nodes for building the next level of the hierarchy will be better downscaled, making the rest of the hierarchy construction even faster. As a reference, for the two datasets and their specific settings used in this paper, building the hierarchy from the MNIST dataset cost about four and half minutes, and from the Aerosol dataset about 30 min.

Although the resultant hierarchy is decided by the algorithm parameters and the distance metric employed, the stability of the result under different settings depends on the dataset. We have seen in the MNIST dataset example that the distance threshold changes between levels are quite small, yet the number of nodes is fast narrowing as levels grow. This means that this dataset could be very sensitive to parameter settings, so that a small variation could lead to big changes in the result structure. In contrast, we also tested on the Aerosol dataset, with some initial distance thresholds a few hundred off from 5000. However, the results are basically the same as shown in Section 5.2, with only small differences on the number of nodes in each level.

Also, our algorithm can be applied to datasets from a wide range of areas, and the result hierarchy can be used in any further analysis requires such information. For example, we can conduct a fast nearest search or hierarchical sampling on the MNIST dataset with the constructed structure, and clustering on the Aerosol dataset. Acquiring such structural information is often a general and important step in big data pre-processing.

Finally, the extended circular dendrogram presented in this paper could be further developed. A richer scope of information, such as the quality metric of each node for instance, could be visualized to make the graph more informative. Interactions can also be added so that users can visually explore and modify the hierarchy to look for new knowledge from data. Based on these developments, a comprehensive visual analytic interface could be well established. This is the subject of future work.

Acknowledgments: This research was partially supported by NSF grant IIS 1527200 and the Ministry of Science, ICT and Future Planning, Korea, under the "IT Consilience Creative Program (ITCCP)" supervised by NIPA. Partial support (for Alla Zelenyuk and Jun Wang) was also provided by the US Department of Energy (DOE) Office of Science, Office of Basic Energy Sciences, Division of Chemical Sciences, Geosciences, and Biosciences. Some of this research was performed in the Environmental Molecular Sciences Laboratory, a national scientific user facility sponsored by the DOE's OBER at Pacific Northwest National Laboratory (PNNL). PNNL is operated by the US DOE by Battelle Memorial Institute under contract No. DE-AC06-76RL0.

Author Contributions: K.M. and J.W. conceived and developed the algorithms. J.W. implemented the system and performed the case studies. Partial data was provided by A.Z. and D.I. The overall work has been done under the supervision of K.M. and A.Z. All authors were involved in writing the paper.

Conflicts of Interest: The authors declare no conflict of interest.

References

1. Big data. *Nature* **2008**, *455*, 1–136.
2. Dealing with data. *Science* **2011**, *331*, 639–806.
3. Muja, M.; Lowe, D.G. Scalable nearest neighbor algorithms for high dimensional data. *IEEE Trans. Pattern Anal. Mach. Intell.* **2014**, *36*, 2227–2240. [CrossRef] [PubMed]
4. Dasgupta, S.; Hsu, D. Hierarchical sampling for active learning. In Proceedings of the 25th International Conference on Machine Learning, Helsinki, Finland, 5–9 July 2008; pp. 208–215.
5. Fahad, A.; Alshatri, N.; Tari, Z.; Alamri, A.; Khalil, I.; Zomaya, A.Y.; Foufou, S.; Bouras, A. A survey of clustering algorithms for big data: Taxonomy and empirical analysis. *IEEE Trans. Emerg. Top. Comput.* **2014**, *2*, 267–279. [CrossRef]
6. Wong, P.C.; Shen, H.W.; Johnson, C.R.; Chen, C.; Ross, R.B. The top 10 challenges in extreme-scale visual analytics. *IEEE Comput. Graph. Appl.* **2012**, *32*, 63–67. [CrossRef] [PubMed]
7. MacQueen, J. Some methods for classification and analysis of multivariate observations. In Proceedings of the Fifth Berkeley Symposium on Mathematical Statistics and Probability, Berkeley, CA, USA, 21 June–18 July 1965 and 27 December 1965–7 January 1966; Lucien, C., Jerzy, N., Eds.; University of California Press: Berkeley, CA, USA, 1967; Volume 1, pp. 281–297.
8. Farivar, R.; Rebolledo, D.; Chan, E. A parallel implementation of k-means clustering on GPUs. In Proceedings of the International Conference on Parallel and Distributed Processing Techniques and Applications, PDPTA 2008, Las Vegas, NV, USA, 14–17 July 2008; Volume 13, pp. 340–345.
9. Bezdek, J. C.; Ehrlich, R.; Full, W. FCM: The fuzzy c-means clustering algorithm. *Comput. Geosci.* **1984**, *10*, 191–203. [CrossRef]
10. Zhang, T.; Ramakrishnan, R.; Livny, M. *BIRCH: An Efficient Data Clustering Method for Very Large Databases*; ACM: New York, NY, USA, 1996; Volume 25, pp. 103–114.
11. Guha, S.; Rastogi, R.; Shim, K. CURE: An efficient clustering algorithm for large databases. *Inf. Syst.* **2001**, *26*, 35–58. [CrossRef]
12. Karypis, G.; Han, E.H.; Kumar, V. Chameleon: hierarchical clustering using dynamic modeling. *Computer* **1999**, *32*, 68–75. [CrossRef]
13. Ester, M.; Kriegel, H.P.; Sander, J.; Xu, X. A Density-Based Algorithm for Discovering Clusters in Large Spatial Databases with Noise. In Proceedings of the International Conference on Knowledge Discovery and Data Mining, Portland, OR, USA, 2–4 August 1996; pp. 226–231.
14. Ankerst, M.; Breunig, M.M.; Kriegel, H.; Sander, J. *OPTICS: Ordering Points to Identify the Clustering Structure*; ACM: New York, NY, USA, 1999; Volume 28, pp. 49–60.
15. Agrawal, R.; Gehrke, J.; Gunopulos, D.; Raghavan, P. *Automatic Subspace Clustering of High Dimensional Data for Data Mining Applications*; ACM: New York, NY, USA, 1998; Volume 27, pp. 94–105.
16. Wang, W.; Yang, J.; Muntz, R. STING: A statistical information grid approach to spatial data mining. *Int. Conf. Very Large Data* **1997**, *97*, 1–18.
17. Fraley, C.; Raftery, A.E. MCLUST: Software for Model-Based Cluster Analysis. *J. Classif.* **1999**, *16*, 297–306. [CrossRef]
18. Dempster, A.P.; Laird, N.M.; Rubin, D.B. Maximum Likelihood from Incomplete Data via the EM Algorithm. *J. R. Stat. Soc. Ser. B* **1977**, *39*, 1–38.

19. Bentley, J.L. Multidimensional binary search trees used for associative searching. *Commun. ACM* **1975**, *18*, 509–517. [CrossRef]
20. Freidman, J.H.; Bentley, J.L.; Finkel, R.A. An Algorithm for Finding Best Matches in Logarithmic Expected Time. *ACM Trans. Math. Softw.* **1977**, *3*, 209–226. [CrossRef]
21. Silpa-Anan, C.; Hartley, R. Optimised KD-trees for fast image descriptor matching. In Proceedings of the IEEE Conference on Computer Vision and Pattern Recognition, Anchorage, AK, USA, 23–28 June 2008; pp. 1–8.
22. Lamrous, S.; Taileb, M. Divisive Hierarchical K-Means. In Proceedings of the International Conference on Computational Inteligence for Modelling Control and Automation and International Conference on Intelligent Agents Web Technologies and International Commerce, Sydney, Australia, 28 November–1 December 2006.
23. Jose Antonio, M.H.; Montero, J.; Yanez, J.; Gomez, D. A divisive hierarchical k-means based algorithm for image segmentation. In Proceedings of the IEEE International Conference on Intelligent Systems and Knowledge Engineering, Hangzhou, China, 15–16 November 2010; pp. 300–304.
24. Nister, D.; Stewenius, H. Scalable Recognition with a Vocabulary Tree. In Proceedings of the IEEE Computer Society Conference on Computer Vision and Pattern Recognition, New York, NY, USA, 17–22 June 2006; Volume 2, pp. 2161–2168.
25. Imrich, P.; Mueller, K.; Mugno, R.; Imre, D.; Zelenyuk, A.; Zhu, W. Interactive Poster: Visual data mining with the interactive dendrogram. In Proceedings of the IEEE Information Visualization Symposium, Boston, MA, USA, 28–29 October 2002.
26. Satish, N.; Harris, M.; Garland, M. Designing efficient sorting algorithms for manycore gpus. In Proceedings of the IEEE International Parallel and Distributed Processing Symposium, Rome, Italy, 23–29 May 2009.
27. Papenhausen, E.; Wang, B.; Ha, S.; Zelenyuk, A.; Imre, D.; Mueller, K. GPU-accelerated incremental correlation clustering of large data with visual feedback. In Proceedings of the IEEE International Conference on Big Data, Silicon Valley, CA, USA, 6–9 October 2013; pp. 63–70.
28. Herrero-lopez, S.; Williams, J.R.; Sanchez, A. Parallel Multiclass Classification Using SVMs on GPUs. In Proceedings of the 3rd Workshop on General-Purpose Computation on Graphics Processing Units, Pittsburgh, PA, USA, 14 March 2010; ACM: New York, NY, USA, 2010; p. 2.
29. Liang, S.; Liu, Y.; Wang, C.; Jian, L. A CUDA-based parallel implementation of K-nearest neighbor algorithm. In Proceedings of the International Conference on Cyber-Enabled Distributed Computing and Knowledge Discovery, Zhangjiajie, China, 10–11 October 2009; pp. 291–296.
30. Jia, Y.; Shelhamer, E.; Donahue, J.; Karayev, S.; Long, J.; Girshick, R.; Guadarrama, S.; Darrell, T. Caffe: Convolutional architecture for fast feature embedding. In Proceedings of the 22nd ACM International Conference on Multimedia, Orlando, FL, USA, 3–7 November 2014; ACM: New York, NY, USA, 2014; pp. 675–678.
31. Kreuseler, M.; Schumann, H. A flexible approach for visual data mining. *IEEE Trans. Vis. Comput. Graph.* **2002**, *8*, 39–51. [CrossRef]
32. Beham, M.; Herzner, W.; Groller, M.E.; Kehrer, J. Cupid: Cluster-Based Exploration of Geometry Generators with Parallel Coordinates and Radial Trees. *IEEE Trans. Vis. Comput. Graph.* **2014**, *20*, 1693–1702. [CrossRef] [PubMed]
33. Van Der Maaten, L.J.P.; Hinton, G.E. Visualizing high-dimensional data using t-sne. *J. Mach. Learn. Res.* **2008**, *9*, 2579–2605.
34. Harris, M. Optimizing Parallel Reduction in CUDA. In *NVIDIA CUDA SDK 2. 2008*; NVIDIA Developer: Santa Clara, CA, USA, August 2008.
35. Zelenyuk, A.; Yang, J.; Choi, E.; Imre, D. SPLAT II: An Aircraft Compatible, Ultra-Sensitive, High Precision Instrument for In-Situ Characterization of the Size and Composition of Fine and Ultrafine Particles. *Aerosol Sci. Technol.* **2009**, *43*, 411–424. [CrossRef]

informatics

MDPI

Article

PERSEUS-HUB: Interactive and Collective Exploration of Large-Scale Graphs

Di Jin *, Aristotelis Leventidis †, Haoming Shen †, Ruowang Zhang †, Junyue Wu and Danai Koutra

Department of Computer Science and Engineering, University of Michigan, Ann Arbor, MI 48109, USA; leventid@umich.edu (A.L.); hmshen@umich.edu (H.S.); jackierw@umich.edu (R.Z.); junyuew@umich.edu (J.W.); dkoutra@umich.edu (D.K.)
* Correspondence: dijin@umich.edu
† These authors contributed equally to this work.

Academic Editors: Achim Ebert and Gunther H. Weber
Received: 19 June 2017; Accepted: 12 July 2017; Published: 18 July 2017

Abstract: Graphs emerge naturally in many domains, such as social science, neuroscience, transportation engineering, and more. In many cases, such graphs have millions or billions of nodes and edges, and their sizes increase daily at a fast pace. How can researchers from various domains explore large graphs interactively and efficiently to find out what is 'important'? How can multiple researchers explore a new graph dataset collectively and "help" each other with their findings? In this article, we present PERSEUS-HUB, a large-scale graph mining tool that computes a set of graph properties in a distributed manner, performs ensemble, multi-view anomaly detection to highlight regions that are worth investigating, and provides users with uncluttered visualization and easy interaction with complex graph statistics. PERSEUS-HUB uses a Spark cluster to calculate various statistics of large-scale graphs efficiently, and aggregates the results in a summary on the master node to support interactive user exploration. In PERSEUS-HUB, the visualized distributions of graph statistics provide preliminary analysis to understand a graph. To perform a deeper analysis, users with little prior knowledge can leverage patterns (e.g., spikes in the power-law degree distribution) marked by other users or experts. Moreover, PERSEUS-HUB guides users to regions of interest by highlighting anomalous nodes and helps users establish a more comprehensive understanding about the graph at hand. We demonstrate our system through the case study on real, large-scale networks.

Keywords: large-scale graphs; visualization; visual analytics; distributions; interaction; distributed computations; anomaly detection; collective analysis

---

1. Introduction

The concept of big data has been deeply rooted among people in both academia and industry. Researchers and analysts from various domains require fast and efficient techniques to mine useful information from large-scale datasets. Among the different types of data, an important category is graphs or networks as they naturally represent entities in different domains and the relationships between them, such as compounds in chemistry [1,2], interactions between proteins in biology [3], symptom relations in healthcare [4], communication between people and behavioral patterns [5,6]. Accordingly, the problem of large-scale graph mining has attracted tremendous interest and efforts and, as a result, many large-scale mining techniques, tools, and systems have been developed.

Interactive visualization (or visual analytics) is commonly used to obtain an overview of the data at hand and to perform preliminary, exploratory analysis. Although such "rough" analysis is not always accurate and could be biased, it is a general technique to handle data from different domains before forming concrete questions about the data and seeking specific patterns in them.

Unlike most distributed algorithms based on abstraction or sophisticated workflows to achieve efficiency, visualization can put the spotlight on the computed patterns to help the user understand both the data and the algorithm. This is particularly important in domains where a researcher or analyst may not have programming expertise. Moreover, beyond static visualization (e.g., plot of a graph statistics distribution which is often used to discover patterns and laws in the data), interaction provides users with the option of exploring patterns of interest with rich side information and discovering relations within the data.

There are many existing tools and systems that support the interactive exploration of graph data [7,8], yet most of them either assume the user has some prior knowledge and knows what she is looking for, or guide the user to patterns that are worth exploring based on predefined interestingness measures (e.g., low degree). Although such techniques could help users discover new knowledge in specific graphs, they cannot be readily used in different domains. For example, low-degree nodes are common in Twitter and other social networks, as most users have a limited number of followers/friends and few interactions. However, if a user applies the techniques designed for analyzing social networks to structural brain graphs, important information could be overlooked. This is because in brain graphs, low-degree nodes could represent neurons that are less active, and could be an indicator of mental diseases (e.g., Schizophrenia). A different system per type of input would avoid this problem, but, at the same time, it would be impractical and time-consuming. Therefore, a graph mining platform that allows users across different domains to collectively analyze datasets, guide the process of exploration, and share the outcomes with others, could help alleviate the above-mentioned shortcomings.

Motivated by this observation, we present PERSEUS-HUB, a general large-scale graph mining framework that computes a set of commonly explored graph properties in a distributed manner, performs ensemble, multi-view anomaly detection to highlight regions that are worth investigating, provides users with uncluttered interactive visualization, and supports collective analysis of data from multiple users. It builds and improves upon a simple, Hadoop-based graph mining framework, PERSEUS [9,10], in terms of speed, depth of analysis, system-guided anomaly detection, interactive features, and collective analysis. Our proposed framework provides tools that support both preliminary and in-depth analysis so that users from different domains can gain insights into their graph data.

Our system consists of three main components:

- **Two-Level Summarization Scheme**: To summarize the input graph efficiently, PERSEUS-HUB interactively visualizes both aggregate and node-specific patterns. It summarizes the graph data in terms of univariate and bivariate distributions—such as the clustering coefficient distribution and the degree vs. PageRank distribution—to reveal different aspects of the input graph. The compliance to or deviation from common laws in the distributions (e.g., power-law pattern in social networks) can guide attention to specific nodes or graph regions. At the aggregate level, points that are close to each other in the plots are aggregated into 'super-points' to efficiently render, visualize, and support interaction with the distribution plots for graphs with even millions of nodes. At the node-specific level, detailed information can be obtained for fine-granularity analysis such as queries for per node statistics.

- **Collective Data Analysis**: To make graph analysis effective and comprehensive for non-experts, PERSEUS-HUB supports publicly sharing processed data and discovered patterns to help other analysts with their exploration. The PERSEUS-HUB service runs on a public AWS (Amazon Web Service) EMR (Elastic MapReduce) Spark cluster and displays datasets and discovered patterns, enabling collective analysis and 'shared' knowledge. Especially non-experts can benefit from interacting with the discoveries of expert users, and this has potential to advance the knowledge in interdisciplinary domains.

- **Ensemble Anomaly Detection**: In addition to collective data analysis that can guide user attention to regions that are identified as interesting by other analysts or experts, PERSEUS-HUB runs multiple scalable anomaly detection algorithms on the summarized statistics distributions to identify outliers of various types in an ensemble manner. For this purpose, the system leverages algorithms such as G-FADD (Grid Fast Anomaly Detection algorithm given Duplicates [11]), a fast, density- based anomaly detection method to find local and global outliers in the distribution plots, as well as ABOD (Angle Based Outlier Detection [12]), an angle-based outlier detection method that is particularly suitable for analyzing highly dimensional data.

The rest of this paper is organized as follows: Section 2 describes the related work. Sections 3 and 4 give the description of PERSEUS-HUB and its components, and an overview of the web application integration, respectively. We showcase the application of PERSEUS-HUB to interactively analyze large graph datasets in Section 5, and conclude with future directions in Section 6.

2. Related Work

Our work is related to graph visualization and interactive analytics, as well as anomaly detection.

2.1. Graph Visualization and Interactive Analytics

There exists a large variety of visualization systems, each of which tries to visualize different types of networks targeting various kinds of users. Refs. [7,13,14] take an exploratory search approach based on the assumption that users are non-experts and unfamiliar with their datasets. Therefore, they guide users to perform analysis by incrementally querying information or browsing through the various relationships in the datasets to discover any unexpected results. FACETS [13] allows users to adaptively explore large graphs by only showing their most interesting parts. The interestingness of a node is based on how surprising its neighbor's data distributions are, as well as how well it matches what the user has chosen to explore. In this way, FACETS builds a user profile that customizes which nodes in a graph are explored. Similarly, Voyager 2 [14] is another graph visualization tool that combines manual and automatic chart specification in one system. More specifically, Voyager 2 allows users to manually vary or specify graph properties through the use of wildcards, which generate multiple charts in parallel for visualization, while the system also automatically recommends relevant visualization based on the user's current focus.

May et al. [15] proposed a system that helps user navigation using signposts to indicate the direction of off-screen regions of the graph that the user would be interested to navigate to. The signposts are generated dynamically depending on the current visible nodes. This visualization system allows the representation of interesting regions of a graph with a small visual footprint. An interesting hybrid visualization system of networks is NodeTrix [16], which uses node-link diagrams to show the global structure of a network and uses adjacency matrices to provide a more detailed view of specific communities in the network. NodeTrix provides users with a large set of interaction techniques, which allows easy modification of network organization. PivotSlice [17] is an interactive visualization tool for faceted data exploration which allows the user to manage a collection of dynamic queries in a multi-focus and multi-scale tabular view. In this way, the user can explore a large dataset by efficiently subdividing it into several meaningful subsets with customized semantics. Refinery [18] is an interactive visualization system that supports bottom-up exploration of heterogeneous networks through associative browsing which allows the user to step iteratively from familiar to novel bits of information. By computing degree-of-interest scores for associated content using a random-walk algorithm, Refinery allows users to visualize sub-graphs of items relevant to their queries. Similarly, GLO-STIX [7] is a graph visualization system that allows a user to interactively explore a graph by applying Graph-Level Operations (GLOs). GLOs are encapsulated manipulations of a graph visualization that allow a user to incrementally modify a graph without committing to a specific graph type, giving the user the choice of fully customizing their visualization through a set of primitive graph-level operations.

In addition to providing the user with some exploratory approaches to interact with their network data, there are some visualization tools designed to help users find anomalies or malicious nodes. Apolo [8] is a graph tool used for attention routing. The user picks a few seed nodes, and Apolo interactively expands their vicinities, enabling sense-making. An anomaly detection system for large graphs, OPAvion [19], mines graph features using Pegasus [20] and spots anomalies by employing OddBall [21] and, lastly, interactively visualizes the anomalous nodes via Apolo. Perseus [9] is a Hadoop-based graph mining framework, which inspired the design of PERSEUS-HUB. Improving on that, PERSEUS-HUB supports faster analytics of a variety of graph statistics on Spark; fast and ensemble, multi-view anomaly detection; collective analysis of data and findings between users enabled by its deployment on AWS; and a wide range of interaction features in the frontend that make the interface easy-to-use. Other works on visualization tackle the issue of how to aggregate large-scale graphs while retaining important information. NetRay [22] focuses on informative visualizations of the spy plots, distribution plots, and correlation plots of web-scale graphs. Dunne and Shneiderman [23] introduce the idea of motif simplification in which common patterns of nodes and links are replaced with compact and meaningful glyphs that can portray relationships within a network more clearly while requiring less screen space to represent the network.

According to Nielsen [24], website response time of 0.1 s gives the feeling of instantaneous response; one second keeps the user's flow of thought seamless, although it is a noticeable delay; and 10 s keeps the user's attention, but faster response time is definitely desired. To support interaction and achieve very short response time, most approaches apply aggregation techniques, clustering, sketches, or other offline pre-processing algorithms which result in a significantly smaller view of the input data that is fast to query. Recent works in the database community that turned to interactive systems include query refinement with feedback [25]; query exploration for WHERE queries [26]; query steering in DBMS (Database Management System) [27]; query recommendations for interactive database exploration [28]; interactive exploration of interesting itemsets and association rules in transaction (supermarket) DB [29]; and visual analytics with data-driven recommendations that support simple aggregation-based queries of relational data [30,31]. In the area of graphs, most methods focus on interactively exploring the graph structure by starting from a set of user-selected nodes [32], and also the idea of community-within-community visualization via multi-resolution [33].

Unlike the above-mentioned systems, PERSEUS-HUB focuses on efficiently summarizing large graph data in terms of several distributions (or correlations) of properties instead of visualizing only the network structure or the vicinities of specific 'interesting' nodes. Moreover, unlike its precursor [9], it enables collective analysis of datasets from multiple users or experts. Table 1 gives a qualitative comparison of PERSEUS-HUB with existing graph visualization tools.

Table 1. Qualitative comparison of PERSEUS-HUB with alternative approaches. (i) **Node analytics:** Does the visualization framework provide node-specific statistics or analytics? (ii) **Mixed Expertise:** Can the visualization tool be used by users with varying levels of expertise (e.g., experts, beginners or users with limited familiarity with the graph data)? (iii) **Automatic Pattern Detection:** Does the visualization tool automatically detect specific patterns and guide users to them? (iv) **Collective Analysis:** Does the visualization tool support information sharing among users? (v) **Scalability:** Can users analyze large-scale, real-world graphs with the tool? (vi) **Domain-independence:** Does the tool support graphs from multiple domains (e.g., social networks, neural networks, etc.)?

	Node Analytics	Mixed Expertise	Auto Detection	Collective An.	Scalability	Domain-Ind.
GLO-STIX [7]	✗	✓	✓	✗	?	✓
FACETS [13]	✓	✓	✓	✗	✓	✓
Voyager 2 [14]	✓	✓	✓	✗	✗	✓
Signposts [15]	✓	✗	✓	✗	✗	✗
NodeTrix [16]	✓	✗	✗	✗	✗	✗
PivotSlice [17]	✓	✗	✗	✗	✓	✓
Refinery [18]	✓	✗	✓	✗	✓	✓
Apolo [8]	✓	✓	✗	✗	✗	✓
OPAvion [19]	✓	✓	✓	✗	✓	✓
NetRay [22]	✗	✗	✓	✗	✓	✓
Motif-Simp [23]	✓	✓	✗	✗	✓	✓
PERSEUS [9,10]	✓	✗	✓	✗	✓	✓
PERSEUS-HUB	✓	✗	✓	✓	✓	✓

2.2. Anomaly Detection

Anomaly detection in static graphs has been studied using various data mining and statistical techniques [11,19,22,34]. Two surveys [35,36] describe various outlier detection methods for static and time-evolving graphs. Here we describe just a small set of approaches. Local outlier factor (LOF) [37] is a popular density-based anomaly detection method that compares the local density of each data point to the density of its neighbors by employing the k-nearest neighbor (kNN) technique to measure how isolated an object is with respect to its neighborhood. Grid-based FADD (G-FADD) [38] improves on the quadratic complexity of LOF by cleverly handling duplicate points and applying a grid on a multi-dimensional space so that only cells that satisfy density-specific criteria need investigation for outliers. AutoPart [39] is a graph clustering algorithm that detects anomalous edges based on the Minimum Description Language (MDL) encoding principle. In AutoPart, an edge whose removal significantly reduces the total encoding cost is considered to be an outlier. Thus, AutoPart finds the block of nodes where the removal of an edge leads to the maximum immediate reduction in cost. Since all edges within that block contribute equally to the cost, all of them are considered outliers. SCAN [40] detects clusters and outliers (vertices that are marginally connected to clusters) in networks by using the structure and connectivity as clustering criteria. SCAN finds the core nodes which share structural similarity with many neighbors and cluster them with their neighbors. Nodes that connect to one or zero clusters are determined as the outliers.

PERSEUS-HUB runs anomaly detection algorithms with and without ensembles in an offline manner and displays the generated results in an aggregate summarization panel. PERSEUS-HUB also provides detailed information about the anomalous nodes through node-specific interactive visualization.

3. PERSEUS-HUB: System Description

PERSEUS-HUB is an interactive, collective graph exploration system deployed in the public AWS EMR-Spark cluster that supports large-scale data analysis. Users can interactively explore their graph data in their own private space, and, if they choose, they can also publish (some of) their data and discoveries so that other experts or users can use them as a starting point and further investigate the data without duplicating the work.

The system overview of PERSEUS-HUB and the workflow of collective analysis is illustrated in Figure 1, while an instance of the 'user view' is given in Figure 2. In a nutshell, PERSEUS-HUB consists of three modules: (1) Distributed Statistics Computation, which computes various graph statistics for the input graph in a distributed manner (distribution and correlation plots of those statistics are shown in Figure 2); (2) Ensemble anomaly detection that discovers 'outlier' nodes in the graph by integrating multiple views (i.e., different statistic distributions) and multiple detection algorithms; and (3) Interactive visualization that provides both aggregated summarization and node-specific statistics to the users. In the following subsections, we describe each module of PERSEUS-HUB (Sections 3.1–3.3) and explain how we integrate them in a system in Section 4.

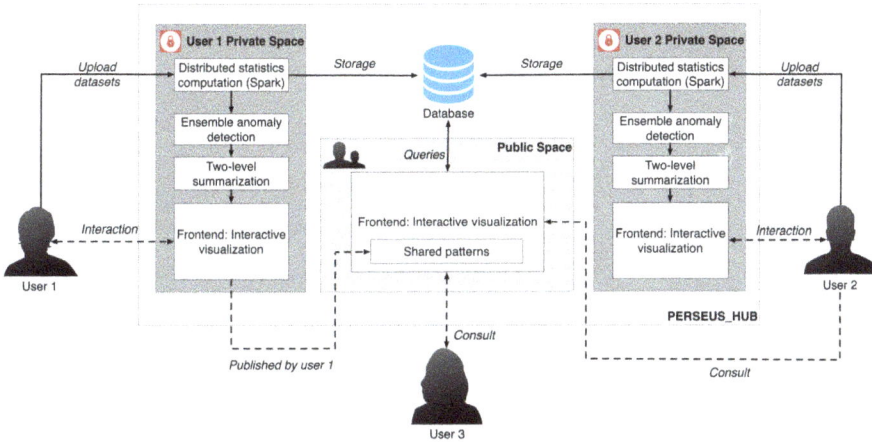

Figure 1. Overview of PERSEUS-HUB. The solid arrows indicate the system behaviors and the dashed arrows indicate the user behaviors. In this example, User 1 uploads the dataset to analyze to the private working space, and publishes it with marked patterns that he thinks are worth exploration. Other users who are interested in the published graph (User 2 and User 3) may refer to the analysis by User 1 in the public working space, or perform analysis in their private working space (e.g., User 2).

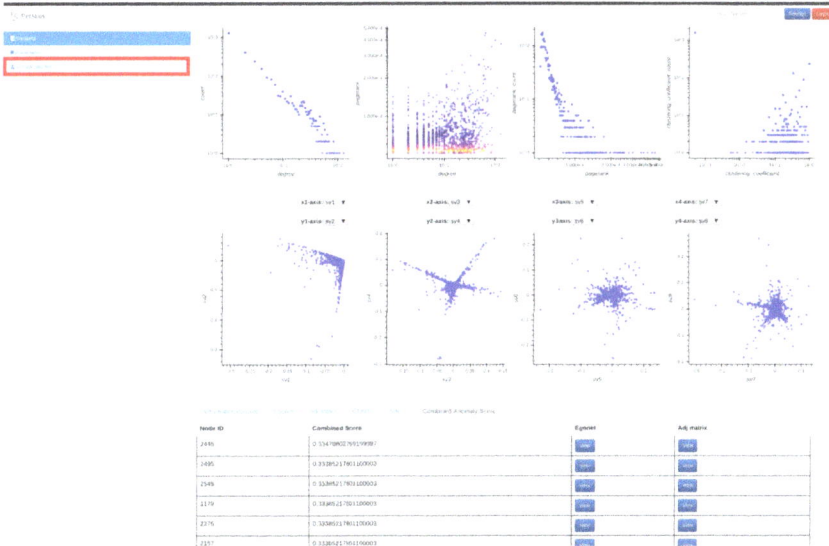

Figure 2. PERSEUS-HUB 'user' view: Node statistic distribution and correlation plots summarize the key structural properties of the input graph, and provide the user with an initial point for exploration and in-depth analysis. The red box annotation indicates the tab for uploading input graph files.

3.1. Module 1: Efficient Computation of Graph Statistics

Univariate and bivariate distributions of node statistics are crucial for graph analysis, and are traditionally used for anomaly or spam detection in networks [35,41]. For example, the degree and PageRank distributions are expected to follow a power law [42], while singular vectors exhibit the "EigenSpokes" pattern [43] that can be used to explore the community structures.

Building on these observations, PERSEUS-HUB computes both weighted and unweighted graph statistics in a distributed manner and visualizes their distributions as an overview for further investigation of the characteristics of an input graph. Specifically, the distributions computed include: degree distribution, PageRank distribution, degree vs. PageRank distribution, top-10 singular vectors, and clustering coefficient distribution. PERSEUS-HUB assumes the format of input file to be <source, destination> for the unweighted graphs and <source, destination, weight> for the weighted graphs. Users can easily upload the input file to PERSEUS-HUB by clicking the tab shown in the red box of Figure 2. In this subsection, we explain how these node properties are computed by using the Spark framework.

3.1.1. Degree Centralities and Degree Distribution

For a node in a graph $\mathcal{G}\,(\mathcal{V}, \mathcal{E})$, its degree centrality is defined as the number of its neighbors. The degree centrality of an arbitrary node $v \in \mathcal{V}$ in a *weighted undirected* graph is the sum of the weights of edges $e_{uv}, u \in \mathcal{V}, u \neq v$ incident upon this node. For an unweighted undirected graph, the weight of each edge is 1. For *weighted directed* graphs, the node degree centralities can be distinguished to in-degree, out-degree, and total degree. In-degree of a node v is defined as the total weights of edges e_{uv} starting from some other node u pointing to v. Similarly, the out-degree of a node v is defined as the total weights of edges e_{vu} starting from this node v pointing to some other node u. The total degree of a node is defined as the sum of in-degree and out-degree.

In Spark, degree centralities can be efficiently computed by 'grouping' the edges of the input graph data that are in the form of source-destination pairs, $< v_{src}, v_{dst} >$, as key-value pairs. The in-degree (out-degree) centrality of an arbitrary node in the graph can be computed by grouping edges with the same destination (source) and then summing up weights of each edge. The total degree centrality can be similarly computed after a union operation being performed on every edge and its reverse.

The degree distribution, defined as the probability distribution of degree over all the nodes, can be obtained by calculating its histogram (i.e., the number of nodes per degree centrality).

3.1.2. PageRank Centralities and PageRank Distribution

PageRank is a way of measuring the importance of website pages introduced by Page et al. [44]. Without loss of generality, we present the definition of PageRank for a weighted graph $\mathcal{G}\,(\mathcal{V}, \mathcal{E})$, with weighted adjacency matrix **W**. To incorporate the weights of each edge into the calculation of PageRank, the weighted transition probability is normalized so that the total probability of leaving node u is one. Thus, the weighted PageRank of an arbitrary node in a graph can be defined as:

$$pr_w(v) = \frac{1-d}{N} + d \sum_{u \in M(v)} \underbrace{\frac{w_{uv} \cdot pr_w(u)}{C_u}}_{wpr_{u \to v}:\text{contribution term}}, \tag{1}$$

where $u, v \in \mathcal{V}$ are nodes in graph; $pr_w(v)$ denotes the weighted PageRank for node v; N represents the total number of nodes in $\mathcal{G}$; d is the damping factor; $M(v) \subseteq \mathcal{V}$ are nodes pointing to v; w_{uv} denotes the weight of the edge from u to v; $C_u = \sum_k w_{uk}$ is the normalizing constant of node u. The contribution term, $wpr_{u \to v}$, denotes the weighted PageRank (wpr) that node u contributes to node v. The unweighted graphs can be seen as a special case with $w_{uv} = 1, \forall u, v \in \mathcal{V}; C_u = L(u), \forall u \in \mathcal{V}$, where $L(u)$ is the number of edges of node u. Therefore, Equation (1) simplifies to:

$$pr(v) = \frac{1-d}{N} + d \sum_{u \in M(v)} \frac{pr(u)}{L(u)}. \tag{2}$$

Algorithm 1 illustrates the Spark procedure for computing the weighted PageRank distribution. The input graph is an edge list organized into Resilient Distributed Dataset (RDD) structure in PYSPARK. To compute the PageRank of the nodes, we employ the power method, which relies on fast spare matrix-vector multiplication. To parallelize the computation of weighted PageRank in Spark, each iteration of the power method can be divided into two parts: (1) computation of the weighted PageRank contribution of node u to every other node $k, k \in \mathcal{V}$ in graph $\mathcal{G}$ through edge e_{uk}; (2) calculation of the total weighted PageRank of node v by summing up the weighted transition probability from every other node $u \in \mathcal{V}, u \neq v$. The correction term $\frac{1-d}{N}$ is added last. The first step can be mapped for every node in the graph; the second step can also be computed in parallel given the result of the first step. The PageRank distribution is attained by computing its histogram after grouping the PageRank value into a number of bins (the default setting is 1000).

Algorithm 1 Weighted PageRank (wpr).

Input: weighted edges RDD **D**: <srcID, (dstID, w)>, total number of iteration: iter_max, dampling factor: d
Output: Weighted PageRank RDD, **wpr**: <nodeID, wpr>

nodeID = {srcID} $\cup$ {dstID}   // Set of all node IDs
N = max(nodeID)   // Getting total number of nodes in the graph
wpr =<nodeID, $\frac{1}{N}$>   // Initializing wpr for each node as $\frac{1}{N}$
while iter $\leq$ iter_max **do**
    temp = group **D** by srcID (key),   // Joining edges with **wpr** of srcID
        join with **wpr** by srcID (key)   // temp: <srcID, ([<dstID$_1$w$_1$>,...,<dstID$_k$, w$_k$>], wpr_{srcID})>
    contribs = map temp to $wpr_{u \rightarrow v}$ in Equation (1)   // calculating contribution term for each node
    wpr = reduce contribs by key,   // adding up all contribution term, $wpr_{u \rightarrow v}$
        multiply the sum by d and add (1-d)/N   // update weighted PageRank for each node **wpr**
    iter ++
end while
return wpr

3.1.3. Clustering Coefficient

In PERSEUS-HUB, we use the local clustering coefficient to measure the extent to which nodes in the input graph tend to cluster together. This measure is widely used in social and other sciences to better understand the connectivity properties of a graph. For example, in connectomics, nodes with high clustering coefficients indicate neurons that have dense connections, which represent areas of the brain that are functionally correlated; in social science, nodes with high clustering coefficients could be an indicator of tightly connected communities that are anomalous.

The clustering coefficient for each node (i.e., Local Clustering Coefficient) in an *undirected* graph is defined as:

$$\text{LCC}(v) = \frac{2 \cdot |\{e_{uk} \in \mathcal{E} | u, k \in \Gamma(v)\}|}{d_v \cdot (d_v - 1)}, \tag{3}$$

where $\Gamma(v)$ is the neighborhood of node v: $\Gamma(v) = \{u \in \mathcal{V} | e_{uv} \in \mathcal{E}\}$, and d_v is the number of nodes in the neighborhood of node v. To measure the average clustering coefficient of an undirected graph, the concept of average clustering coefficient is introduced:

$$\text{ACC}(\mathcal{G}) = \sum_{v \in \mathcal{V}} \frac{\text{LCC}(v)}{N}. \tag{4}$$

The Spark procedure, CCFinder [45], for computing the local clustering coefficient is given in Algorithm 2. Our implementation of local clustering coefficient uses the FONL (Filtered Ordered Neighbor List) data structure [45]. FONL is designed to increase the computation speed by efficiently storing necessary data and exploiting the parallel computing framework. Given $< srcID, dstID >$ pairs, the input graph data is organized into $< key, value >$ form: $< v_i, (d_{v_i}, [v_{i1}, v_{i2}, \cdots v_{im}]) >$, where v_i is the ID of node i; $\forall 1 \leq k \leq m, v_{ik} \in \Gamma(v_i)$ are neighbors of v_i with a larger degree ($d_{v_{ik}} > d_{v_i}$) or having a greater node ID if their degrees are the same ($v_{ik} > v_i$ and $d_{v_{ik}} = d_{v_i}$). Moreover, v_{ik} are arranged by their degree in ascending order, i.e., $d_{v_{i1}} \leq d_{v_{i2}} \leq \ldots \leq d_{v_{ik}}$. The degree of v_i, d_{v_i}, is appended after v_i and will be used to calculate the denominator in Equation (3). For each node in the graph, a $<$ key, value $>$ pair with the above form is generated.

In Equation (3), the term $|\{e_{uk} \in \mathcal{E} | u, k \in \Gamma(v)\}|$ denotes the number of edges linking neighbors of node v. Since every such edge indicates the existence of a triangle involving v, solving for $|\{e_{uk} \in \mathcal{E} | u, k \in \Gamma(v)\}|$ is equivalent to counting the number of triangles in the neighborhood of v. Thus, to find triangles in a graph, we check whether the neighbors of v_i, i.e., v_{ik} have matching neighbors or not. Therefore, for each v_{ik}, the FONL structure is organized into a candidate list of the form $< v_{ik}, (v_i, [v_{i(k+1)}, \ldots, v_{im}]) >$, in which only nodes with a higher or equal degree than v_{ik} are kept, i.e., $\forall \, v_{il} \in \{v_{i(k+1)}, \ldots, v_{im}\}, d_{v_{ik}} \leq d_{v_{il}}$. Next, a join operation is performed on FONL and the above candidate list. Taking the intersection of $[v_{i(k+1)}, \ldots, v_{im}]$ of v_{ik}'s candidate list and $[v_{(ik)1}, v_{(ik)2}, \cdots v_{(ik)m'}]$ of v_{ik}'s FONL structure, we return the matching node ID together with their common neighbors, v_i and v_{ik}, identify the triangles in the graph. Due to the ascending ordering of degrees, every triangle in the graph can only be counted once, and thus counting duplicates is avoided. Now, the local clustering coefficient for each node v_i can be computed using the number of triangles it involves in and its degree, d_{v_i}.

As in the case of PageRank, the distribution of local clustering coefficients can also be obtained by computing its histogram after appropriate binning.

Algorithm 2 CCFinder [45].

Input: edge_list: <srdID, dstID>
Output: Local Clustering Coefficient for every node, lccRdd: <nodeID, lcc>
nodeID = {srcID} ∪ {dstID}
N = max(nodeID)   // Getting the total number of nodes in the graph
edge_list = <srdID, dstID> ∪ <dstID, srdID>   // Getting reversed edges
edge_list = group edge_list by nodeID (key),   // Appending nbrDeg to <nodeID, [(nbrIDs)]>
      append neighbor degree after nbrID   // <nodeID, [(nbrID, nbrDeg)]>
fonlRdd = CREATEFONL(edge_list)   // Creating FONL Structure
save fonlRdd both in memory and on disk
candRdd = map fonlRdd to <nbrID, (nodeID, [(nbrIDs)])>   // Creating Candidate List
     $\forall$ nbrID$_i$ ∈ [(nbrIDs)], nbrDeg$_i$ ≥ nbrDeg
triRdd = join candRdd with fonlRdd by nodeID,   // Counting the number of triangles
     compute triCounts = | fonlRdd.[(nbrIDs)] ∩ candRdd.[(nbrIDs)] |
     map to the form <nodeID, triCounts> and reduce by nodeID
lccRdd = map triRdd to get local clustering coefficient using Equation (3)   // Computing lccRdd
return lccRdd

function CREATEFONL(edge_list)
  compute nodeDeg, $\forall$ nodeID ∈ edge_list
  [(nbrIDs)] = sort {(nbrID,nbrDeg) ∈ edge_list : (nbrDeg > nodeDeg) or
     (nbrID > nodeID and nbrDeg = nodeDeg)}   // Filtering and sorting edge_list
     by (nbrDeg and nbrID) in ascending order
  return <nodeID, nodeDeg, [(nbrIDs)]>
  end function

3.1.4. Execution Time Analysis

In this subsection, we provide runtime analysis of the graph statistics module. The AWS EMR-Spark cluster that PERSEUS-HUB runs on has one master node and two slave nodes, each of which is deployed on the general purpose m4.large instance (two CPU cores and 8 GB memory). We also measure the average response time by performing 1000 queries in the frontend. Specifically, we consider the two graphs used in Section 5, the size of which is briefly described in Table 2.

Table 2. Execution time: computation of graph statistics.

Dataset	# Nodes	# Edges	Statistics Computation (s)	Average Response (ms)
cit-HepTh	27,770	352,807	790.014	7.320
Twitter_retweet	35,366	78,058	711.592	8.311

From Table 2, we find that the time for computing these two real-world graphs takes about 10 min on three machines. Using additional machines could help reduce the computation time further, and support the analysis of larger graphs at a relatively short time. We also note that the average response time in both cases is very short and less than 0.1 s, which, according to Nielsen [24] gives the feeling of instantaneous response.

3.2. Module 2: Multi-View, Ensemble Anomaly Detection

The second (offline) module of PERSEUS-HUB consists of a system-based detection of anomalous nodes in the input network, which can guide the exploration process by highlighting 'abnormal' node behaviors. Discovery and evaluation of outliers from a multi-dimensional set of points is a difficult task. Moreover, different anomaly detection algorithms calculate outlier scores under different criteria (e.g., distance from a trend-line [21] vs. density differences from surrounding points [37] vs. latent node properties), and thus their scores are vastly different. To benefit from various methods and detect outliers with higher confidence, PERSEUS-HUB uses a multi-view, ensemble approach: The approach is multi-view because outlier detection methods are applied on all the distribution plots, which capture different aspects and normality patterns of the input data. Moreover, it is also considered as an ensemble because it normalizes multi-dimensional outlier scores from different anomaly detection methods to form a unified combined score for each multi-dimensional point.

3.2.1. Multi-View, Ensemble Anomaly Detection: Approach

Given a set of plots that portray the relationship between different attributes of a multi-dimensional dataset (e.g., the distribution plots computed by the first module of PERSEUS-HUB), we apply an anomaly detection method to calculate the anomalousness score for each point (or node). PERSEUS-HUB leverages two anomaly detection methods (an angle- and a density-based). Specifically, we use:

- ABOD (Angle Based Outlier Detection [12]), an anomaly detection method that assesses the variance in the angles between the different vectors of a point to other points. ABOD is particularly effective for high-dimensional data since effects from the "curse of dimensionality" are alleviated compared to purely distance-based approaches.
- G-FADD (Grid-Based Fast Outlier Detection algorithm given Duplicates [38]), a density-based anomaly detection method that is remarkably fast in determining outliers from multi-dimensional points with a large number of (near-) duplicates (i.e., points that have almost the same coordinates). Specifically, by considering identical coordinates in an n-dimensional space as a super node with their duplicate count information, G-FADD can compute anomalousness scores much faster than other density-based approaches such as LOF [37] (Local Outlier Factor), which do not treat duplicate nodes specially, and are inefficient at handling real datasets with many near-duplicate points. The time complexity of LOF is near-quadratic with respect to the number of duplicates, whereas for G-FADD it is near-linear.

Although we used two anomaly detection methods, the methodology that we describe below can be applied to any number of approaches. Since the anomaly detection module is offline, its runtime does not burden the user significantly. Nevertheless, choosing fast methods that scale well with the size of the input dataset is important when handling large graph datasets.

Each method returns a list of real-valued anomalousness scores for all the nodes in each distribution plot computed by Module 1. However, scores computed by different methods and on different plots have widely varying scales, and combining them to get a uniform score is not straightforward. In our implementation, we compare the anomalousness scores of different methods by following the methodology in [46] and normalizing all the anomalousness scores into probabilistic scores that represent the probability that node v in a certain plot is anomalous according to a certain anomaly detection method. Such a probability can be calculated in the following way:

$$P(v) = max\left(0, \ erf\left(\frac{S(v) - \mu_S}{\sigma_S \cdot \sqrt{2}}\right)\right), \tag{5}$$

where $S(v)$ is the anomalousness score of node v, erf is the error function, and μ_S and σ_S are the mean and standard deviation of anomalousness score for the method within the plot of interest.

In our case, because we use ABOD where a high score for a node means a low probability of being an outlier, before normalizing the ABOD scores, we first need to invert them so that a high score for a node maps to a larger probability for being an outlier. To enhance the contrast between inliers and outliers, we use logarithmic inversion:

$$Reg_S^{loginv}(v) = -\log(S(v)/S_{max}), \tag{6}$$

where $S(v)$ is the original unregularized ABOD score and S_{max} is the maximum ABOD score from all the ABOD scores in a given distribution plot. The inversion of ABOD scores allows the proper computation of probabilistic scores based on Equation (5).

At this point, for each method j and each node v in each plot p_i, we have normalized the probability score, which we can denote as $P_{p_i}^j(v)$. In our case, $j = \{ABOD, G\text{-}FADD\}$ and p_i ranges from p_1 to p_9:

- p_1: Degree distribution plot,
- p_2: Two-dimensional distribution (or correlation plot) of degree vs. PageRank,
- p_3: PageRank distribution plot,
- p_4: Clustering coefficient distribution plot,
- p_5 to p_9: Five pairwise singular vector (sv) plots—sv_1 vs. sv_2, sv_3 vs. sv_4, sv_5 vs. sv_6, sv_7 vs. sv_8 and sv_9 vs. sv_{10}, respectively (where sv_i is the ith singular vector of a graph's adjacency matrix). By default, the frontend shows the first four pairs of singular vectors. However, when PERSEUS-HUB computes the combined anomalousness scores, all plots are considered.

To obtain a combined anomalousness score, we compute a weighted combination of the normalized anomalousness scores from the various views (or distributions) of our data in the following way:

$$P(v) = \frac{\sum\limits_{i=1}^{9}\sum\limits_{j} w_{p_i} \cdot P_{p_i}^j(v)}{\sum\limits_{i=1}^{9}\sum\limits_{j} w_{p_i}} \quad , \quad \forall \ node \ v, \tag{7}$$

where w_{p_i} is the contribution of plot p_i to the total anomalousness score of each node. These weights can be modified as desired by the user based on domain knowledge. For example, in neuroscience, when studying connectomes of subjects with Alzheimer's, a scientist may choose to set higher weight for the clustering coefficient distribution plot (based on research that has shown that the clustering coefficient is a discriminative feature for patients and control subjects [47]), and minimize the weight for the PageRank distribution. For simplicity, we provide default values of equal contribution per data aspect, where we consider four main aspects: degree, PageRank, clustering coefficient, and singular vector (i.e., $w_{p_i} = 1$ for $i \in \{1,2,3,4\}$ and $w_{p_i} = \frac{1}{5}$ for $i \in \{5,6,7,8,9\}$). Additionally, the weights can be further customized if we introduce a $w_{p_i}^j$ notation so that different methods and plots can be weighted differently.

3.2.2. Scalability

To improve the performance of calculating the anomalousness scores using ABOD, parallelization was used. In particular, a separate process was used for the calculation of ABOD scores for each plot. The runtime of G-FADD and combining anomalousness scores is negligible when compared to the runtime of ABOD, and thus no further speedup was required.

To analyze the scalability of the two anomaly detection routines that are integrated in PERSEUS-HUB, we varied the number of nodes (or points) from a few hundreds to one million, and applied the methods of the PageRank vs. degree correlation plot. As seen from Figure 3, the runtime of ABOD grows quadratically as a function of the number of nodes, whereas G-FADD runs in near-linear time and is significantly faster even on datasets with millions of nodes. The outlier detection analysis is offline, and the precomputed results are stored and used for interactive visualization. For larger graphs, ABOD can be 'disabled' and G-FADD (in combination with other fast approaches) can be used for multi-view anomaly detection.

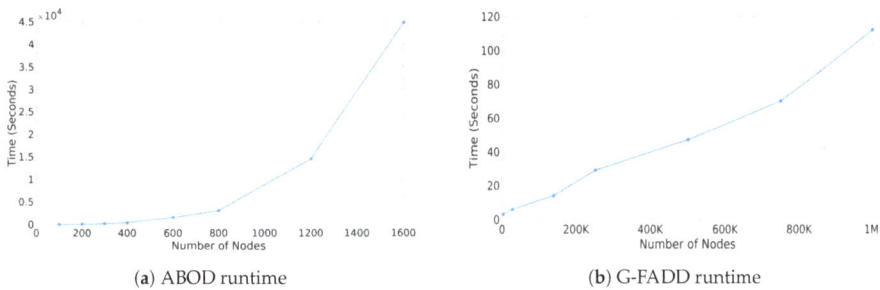

(a) ABOD runtime (b) G-FADD runtime

Figure 3. Scalability of ABOD and G-FADD for varying graph size. *x*-axis: number of nodes; *y*-axis: runtime in seconds.

3.3. Module 3: Frontend Visualization

The frontend visualization consists of two parts: (1) aggregated summarization and (2) node-specific summarization. The former displays the linked statistics distributions, and the latter provides users with three components to analyze the node statistics and local connectivity with its neighbors.

3.3.1. Aggregated Summarization

Given a dataset, PERSEUS-HUB plots the distributions of the node-specific statistics (e.g., degree vs. PageRank, pairwise singular vector plots) and tracks specific nodes through linking and brushing. The layout of the 'aggregated summarization' is shown in Figure 4. In particular, the interactive visualization consists of eight linked distribution plots, where each point in the distributions is trackable: when the user clicks on it in one of the distributions, the corresponding point in other statistics distributions will be highlighted as well. The distribution plots p_1 to p_4 (described in Section 3.2.1) are aligned in the first row, and p_5 to p_8 in the second row. For p_5 to p_8, users can change the axes of each of the four plots accordingly to portray the visualization they want.

Since the input graphs in PERSEUS-HUB may consist of thousands or millions of distinct nodes, directly plotting distributions of *node-wise* graph statistics—using identical amount of points to the number of nodes—is significantly memory-consuming and inefficient. To overcome such difficulties, in each distribution, PERSEUS-HUB reduces the amount of points required for plotting by dividing the plot into a $k \times k$ grid (e.g., k = 1000 by default) and using the center to represent points falling in each grid. To show the density of each grid, the number of points in each grid is marked with different colors: each non-empty grid is initialized as blue and becomes closer to red with more nodes.

The distribution plots are created using Bokeh, a Python visualization library that allows functionality such as linking, selecting, taping, and zooming. The Bokeh toolbar, which is illustrated in Figure 5, is given at the top-right of each row of plots and supports the following interactions starting from left to right:

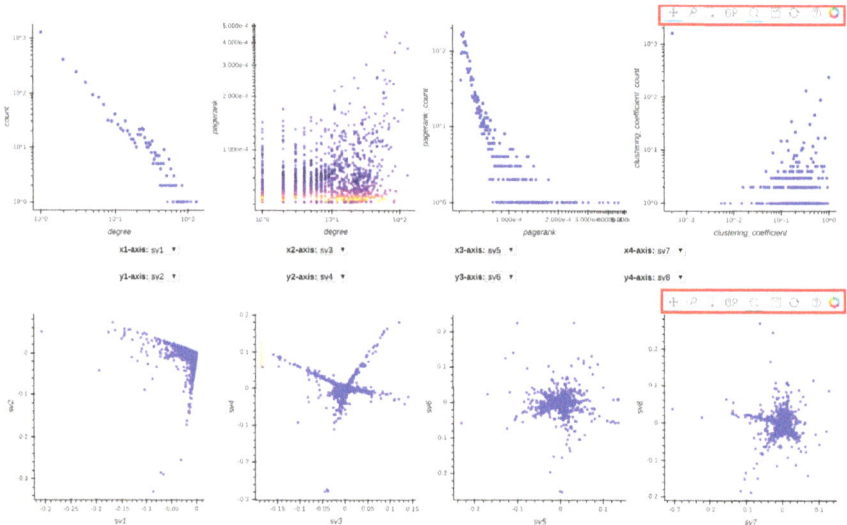

Figure 4. Layout of the graphs displayed in the frontend. The plots in the first row are degree vs. count, degree vs. PageRank, PageRank vs. PageRank count and ClusteringCoefficient vs. ClusteringCoefficient count. The plots in the second row are singular vector (sv) relationships, specifically they are sv1 vs. sv2, sv3 vs. sv4, sv5 vs. sv6, sv7 vs. sv8. Note that each node in the degree vs. PageRank plot is colored according to the PageRank count value of each node, a high count is red and a low count is blue. The red box annotations indicate the Bohek toolbar.

Figure 5. Bokeh Toolbar (the last two buttons at the rightmost end of the toolbar are just links for more information about Bokeh and they don't serve any interaction purpose).

- pan: When the pan tool is selected, the user can move the cloud of points in each distribution plot by holding the left mouse button and moving the cursor in the direction they want to move the graph.
- box-zoom: When the box-zoom tool is selected, the user can box-select a region in the distribution plot they would like to zoom in.
- box-select: When the box-select tool is selected, the user can box-select and highlight a region of points in the selected distribution plot. Since all plots are linked, the corresponding points in the other plots are highlighted at the same time. This can be useful when a user wants to track micro-clusters of points across different 'views' (or graph properties).

- wheel-zoom: When the wheel-zoom tool is selected, the user can scroll over a given distribution plot to zoom in or out.
- tap: When the tap tool is selected, the user can left click on a specific single point. This is useful when the user wants to explore a single node (via the information console, egonet, or adjacency matrix) instead of using the box-select tool.
- save: When the user clicks on the save button, the corresponding distribution plots are saved in png format.
- reset: When the reset button is clicked, the distribution plots are re-positioned to their original setup and the previously selected points are unselected.

3.3.2. Node-Specific Summarization

PERSEUS-HUB provides three components to explore a specific point in a distribution (or correlation) plot: (1) an information console revealing the details of a collection of nodes that are similar to the a user-selected node in terms of at least one statistic (e.g., similar degree); (2) a dynamic egonet representation; and (3) an adjacency matrix centering a specific node. In addition, users who are interested in system-detected anomalies can explore the corresponding nodes using the provided anomaly detection methods.

Walk-through of a user interaction case. When a user selects a point in any of the plots, the point's coordinates are sent to the backend server through an Ajax request. In the backend, the database is queried with the coordinates and attributes of the selected point and it returns at most 10 non-aggregate nodes that match the selected aggregate node. For example, if the user selects the point, *degree = 130* and *PageRank = 0.00037250011576710077* in the PageRank vs. degree graph, then the database will be queried so that it finds at most ten distinct nodes that have the same PageRank and degree values. Once all that information is retrieved, the backend will respond back and display the results in the information console as shown in Figure 6.

Node ID	Degree	Pagerank	Singular Vector 1	Egonet	Adj matrix	Bookmark / Unbookmark
1247	82	0.0000091296137563473432	-0.00507913742506	View	View	
1287	82	0.000006244820013151303	-0.00522599343256	View	View	
1291	82	0.000008970807903345281	-0.00628850203511	View	View	
130	82	0.0000058606409794742254	-0.00000136337323682	View	View	
1303	82	0.000007366313963398762694	-0.0114327090978	View	View	
1425	82	0.0000128114311016674436	-0.00302712913573	View	View	
14995	82	0.0000302338063720762	-0.00461157279422	View	View	
1510	82	0.0000813219377624271487	-0.000785894379443	View	View	
15668	82	0.0000289925678021800302	-0.00794544743959	View	View	
1627	82	0.00000546835705202722795	-0.000263239988712	View	View	

Figure 6. Layout of the information console.

For each retrieved node in the information console, the user has the option to display the egonet and the adjacency matrix (shown in Figure 7) of that node by clicking the appropriate button. Then, the user can navigate to the egonet and adjacency matrix tabs to view their respective interactive graphs. The egonet graph shows at most 10 other nodes adjacent to the selected node and all the connections between them. The user also has the ability to expand the egonet by clicking on any desired node from which they want to expand the network as well as deleting any desired node from the egonet. For efficiency, when a user clicks on a node, it expands to a *random* set of ten neighbors. However, this choice can be adjusted to better serve the exploration needs for different domains. The egonet (which is displayed using a force-directed layout) is interactive and can be re-positioned as desired by the user in order to provide a better visualization.

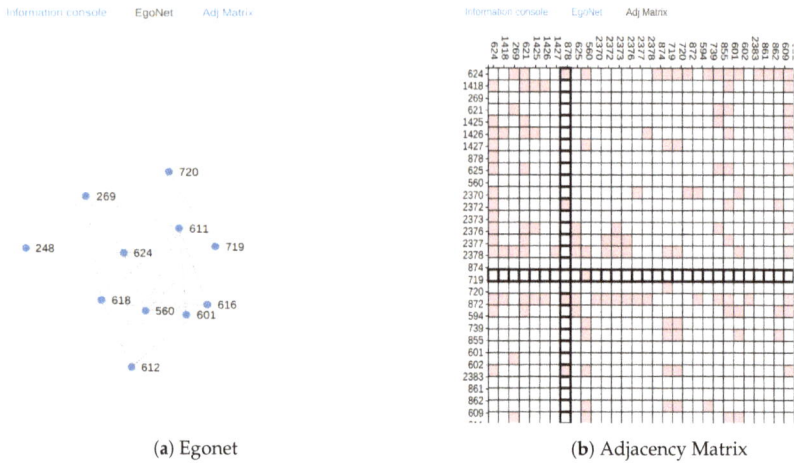

(a) Egonet   (b) Adjacency Matrix

Figure 7. Node-specific summarization: representation of the egonet and adjacency matrix of a selected node.

To explore nodes detected as anomalies, users can click on the *G-FADD*, *ABOD*, and *Combined Anomaly Score* buttons at the bottom of the screen to display the respective anomaly scores (Figure 8). Once the button is clicked, the top ten most anomalous nodes from the input graph will be presented in the table format ordered by their scores. Simultaneously, the top ten anomalous nodes are automatically highlighted in all the distribution plots to give the user a sense of the most anomalous nodes in their dataset (Figure 9). Node-specific interaction via egonet and adjacency matrix representation is supported for the anomalous nodes as well (through the tab for anomaly detection, as shown in Figure 8).

Figure 8. Layout of the anomaly scores table.

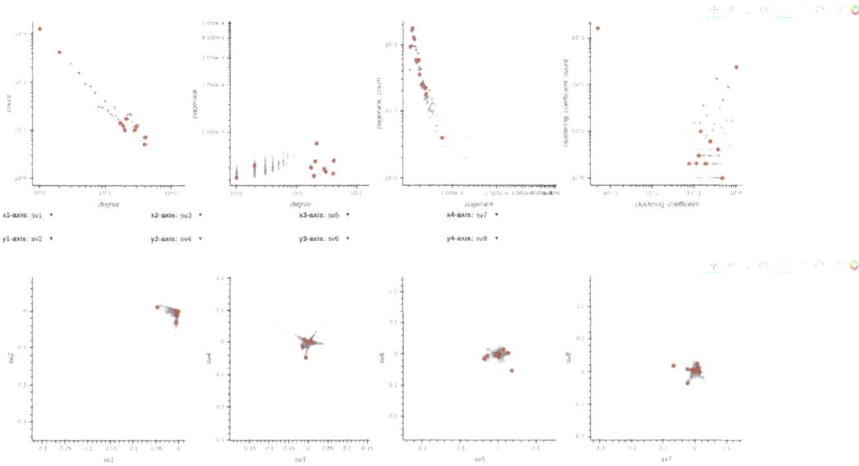

Figure 9. Top ten anomalous nodes are highlighted in red.

4. PERSEUS-HUB: Web Application Integration

In order to combine the above visualization features into one cohesive user interface, we wrap PERSEUS-HUB inside a web application interface, which is accessible universally within a browser. The application backend is hosted on Amazon Web Service (AWS), consisting of an Apache Spark cluster for processing user-uploaded raw datasets, an application architecture powered by an uWsgi server running Django, a Nginx reverse proxy server for IP resolution and static file serving, and a MySQL server for application data storage, while the frontend user interface is powered by the Bootstrap 3 framework.

The data visualization process is as follows: first, the application lets users upload raw datasets in the form of a text file from the upload page. The uploaded raw data is then passed to the master node of Apache Spark cluster to compute the graph statistics. Due to the potentially long time needed, the system is set to send the user an email when the processing is done, rather than providing an unresponsive interface which requires the user to wait.

After the raw dataset is processed, the computed node-specific statistics is stored in the MySQL table for dynamic querying. The dataset is then made available for viewing through a templated URL. When the templated dataset webpage is loaded, the dataset's respective data files are loaded to plot statistics distributions with the Bokeh visualization library. One important goal of PERSEUS-HUB is to create a multi-user environment for sharing and analyzing a variety of user data sourced from the crowd. For this purpose, PERSEUS-HUB provides four user-related modules:

- **Authentication module**: The web application includes a user registration and authentication module where new users can register and start processing and sharing their data immediately. This module is also the foundation of some other user-controlled features, which are elaborated on below.
- **Exploration module**: The dashboard page of the web application is filled with a collection of recently added datasets. This data "feed" will be enhanced with suggestion algorithms in the future for better personalization.
- **Search module**: When a user uploads her raw data, she is asked to enter a descriptive title for them. This piece of metadata provides a simple indexing method for the users to either search for a specific dataset or browse for datasets with a similar title.
- **Bookmark module**: The bookmark feature allows users to mark a specific node within a dataset by clicking on the "star" button next to the statistics in the information console. This feature

gives users a finer-grained choice to mark their interest, e.g., a specific point in social networking datasets. Moreover, users can set a priority flag for their bookmarks, enabling better categorization and recognition for the creator. For collective analysis, users can choose to share their bookmarks publicly so that other analysts can incorporate these findings in their exploration process. Screenshots of the bookmark button and bookmark creation are shown in Figure 10.

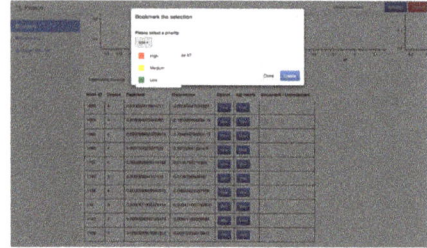

(a) Bookmark in information console

(b) Bookmark creation notification

Figure 10. Illustration of the bookmark functionality.

5. Case Study

PERSEUS-HUB can be used to analyze various graph patterns, and in this section, we run PERSEUS-HUB on two real-world datasets from different domains to showcase how to analyze the spike patterns that appear in singular vector distributions.

5.1. The Citation Network

The first dataset we use is the Arxiv Hep-Th (high energy physics theory) citation graph [48] which contains 27,770 nodes and 352,807 edges. The frontend is shown in Figure 11. To analyze the spike patterns in the singular vector relationship plots, we study the points that locate at the ends of the spikes (extreme points). Specifically, we search for detailed information about the papers denoted by these points under the physics category in Arxiv (https://arxiv.org). The points explored are shown in Figure 12, and their detailed information is listed in Table 3.

Table 3. Detailed information of selected points in singular vector plots.

Color	Paper ID	Paper title
Red	9905111	"Large N Field Theories, String Theory and Gravity"
Cyan	0201253	"Supersymmetric Gauge Theories and the AdS/CFT Correspondence"
Orange	9710046	"Black Holes and Solitons in String Theory"
Green	0109162	"Quantum Field Theory on Noncommutative Spaces"
Purple	0102085	"A Review on Tachyon Condensation in Open String Field Theories"
Dark green	0302030	"Brane World Dynamics and Conformal Bulk Fields"
Yellow	0101126	"M(atrix) Theory: Matrix Quantum Mechanics as a Fundamental Theory"
Blue	0111208	"Noncommutative Field Theories and (Super)String Field Theories"
Light brown	211178	"Supersymmetric D3 brane and N=4 SYM actions in plane wave backgrounds"
Brown	9911022	"Tests of M-Theory from N=2 Seiberg-Witten Theory"
Light pink	0104127	"Thermodynamic properties of the quantum vacuum"

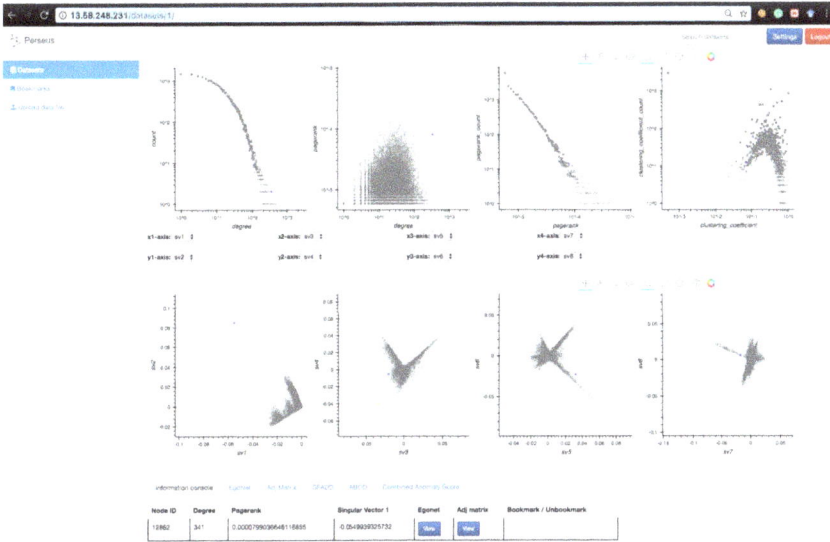

Figure 11. Illustration of running PERSEUS-HUB on the cit-HepTh (citation in high energy physics theory) graph. To explore the spike patterns in the singular vector plots, extreme points on the spikes are selected and explored. Only one is shown in this case to obtain the node ID.

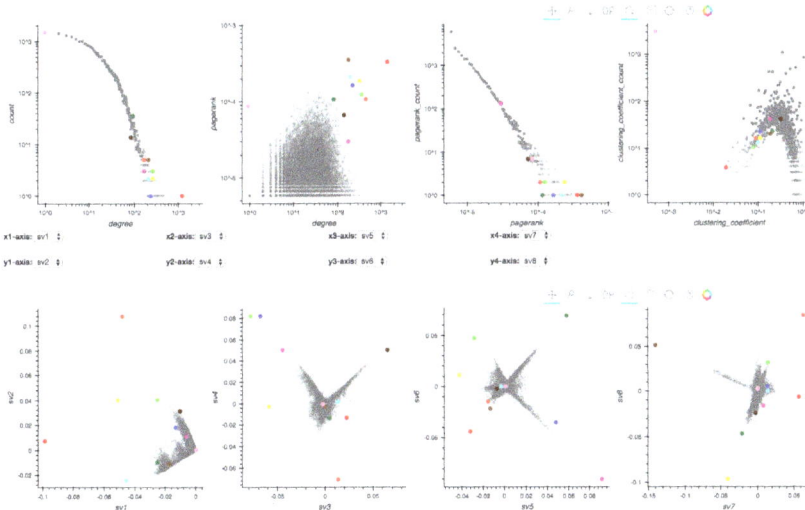

Figure 12. Illustration of points explored, marked with different colors.

Although singular vectors are known to capture structural information in graphs, there is no solid theory about what exact information is encoded and how to use it. By using the correspondence of extreme points in different singular vector relationships in PERSEUS-HUB, we find correlation between the paper topics and the structural information in the citation graph. Specifically, we make the following empirical observation: (1) in the citation graph, papers that are rarely cited or citing others are located at the "centers" of all singular vector relationship plots; (2) the similarity and difference between topics can be captured in different singular vector relationships; and (3) different spikes reflect different topics captured by the structural information in the citation graph.

To support our observation (1), we explored the point marked in light pink and found several papers that have only one total degree (degree distribution plot) in the dataset. In addition, according to the clustering coefficient distribution plot, the low value indicates that the corresponding papers are unlikely to be clustered with others, which demonstrates the fact that these papers are rarely cited by or citing other papers in this dataset. To support (2), we compare the paper titles of several extreme points. For example, the point marked in green and point marked in purple are located closely in the 3rd vs. 4th singular vector plot, but far away in the 5th vs. 6th singular vector plot. From the titles, we find that both green and purple discuss the topic of "field theory", but, more specifically, one is about "quantum field theory" and the other is about "string field theory". This explains why they are similar in terms of some singular vectors but different in terms of the others. Similarly, we observe this from the points marked in yellow and green. This similarity and difference in topic captured by the graph connectivity also demonstrates (3), since most papers discuss different specific areas in high energy physics theory. Nearby extreme points, such as the blue and green points in sv3 vs. sv4, show that papers about the same topic are likely to cite each other and share the same statistics.

5.2. The Social Network

The second dataset we use is a Twitter dataset, which contains 35,366 accounts and 78,058 retweet records, ranging from October 2015 to January 2016. Similar to the analysis in Section 5.1, we illustrate the frontend of PERSEUS-HUB in Figure 13 with patterns of interest marked in red ellipses.

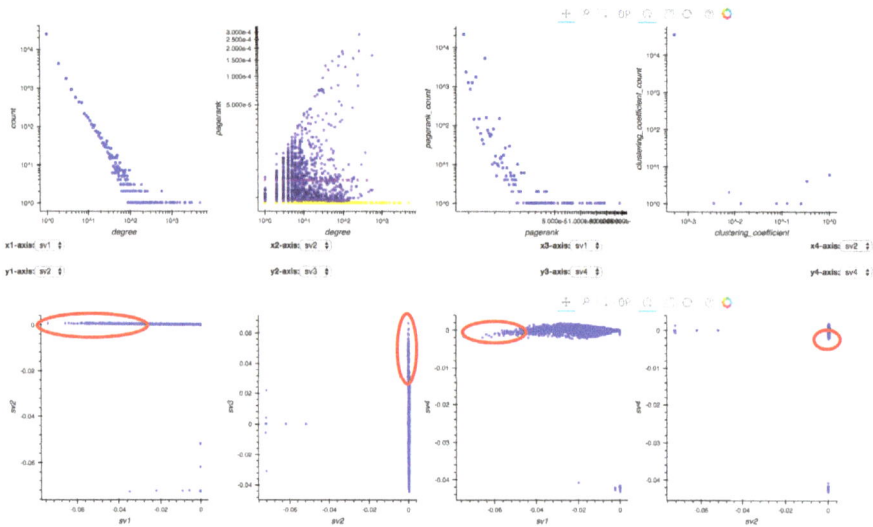

Figure 13. Analysis of Twitter retweet network: suspicious patterns are marked in red ellipses.

In Figure 13, we manually tune the x- and y-axes of four singular vector relationship plots, which exhibit clear "EigenSpokes" patterns [43]. The spokes, or the straight lines that are aligned with the axes, represent the existence of groups of tightly connected nodes, and, in this case, accounts that often retweet or are being retweeted by other accounts. To further explore these suspicious groups of users (marked in red ellipses), we investigate their detailed screen names and their behaviors in Twitter. We find there are accounts that post tremendous amounts of advertising tweets such as "bmstores" (B&M online store), "@ProtecDirect" (UK supplier of personal protective equipment), "@CartridgeSave" (printer supplies) and so forth. On the other hand, there are accounts which mainly retweet those ads such as "@POY_UK" (product of the year) and "@harrisonsdirect" (product direct delivery). This finding explains why there are accounts with frequent retweet behaviors: Online sales are always trying to promote their products through either their official accounts (e.g., "@bmstores") or proxies (e.g., "@POY_UK"). Similarly, we can run PERSEUS-HUB to detect bots in Twitter that spread rumors or fake news by evaluating the singular vector plots.

In a nutshell, we illustrate that PERSEUS-HUB provides analysts a way to distinguish normal users from bots or spam accounts, which further demonstrates the usage of PERSEUS-HUB in the domain of social networks.

6. Conclusions

In this paper, we introduce PERSEUS-HUB, an efficient large-scale graph mining tool with interactive visualization to support preliminary analysis of graph data. PERSEUS-HUB adopts a two-level summarization scheme that visualizes both aggregate and node-specific patterns. Our system also enables collective analysis of data from multiple users, as well as machine-driven, ensemble, multi-view outlier detection analysis that guides the data exploration process to anomalous patterns. We believe that PERSEUS-HUB provides a new way to analyze large, complex graphs that could be beneficial to users across various domains.

Most analysis performed in PERSEUS-HUB is based on the node-specific statistics of the input graph; therefore, a natural direction to consider as future work is the integration of side node information (e.g., node attributes). Another direction is to extend the anomaly detection functionality with the integration of more detection algorithms and the support of additional frontend options for the user (e.g., selection of weight contributions for various views to the ensemble scores). More importantly, for domain-specific summarization, it would be beneficial to devise efficient computational methods for 'customized' summarization of graph data by leveraging the domain knowledge, i.e., automatic selection of the 'summarized' statistics to be shown to the user, instead of providing a set of predefined statistics as in PERSEUS-HUB. At this stage, we consider our work to be the first step towards a 'collaborative' tool that supports large-scale graph analysis.

Acknowledgments: The authors would like to thank Christos Faloutsos, Yuanchi Ning, and Esther Wang for their feedback and contributions during the early steps of this project. This work is supported by the University of Michigan and AWS Cloud Credits for Research.

Author Contributions: Danai Koutra had the original idea for the system, which was further developed with Di Jin. Di Jin led the implementation of the system and managed a team of undergraduate and Masters students, who contributed to various parts of PERSEUS-HUB: Aristotelis Leventidis, Haoming Shen, Ruowang Zhang, and Junyue Wu. Di Jin performed the analysis of the case studies. Di Jin, Danai Koutra, Aristotelis Leventidis and Haoming Shen were involved in writing the paper.

Conflicts of Interest: The authors declare no conflict of interest.

References

1. Kuramochi, M.; Karypis, G. Frequent Subgraph Discovery. In Proceedings of the 2001 1st IEEE International Conference on Data Mining (ICDM), San Jose, CA, USA, 29 November–2 December 2001; pp. 313–320.
2. Leardi, R. Multi-way analysis with applications in the chemical sciences, age smilde, Rasmus Bro and Paul Geladi, Wiley, Chichester, 2004, ISBN 0-471-98691-7, 381 pp. *J. Chemometr.* **2005**, *19*, 119–120.
3. Tong, H.; Faloutsos, C. Center-piece subgraphs: Problem definition and fast solutions. In Proceedings of the 12th ACM SIGKDD International Conference on Knowledge Discovery and Data Mining (KDD '06), New York, NY, USA, 20–23 August 2006; ACM: New York, NY, USA, 2006; pp. 404–413.
4. Sondhi, P.; Sun, J.; Tong, H.; Zhai, C. SympGraph: A framework for mining clinical notes through symptom relation graphs. In Proceedings of the 18th ACM SIGKDD International Conference on Knowledge Discovery and Data Mining (KDD '12), Beijing, China, 12–16 August 2012; pp. 1167–1175.
5. Backstrom, L.; Kumar, R.; Marlow, C.; Novak, J.; Tomkins, A. Preferential behavior in online groups. In Proceedings of the International Conference on Web Search and Web Data Mining (WSDM '08), New York, NY, USA, 11–12 February 2008; ACM: New York, NY, USA, 2008; pp. 117–128.

6. Barabási, A.L.; Jeong, H.; Néda, Z.; Ravasz, E.; Schubert, A.; Vicsek, T. Evolution of the social network of scientific collaborations. *Physica A* **2002**, *311*, 590–614.
7. Stolper, C.D.; Kahng, M.; Lin, Z.; Foerster, F.; Goel, A.; Stasko, J.; Chau, D.H. Glo-stix: Graph-level operations for specifying techniques and interactive exploration. *IEEE Trans. Vis. Comput. Graph.* **2014**, *20*, 2320–2328.
8. Chau, D.H.; Kittur, A.; Hong, J.I.; Faloutsos, C. Apolo: Making Sense of Large Network Data by Combining Rich User Interaction and Machine Learning. In Proceedings of the 17th ACM International Conference on Knowledge Discovery and Data Mining (SIGKDD), San Diego, CA, USA, 21–24 August 2011.
9. Koutra, D.; Jin, D.; Ning, Y.; Faloutsos, C. Perseus: An Interactive Large-Scale Graph Mining and Visualization Tool. *Proc. VLDB Endow.* **2015**, *8*, 1924–1927.
10. Jin, D.; Sethapakdi, T.; Koutra, D.; Faloutsos, C. PERSEUS3: Visualizing and Interactively Mining Large-Scale Graphs. In Proceedings of the WOODSTOCK '97, El Paso, TX, USA, July 1997.
11. Lee, J.Y.; Kang, U.; Koutra, D.; Faloutsos, C. Fast Anomaly Detection Despite the Duplicates. In Proceedings of the 22nd International Conference on World Wide Web (WWW Companion Volume), Rio de Janeiro, Brazil, 13–17 May 2013; pp. 195–196.
12. Kriegel, H.P.; Zimek, A.; Hubert, M.S. Angle-based outlier detection in high-dimensional data. In Proceedings of the 14th ACM SIGKDD International Conference on Knowledge Discovery and Data Mining, Las Vegas, NV, USA, 24–27 August 2008; pp. 444–452.
13. Pienta, R.; Kahng, M.; Lin, Z.; Vreeken, J.; Talukdar, P.; Abello, J.; Parameswaran, G.; Chau, D.H. FACETS: Adaptive Local Exploration of Large Graphs. In Proceedings of the 2017 SIAM International Conference on Data Mining. Society for Industrial and Applied Mathematics, Houston, TX, USA, 27–29 April 2017.
14. Wongsuphasawat, K.; Qu, Z.; Moritz, D.; Chang, R.; Ouk, F.; Anand, A.; Mackinlay, J.; Howe, B.; Heer, J. Voyager 2: Augmenting Visual Analysis with Partial View Specifications. In Proceedings of the 2017 CHI Conference on Human Factors in Computing Systems, Denver, CO, USA , 6–11 May 2017; ACM: New York, NY, USA, 2017; pp. 2648–2659.
15. May, T.; Steiger, M.; Davey, J.; Kohlhammer, J. Using signposts for navigation in large graphs. *Comput. Gr. Forum* **2012**, *31*, 985–994.
16. Henry, N.; Fekete, J.D.; McGuffin, M.J. NodeTrix: A hybrid visualization of social networks. *IEEE Trans. Vis. Comput. Graph.* **2007**, *13*, 1302–1309.
17. Zhao, J.; Collins, C.; Chevalier, F.; Balakrishnan, R. Interactive exploration of implicit and explicit relations in faceted datasets. *IEEE Trans. Vis. Comput. Graph.* **2013**, *19*, 2080–2089.
18. Kairam, S.; Riche, N.H.; Drucker, S.; Fernandez, R.; Heer, J. Refinery: Visual exploration of large, heterogeneous networks through associative browsing. *Comput. Gr. Forum* **2015**, *34*, 301–310.
19. Akoglu, L.; Chau, D.H.; Kang, U.; Koutra, D.; Faloutsos, C. OPAvion: Mining and Visualization in Large Graphs. In Proceedings of the 2012 ACM International Conference on Management of Data (SIGMOD), Scottsdale, AZ, USA, 20–24 May 2012; ACM: New York, NY, USA, 2012; pp. 717–720.
20. Kang, U.; Tsourakakis, C.E.; Faloutsos, C. PEGASUS: A Peta-Scale Graph Mining System—Implementation and Observations. In Proceedings of the 9th IEEE International Conference on Data Mining (ICDM), Miami, FL, USA, 6–9 December 2009.
21. Akoglu, L.; McGlohon, M.; Faloutsos, C. OddBall: Spotting Anomalies in Weighted Graphs. In Proceedings of the 14th Pacific-Asia Conference on Knowledge Discovery and Data Mining (PAKDD), Hyderabad, India, 21–24 June 2010.
22. Kang, U.; Lee, J.Y.; Koutra, D.; Faloutsos, C. Net-Ray: Visualizing and Mining Web-Scale Graphs. In Proceedings of the 18th Pacific-Asia Conference on Knowledge Discovery and Data Mining (PAKDD), Tainan, Taiwan, 13–16 May 2014.
23. Dunne, C.; Shneiderman, B. Motif Simplification: Improving Network Visualization Readability with Fan, Connector, and Clique Glyphs. In Proceedings of the SIGCHI Conference on Human Factors in Computing Systems (CHI), Paris, France, 27 April–2 May 2013; ACM: New York, NY, USA, 2013; pp. 3247–3256.
24. Nielsen, J. Website Response Times. 21 June 2010. Available online: http://www.nngroup.com/articles/website-response-times/(accessed on 17 November 2015)
25. Mishra, C.; Koudas, N. Interactive query refinement. In Proceedings of the 12th International Conference on Extending Database Technology (EDBT 2009), Saint Petersburg, Russia, 24–26 March 2009; pp. 862–873.

26. Jiang, L.; Nandi, A. SnapToQuery: Providing Interactive Feedback during Exploratory Query Specification. *PVLDB* **2015**, *8*, 1250–1261.

27. Çetintemel, U.; Cherniack, M.; DeBrabant, J.; Diao, Y.; Dimitriadou, K.; Kalinin, A.; Papaemmanouil, O.; Zdonik, S.B. Query Steering for Interactive Data Exploration. In Proceedings of the Sixth Biennial Conference on Innovative Data Systems Research (CIDR 2013), Asilomar, CA, USA, 6–9 January 2013.

28. Chatzopoulou, G.; Eirinaki, M.; Polyzotis, N. Query Recommendations for Interactive Database Exploration. In Proceedings of the 21st International Conference on Scientific and Statistical Database Management (SSDBM 2009), New Orleans, LA, USA, 2–4 June 2009; Winslett, M., Ed.; Springer: New Orleans, LA, USA, 2009; Volume 5566, pp. 3–18.

29. Goethals, B.; Moens, S.; Vreeken, J. MIME: A Framework for Interactive Visual Pattern Mining. In Proceedings of the 17th ACM International Conference on Knowledge Discovery and Data Mining (SIGKDD), San Diego, CA, USA, 21–24 August 2011; ACM: New York, NY, USA, 2011; pp. 757–760.

30. Vartak, M.; Rahman, S.; Madden, S.; Parameswaran, A.; Polyzotis, N. SeeDB: Efficient Data-driven Visualization Recommendations to Support Visual Analytics. *Proc. VLDB Endow.* **2015**, *8*, 2182–2193.

31. Shahaf, D.; Yang, J.; Suen, C.; Jacobs, J.; Wang, H.; Leskovec, J. Information cartography: Creating zoomable, large-scale maps of information. In Proceedings of the 19th ACM SIGKDD International Conference on Knowledge Discovery and Data Mining (KDD 2013), Chicago, IL, USA, 11–14 August 2013; pp. 1097–1105.

32. Chau, D.H.; Akoglu, L.; Vreeken, J.; Tong, H.; Faloutsos, C. TOURVIZ: Interactive Visualization of Connection Pathways in Large Graphs. In Proceedings of the 18th ACM International Conference on Knowledge Discovery and Data Mining (SIGKDD), Beijing, China, 12–16 August 2012; ACM: New York, NY, USA, 2012.

33. Rodrigues, J.F., Jr.; Tong, H.; Traina, A.J.M.; Faloutsos, C.; Leskovec, J. GMine: A System for Scalable, Interactive Graph Visualization and Mining. In Proceedings of the 32nd International Conference on Very Large Data Bases, Seoul, Korea, 12–15 September 2006; pp. 1195–1198.

34. Khoa, N.L.D.; Chawla, S. Robust Outlier Detection Using Commute Time and Eigenspace Embedding. In Proceedings of the 14th Pacific-Asia Conference on Knowledge Discovery and Data Mining (PAKDD), Hyderabad, India, 21–24 June 2010; Springer: Berlin, Germany; Volume 6119, pp. 422–434.

35. Akoglu, L.; Tong, H.; Koutra, D. Graph-based Anomaly Detection and Description: A Survey. *Data Min. Knowl. Discov. (DAMI)* **2014**, *29*, 626–688.

36. Ranshous, S.; Shen, S.; Koutra, D.; Harenberg, S.; Faloutsos, C.; Samatova, N.F. Anomaly detection in dynamic networks: A survey. *WIREs Comput. Statist.* **2015**, *7*, 223–247.

37. Breunig, M.M.; Kriegel, H.P.; Ng, R.T.; Sander, J. LOF: Identifying density-based local outliers. In Proceedings of the ACM SIGMOD 2000 International Conference on Management of Data, Dalles, TX, USA, 15–18 May 2000; ACM: New York, NY, USA, 2000; Volume 29, pp. 93–104.

38. Lee, J.Y.; Kang, U.; Koutra, D.; Faloutsos, C. Fast Outlier Detection Despite the Duplicates. In Proceedings of the WWW 2013 Companion, Rio de Janeiro, Brazil, 13–17 May 2013.

39. Chakrabarti, D. Autopart: Parameter-free graph partitioning and outlier detection. In Proceedings of the 8th European Conference on Principles of Data Mining and Knowledge Discovery, Pisa, Italy, 20–24 September 2004; Springer: Berlin, Germany, 2004; pp. 112–124.

40. Xu, X.; Yuruk, N.; Feng, Z.; Schweiger, T.A. Scan: A structural clustering algorithm for networks. In Proceedings of the 13th ACM SIGKDD International Conference on Knowledge Discovery and Data Mining, San Jose, CA, USA, 12–15 August 2007; ACM: New York, NY, USA, 2007; pp. 824–833.

41. Jiang, M.; Cui, P.; Beutel, A.; Faloutsos, C.; Yang, S. Catchsync: catching synchronized behavior in large directed graphs. In Proceedings of the 20th ACM SIGKDD International Conference on Knowledge Discovery and Data Mining, New York, NY, USA, 24–27 August 2014; ACM: New York, NY, USA, 2014; pp. 941–950.

42. Faloutsos, M.; Faloutsos, P.; Faloutsos, C. On power-law relationships of the internet topology. In *ACM SIGCOMM Computer Communication Review*; ACM: New York, NY, USA, 1999; Volume 29, pp. 251–262.

43. Prakash, B.A.; Sridharan, A.; Seshadri, M.; Machiraju, S.; Faloutsos, C. EigenSpokes: Surprising Patterns and Scalable Community Chipping in Large Graphs. In Proceedings of the Pacific-Asia Conference on Knowledge Discovery and Data Mining (PAKDD), Hyderabad, India, 21–24 June 2010; pp. 435–448.

44. Page, L.; Brin, S.; Motwani, R.; Winograd, T. The PageRank Citation Ranking: Bringing Order to the Web; Stanford Digital Library Technologies Project. In Proceedings of the 7th International World Wide Web Conference, Brisbane, Australia, 14–18 April 1998.

45. Alemi, M.; Haghighi, H.; Shahrivari, S. CCFinder: Using Spark to find clustering coefficient in big graphs. *J. Supercomput.* **2017**, 1–28, doi:10.1007/s11227-017-2040-8.
46. Kriegel, H.P.; Kroger, P.; Schubert, E.; Zimek, A. Interpreting and unifying outlier scores. In Proceedings of the 2011 SIAM International Conference on Data Mining, Phoenix, AZ, USA, 28–30 April 2011.
47. Wang, J.; Zuo, X.; Dai, Z.; Xia, M.; Zhao, Z.; Zhao, X.; Jia, J.; Han, Y.; He, Y. Disrupted functional brain connectome in individuals at risk for Alzheimer's disease. *Biol. Psychiatry* **2013**, *73*, 472–481.
48. Leskovec, J. Stanford Large Network Dataset Collection. Available online: http://snap.stanford.edu/data/cit-HepTh.html (accessed on 17 November 2015)

informatics

MDPI

Article

Visual Exploration of Large Multidimensional Data Using Parallel Coordinates on Big Data Infrastructure

Joris Sansen *, Gaëlle Richer , Timothée Jourde , Frédéric Lalanne, David Auber and Romain Bourqui

LaBRI, UMR 5800, Université de Bordeaux, 351, cours de la Libération F-33405 Talence Cedex, France; gaelle.richer@u-bordeaux.fr (G.R.); itim.lcf@gmail.com (T.J.); frederic.lalanne@u-bordeaux.fr (F.L.); david.auber@labri.fr (D.A.); romain.bourqui@labri.fr (R.B.)
* Correspondence: joris.sansen@u-bordeaux.fr

Academic Editors: Achim Ebert and Gunther H. Weber
Received: 31 May 2017; Accepted: 10 July 2017; Published: 12 July 2017

Abstract: The increase of data collection in various domains calls for an adaptation of methods of visualization to tackle magnitudes exceeding the number of available pixels on screens and challenging interactivity. This growth of datasets size has been supported by the advent of accessible and scalable storage and computing infrastructure. Similarly, visualization systems need perceptual and interactive scalability. We present a complete system, complying with the constraints of aforesaid environment, for visual exploration of large multidimensional data with parallel coordinates. Perceptual scalability is addressed with data abstraction while interactions rely on server-side data-intensive computation and hardware-accelerated rendering on the client-side. The system employs a hybrid computing method to accommodate pre-computing time or space constraints and achieves responsiveness for main parallel coordinates plot interaction tools on billions of records.

Keywords: big data; multidimensional data; parallel coordinates; interactive data exploration and discovery; distributed computing

---

1. Introduction

Recent years have seen a striking increase in the amount of collected and generated data. For instance, in web analytics, visitors' behavior is analyzed to the event-level to improve services and marketing strategies. Naturally, such amounts bring about new challenges for storing, querying and analyzing these ever-growing datasets. Among other approaches, information visualization enables the exploration and understanding of complex data without prior knowledge of the patterns and trends to identify. However, traditional information visualization systems are not fitted for large-scale data exploration and have to be adapted to tackle the new challenges that arise from the surge of dataset sizes. The first consequence of this growth is the increase of the time needed to process the data. Another consequence is the separation of the data storage from the visualization client. Often, large data are stored on distant repositories and cannot reasonably be moved since the time (and storage on the destination computer) necessary for such operation is too important to be considered. Finally, the screen space is physically limited by its number of available pixels and large amounts of visual elements either cannot fit or become indistinguishable. Therefore, scalable visualization systems have to address three main challenges: perceptual scalability, interactive scalability and remoteness. *Perceptual scalability* refers to screen space and user capabilities limitations when depicting large data. *Interactive scalability* encompasses the latencies incurred by processing and querying large data for interaction. *Remoteness* corresponds to data processing and visualization being performed at separate locations which induces data transfer. To limit latencies during interaction, the visualization system has to be designed with this major bottleneck in mind.

Various visualization techniques are dedicated to quantitative multivariate data: scatterplot matrices [1], hyperboxes [2], star coordinates [3], Andrew curves [4] and parallel coordinates [5] are well-known examples. In the parallel coordinates technique, all dimensions of the multidimensional data are represented as parallel axes and each record as a polyline. A record's polyline intersects each axis at its corresponding dimensional value (see Figure 1a). The strength of this plot is that it offers an overview of the multidimensional data since each dimension is displayed uniformly. However, the patterns revealed by a plot strongly depend on the arbitrary placement of its axes [6]. For this reason, interactive axis reordering and flipping are necessary to grasp all pairwise relations between dimensions. *Brushing* is another fundamental interaction as it makes possible to select and enhance a subset of multidimensional items on top of the original view, following a focus+context approach. For hundred of thousands of items, traditional line-based parallel coordinate plot induces substantial visual clutter [7] (see Figure 1b) and interaction latencies. Several approaches have been proposed to deal with these challenges, ranging from using visual enhancement [8], via interactive tools [9], to joint representations of data subsets [10]. These techniques, either data-driven or screen-based, successfully handle up to millions of records. However, some may conceal patterns and most techniques are not adapted for remote visualization constraints.

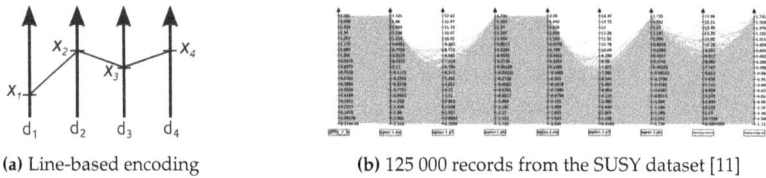

(a) Line-based encoding   (b) 125 000 records from the SUSY dataset [11]

Figure 1. Traditional parallel coordinates. (a) Representation of a four-dimensional record X with a polyline (b) Example of clutter produced by the crossing and overlapping of many lines.

Adapting parallel coordinates for the interactive visualization of large and remote data requires schemes to bound the size of the transferred data from the storing location to the visualization point. One solution lies in using a form of data reduction and representing an abstract version of the data, of a bounded size, regardless of the original dataset size. A second requirement is having a similar adaptability (or scalability) in the processing capabilities of the system, relative to the workload induced by interactive manipulation like *brushing*. This can be achieved using horizontal scalable systems which seamlessly parallelize processing over a network of computing and storage units where data is replicated and partitioned. While vertical scalability denotes the addition of extra resources on the same unit (e.g., expanding storage or adding a CPU), horizontal scalability is the concept of using multiple units as a single one. Systems that scale horizontally can theoretically satisfy any increase in processing demand with the expansion of their network of units.

The main contribution of this paper is a scalable system suited for interactive visual exploration of large multidimensional data with an abstract parallel coordinates representation. The novelty of this approach is the horizontal scalability of the system which relies on distributed processing in the form of pre-computation and on-demand computation, and on data aggregation. The use of distributed processing using scalable components for interaction allows performance to be tailored seamlessly to achieve responsiveness. The use of data aggregation provides boundaries for: (1) storage for pre-computed data, (2) data transfer between a distributed storage and computing infrastructure and a rendering client and (3) displayed items.

In the following section, we present two topics of related work: perceptual scalability on parallel coordinates and large-scale visualization systems. Next, in Section 3, we describe the work-flow as well as the main components of our system. We then describe the abstract parallel coordinates representation implemented (Section 4) and discuss interactive scalability for panning, zooming, axis rearrangement and selection in Section 5. In Section 6 we demonstrate the effectiveness of the abstract visualization and in Section 7, the system performances. Finally, we present the conclusions and discuss the possible future works.

2. Related Work

On line-based parallel coordinate representations, each multidimensional data item is depicted with a polyline and the multiplication of such polylines produces a rapid increase in the usage of pixels as the dataset gets larger. Such consumption of pixels is not practical for large datasets: the plot rapidly becomes overcrowded as polylines cross and overlap. Moreover, rendering and interaction complexity generally depend on the dataset size which requires fast data querying techniques to ensure responsiveness for large datasets. In this context, we discuss two topics of related work. First, existing methods for effectively avoiding clutter and over-plot in parallel coordinates and secondly, previous works on scalable visualization systems for parallel-coordinate representations and others.

2.1. Overcoming Clutter in Parallel Coordinates

A first approach to deal with clutter in parallel coordinates lies in visual enhancement for clarifying dense areas and facilitating pattern recognition. This can be accomplished using density-based methods [8,12,13] or curves instead of lines [14–16]. However, such methods do not scale as they are, since they require to draw every item. According to Ellis and Dix [17], clutter reduction techniques that directly meet the scalable criteria are sampling, filtering, and clustering. In Ellis and Dix [9], sampling is applied locally, with an interactive lens, to unclutter the plot. Similarly, filtering is used by numerous prior works [18–20] as an interactive and user-controlled tool. Johansson and Cooper [21] introduced a screen-space measure to filter items while preserving significant features with better results than data-space measures. Using sampling or filtering for data reduction has the advantage of keeping a consistent visual representation of items, regardless of the chosen level of detail. However, these approaches have limitations as a general methods to avoid clutter. For example, filtering often requires prerequisite knowledge of the data, otherwise meaningful structures or outliers may be unintentionally hidden.

Clustering the data is another possible approach adopted in many previous works. Cluster-based enhancement on traditional representations facilitates identification of multidimensional groups of items by using visual cues such as color, opacity or bundling. For example, in Johansson et al. [22] transfer functions are applied independently to high-precision textures generated for each computed cluster. Luo et al. [23] use curved bundles which both help to trace lines trough axes and facilitate the identification of groups. Representing aggregates instead of their covered subset of items/polylines further reduces visual clutter and may speed up the rendering. Aggregates have been represented using statistical metrics [24], envelopes or bounding-boxes [5,10,18,25]. Fua et al. [18] render both multidimensional clusters as polygons and mean values as dense lines and McDonnell and Mueller [10] draw bundled envelopes. This approach has been generalized to hierarchical clustering [18,26] to support multiscale exploration.

As stated by Palmas et al. [27], multidimensional data reduction (sampling or clustering) enhances global trends to the cost of potentially concealing pairwise relationships between dimensions. Novotny and Hauser [28] propose a hybrid solution which separates the rendering of outliers from the rest. On the one hand, data are clustered using a binning clustering technique and two-dimensional bins are represented using parallelograms. On the other hand, outliers are rendered using polylines. Doing so, in-between axis information is not degraded and clusters are sharper due to outliers having been primarily filtered out. The bundled parallel coordinates presented by Palmas et al. [27] uses one-dimensional clustering which improves visual continuity on axes compared to [28]. One-dimensional clustering creates meta-link between axis aggregates, similar to Kosara et al. [29]'s Parallel Sets which deals with nominal data inducing natural groups on axes. Matchmaker [30] presents comparable weighted and curved meta-links between axes, however, the clustering technique employed is applied on groups of dimensions which relates to the specific case of inherent groupings between dimensions.

Most of these techniques efficiently handle up to dozens of thousands of items for rendering. However, they do not scale to millions or billions of items. To the best of our knowledge, it is mainly due to their computational and memory costs with standard computers. This challenge is precisely the scope of this paper.

2.2. Scalable Visualization Systems

When data becomes too massive to fit inside a computer's memory, input/output communication with slower, external or remote, memory and data-processing time create a substantial bottleneck for interactive visualization. Solutions arise from hardware upgrades (vertical scalability) with parallel processing, distributed methods (horizontal scalability) and other strategies (e.g., out-of-core methods, data abstraction, tailored indexing).

Hardware acceleration of modern GPU is now frequently used [18,22,25,31] to render millions of items in real-time. Parallel processing can also serve data processing with dedicated multi-core computing units [32], distributed systems of servers running on commodity hardware [33,34] or multi-threading [35]. Other mechanisms to enable interactivity at large scale include pre-computing [31,33] and pre-fetching, incremental approaches, and data abstraction with multi-resolution representations [18,33,36]. For instance, Rübel et al. [32] and Perrot et al. [33] propose systems associating several of these solutions to scale interactive representations (respectively parallel coordinate and heatmaps) to extremely large datasets. Rübel et al. [32] present a system combining the multi-resolution binning technique of Novotny and Hauser [28] and FastBit [37], an index/query tool, on a supercomputer. Perrot et al. [33] use Canopy clustering for multilevel heatmaps computed over a distributed infrastructure, most similar to our setting.

Ideally, a visual exploration system should support interaction following the user's flow of thoughts [38] i.e., should operate in less than one second. Generally, bringing interactivity on modest systems operating on large data is a trade-off between pre-computation, approximate computation [39] and the cost of the hardware used. For instance, Rübel et al. [32] use a supercomputer system to achieve 0.15 s response time for tracing 500 items, over a dataset of almost 200 million items, with 100 computing units. In this paper, we target common-hardware and use an infrastructure where the distributed storage is leveraged for computation. Compared to supercomputers, these infrastructures do not offer as efficient communication between units and generally less computing power. Nevertheless, they are much more accessible and affordable and quite common nowadays in various domains (e.g., web analytics). To address data transfer, we use data reduction in the form of aggregation, with per-dimension clustering, as described in the following section.

3. System Overview

Our system lies on two major parts, (1) an abstract representation which addresses the transfer bottleneck and the perceptual scalability, (2) all interactions requiring the full data to be computed occur in a distributed manner on a data-intensive back-end platform. This way, two levers are available for supporting increasingly large data sets and/or improving interactivity: the level of detail which, among other things, impacts the client-server transfer time, and the computing and storage resources of the back-end by expanding the network of computing units. The main components of our system are laid out on Figure 2: the back-end is composed of distributed components that store and process the data, as well as a server acting as its interface. The client renders and animates the representations and interacts with the server for completing interactions that involve the full data. First, we describe the overall work-flow that takes place on the data-intensive platform. Then, we motivate and precise the abstraction used.

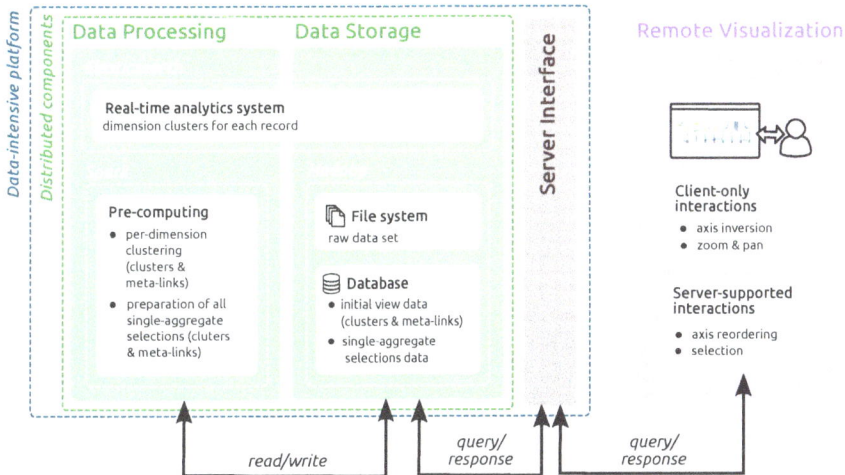

Figure 2. System components. Data processing steps using Spark occur once and consist in the clustering of dimension values, forming the abstraction (clusters and meta-links) for the initial view and all axis ordering, as well as preparing all single-aggregate selections. The server interface communicates with two types of storage system to answer queries received from the rendering client, one holding prepared data, the other processing aggregation on-demand.

3.1. Distributed Processing Work-Flow

The platform is used for hosting the raw data and computing abstract representations for different states of the visualization: corresponding to the initial view or resulting from user interactions. The different types of queries performed by the client are: display of a dataset for a given axis ordering and brushing data given a set of selected aggregates. The back-end computes a per-dimension clustering of the raw dataset and the aggregates (clusters and meta-links) corresponding to those two types of query. While the aggregates formed from the full data are always computed in a preparatory step, the one formed from subsets (used in brushing), result from both beforehand and on-demand computed data depending on the query type.

The pre-computing step (see Figure 2) encompasses computing an abstraction of the raw data and the results of several queries which are inserted into the distributed database. The abstraction, resulting from per-dimension clustering, is subsequently indexed into the real-time analytics system in the form of n vectors (one for each record) where each components gives the cluster identifier of the corresponding dimension. Queries are forwarded by the server, depending on their type, to the prepared data storage or to the analytics system for on-demand filtering and aggregation. On the latter system, cost is related to the number of records involved whereas in the former it is constant.

3.2. Bounding Data Transfer

We use per-dimension clustering as an aggregation strategy. The number of produced clusters for each dimension (called resolution parameter k) together with the number of dimensions (noted d), determines the number of rendered elements. Consequently, controlling and tuning this resolution parameter or degree of reduction conditions the transfer data size, the client memory footprint and the amount of displayed items. Moreover, rendering and interactions that can be performed on the abstract data have complexity independent of the underlying data set size, meeting the interactive scalability criterion. Thus, the performances of the abstract visualization become solely dependent on k and the number of displayed dimension rather than the actual number of records.

Clustering algorithms group items based on a measure of similarity which creates a simplified version of the original data, ideally with meaningful groups (clusters). No omnipotent clustering algorithm is suited for all type of data. Various algorithm have been used in state-of-the-art techniques (Fua et al. [18] uses Birch algorithm, Van Long and Linsen [26] uses a grid-based algorithm, Palmas et al. [27] uses kernel density estimation). In this article, any clustering algorithm can be chosen as long as it produces a strict partitioning of the interval of dimension values, i.e., all resulting clusters should form non-overlapping intervals. Among the various possibilities (k-means [40], DBSCAN [41], binning or adaptive binning as used in [28]). In the performance evaluation of our system, we used Perrot et al. [33]'s Canopy clustering since our prerequisites were similar: a small number of passes over the data to limit processing time and an efficient distributed implementation.

For a fixed resolution parameter k, defined prior to pre-computation, we expect a maximum of k clusters per-dimension. An abstraction is composed of those $d \cdot k$ clusters and the meta-links induces between axes, amounting to at most k^2 between each pair of axes. Clusters have, as properties, their extrema values, their cardinality/weight and a distribution of their values, meta-links have their size. As a result, the total number of items to be transferred between server and client is effectively bounded by $k \cdot d + k^2 \cdot d$, where d is the number of displayed dimensions. Additionally, exchanges between the server and the two different storage components benefit from the same bound. Indeed, in both cases the transferred data have been aggregated beforehand.

4. Abstract Parallel Coordinates Design

In this section, we focus on the abstract representation design. Similar to Palmas et al. [27], we cluster values on each dimension and bundles lines between pairs of clusters into meta-links. We explored visual metaphors for both types of aggregates (clusters and meta-links) inspired by Palmas et al. [27] and Kosara et al. [29]'s Parallel Sets while striving to reduce occlusion and retain the general overview. The overall design also resembles Sankey diagrams (e.g., [42]), which are specifically designed to represent flow data.

Our abstract parallel coordinates design uses a *distribution visual encoding* of clusters. With such encoding, the size of cluster representations depends on the number of elements in the corresponding subset (see Figures 3 and 4b). This approach has been widely used in many visualization techniques for rendering abstract elements (e.g., matrix based diagrams). Using this weight depiction, one can easily find which clusters are the most important for a given dimension and if they are connected to many small clusters or rather to a few large ones. Each cluster of a dimension covers both an interval of values and a subset of records/items, with intervals being non-overlapping as stated in Section 3. Using this visual encoding does not natively provide the information of the interval covered by the subset. To solve this issue, we added an inter-clusters spacing to convey an approximation of their covered interval (see Figure 4b). This way, the relative distance between clusters can be assessed and compared. To provide a better insight of this information we also implement an *interval visual encoding* similar to Palmas et al. [27]'s representation (see Figures 3 and 4c). These two different encoding strategies reflect the dual interpretation of clusters: either as a sub-space of a dimension or as set of close items. In both case, clusters surfaces are leveraged with the display of a smoothed mirrored histogram providing an overview of the values distribution (see Figure 3). This histogram is computed using 10 regular bins (equal-size intervals) for each cluster, with all bin sizes normalized per axis. For each attribute, the larger the bin is, the closer the histogram is to the edges of the cluster.

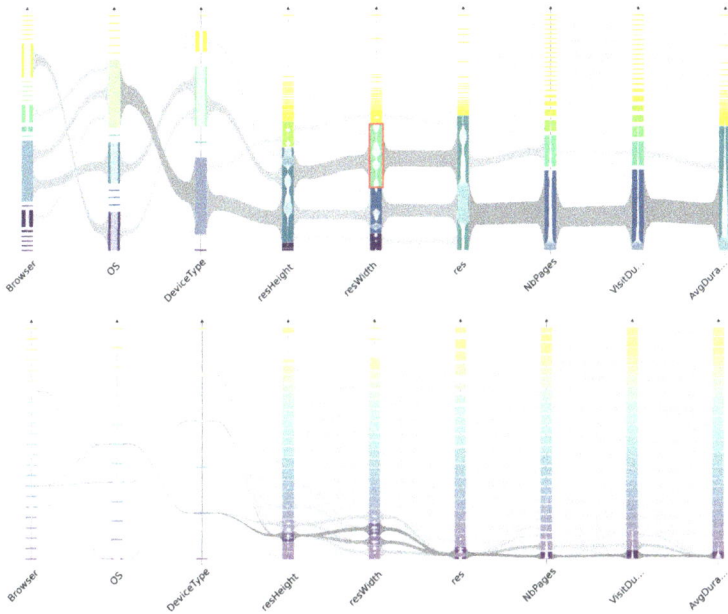

Figure 3. Both implemented visual encodings for the same abstraction ($k = 30$). On top, the *distribution encoding*; on the bottom, the *interval encoding*. Surrounded in red: the inner-cluster smoothed histogram view.

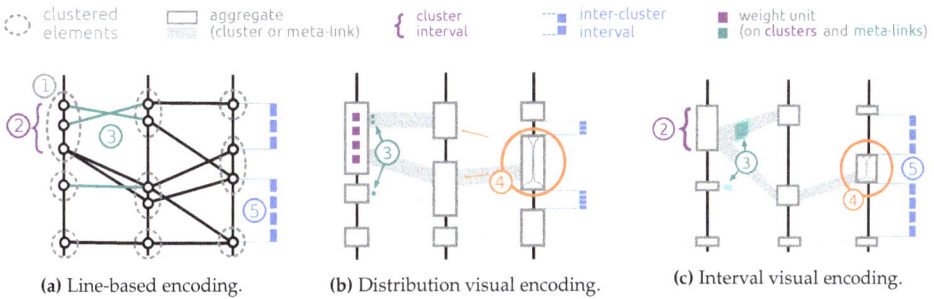

(a) Line-based encoding.　　**(b)** Distribution visual encoding.　　**(c)** Interval visual encoding.

Figure 4. Comparison of the two visual encodings proposed for abstract parallel coordinates compared to the line-based version displayed on (**a**). On (**a**), ① four elements forming the example cluster and ② two sets of connections forming two example meta-links. On (**b**), cluster and meta-links size encodes their weight. Meta-links anchor points on clusters are sorted relative to their destination to limit crossings as shown on ③. On (**c**), cluster sizes correspond to their interval size. Meta-link colors and size respectively depends on their weight and ends. They are depth sorted by weight and attached on each cluster ends to the highest density point as represented on ③. On both (**b**) and (**c**), the inner-cluster smoothed histogram is represented on ③. Finally, ④ shows that inter-cluster intervals can be compared, per-axis, on both encodings.

Meta-links are two-dimensional aggregates, and as such, bear an analogous encoding to clusters. They are represented as ribbons with the size of the subset they cover mapped to their thickness or ends sizes depending on the encoding. Several methods are used to reduce in-between axes clutter. In the distribution visual encoding, meta-links that end on the same cluster are vertically spaced and sorted to minimize crossings as in [29,30] which highly reduces clutter. In both encodings, meta-links are rendered as Bézier curves thinned down in their middle part. Additionally, color intensity and depth sorting are used to further enhance the largest meta-links.

All these features make possible, analyzing both aggregates (clusters and meta-links), to compare at a dimension scale the sizes, densities, separations and distributions on axes. Additionally, using the interval encoding, the slope of meta-links can be used to draw, to some extent, conclusion on the relationship between neighbouring dimensions. The distribution encoding emphasizes larger groups, and consequently major trends, while also tremendously reducing in-between axes clutter but does not provide the ability to assess relationship between dimensions due to the cluster positioning and meta-link positioning schemes. This illustrates the complementarity of both visual encodings.

5. Enabling Interactivity

With usual parallel coordinates tasks in mind, we first identify a set of interactions adapted to abstract parallel coordinates and complying with data processing and transfer constraints. We expose their pre-computation costs and the corresponding chosen strategy (pre-computing or on-demand processing).

5.1. Tasks & Interactions

Common tasks for parallel coordinates include gaining an overview, evaluating correlation between dimensions, identifying subsets of items presenting similar features, and searching for item with a particular multidimensional profile. Gaining a complete overview of multidimensional data implies being able to see every pairs of dimensions. However, without interaction, the ordering of axes only represents a fraction of all the information the dataset contains (non-contiguous dimension relationships are not represented). Bringing axes side-by-side allows an analysis of the meta-links and eases clusters distribution comparison. Using the interval encoding, it also allows the identification of correlation, recognizable by parallel meta-links between the two considered axes. Various tools exist to perform this operation: axis replacement, swap or move.

In abstract parallel coordinates, the aggregation of close dimension values facilitates the distinction of groups of items exhibiting similar features on a given dimension. Highlighting a subset of items permits tracing them across all dimensions. To find items falling into a specific range of values on different dimension, we implement a single-aggregate selection (see Figure 5a) triggered by click on clusters and meta-links and a compound aggregate selection triggered by axis sliders in a filtering fashion (see Figure 5b). Subset highlighting acts as a brushing interaction where selection only operates on aggregates (both clusters and bi-dimensional aggregates, also called meta-links). A selected subset is represented over meta-links and clusters with a gauge showing the proportion of the selected items they contain (see Figure 5a,b) in the distribution encoding and with color intensity in the interval encoding. Additionally, inner-cluster distributions are updated to reflect the distribution of the selected subset only. Making possible to select a subset and emphasize its distribution on the overall representation to compare and analyze the dataset is the core of the focus+context visualization step. The two following sections will describe the interaction tools implemented to allow these tasks. Other solutions that this work will not cover are the grand tour [43], optimum axis placement, and dimensionality reduction.

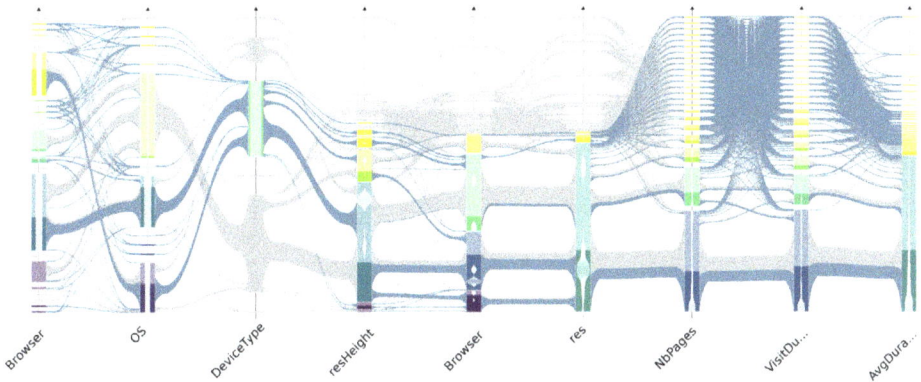

(a) Example of cluster selection (on the *DeviceType* axis)

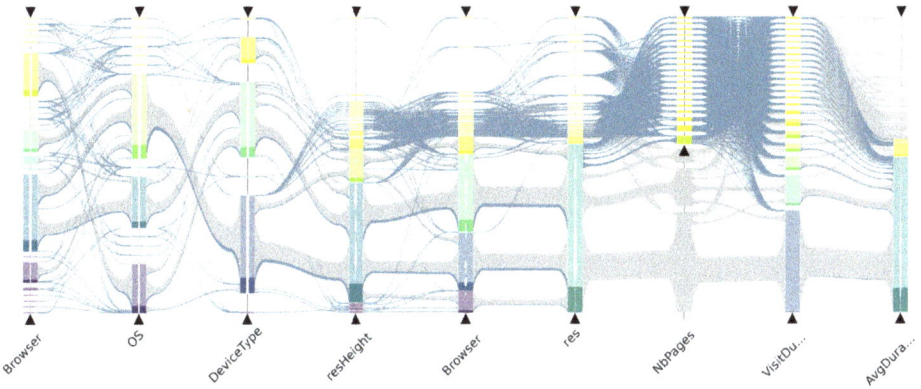

(b) Range selection on the highest values of the *NbPages* axis

Figure 5. Two selection views on a C2C dataset described in the next section. Here, items are visiting session on a website. Selecting an aggregate (cluster or meta-link) triggers the highlight of the selected subset through all the displayed axes. The inner-cluster histograms are also updated according to the selection. (**a**) Selection of the smartphone category (on *DeviceType*); (**b**) Selection of the higher range of values (between 37 and 300) on the *NbPages* axis which relates to the number of visited pages.

5.2. Client-Only Interaction & Parameters

Our system supports various interaction tools to help the user in its exploration. Several rendering parameters can be tweaked to instantly obtain different views as they are handled on the client side solely:

- Zoom and pan: the most classical interaction tool to explore and navigate within a representation.
- Axis height: used to tune the aspect ratio of the representation by increasing or reducing the height of the axes.
- Cluster width: can help the user by emphasizing or reducing the focus on the clusters (and the histogram within).
- Meta-link thickness: changing the thickness makes possible to emphasize the meta-links between clusters rather than the clusters themselves.

- Meta-link curvature: curving and bundling the meta-link is often used to reduce the clutter, tuning the degree of curvature makes possible to optimize the clutter reduction and Meta-link visibility.
- Inter-axis spacing: increasing (or reducing) the space between axes makes possible to increase the focus either on clusters or on meta-links and changes the aspect ratio of the representation.
- Intra-axis spacing: the percentage of empty space allocated to represent the relative distance between clusters (as presented in Section 4) can be reduced at no space (results in displaying a stacked histogram, see Figure 6) or increased to focus on the relative distance between clusters.
- Axis inversion: inverting an axis may help reducing unnecessary clutter by decreasing the number of crossings.

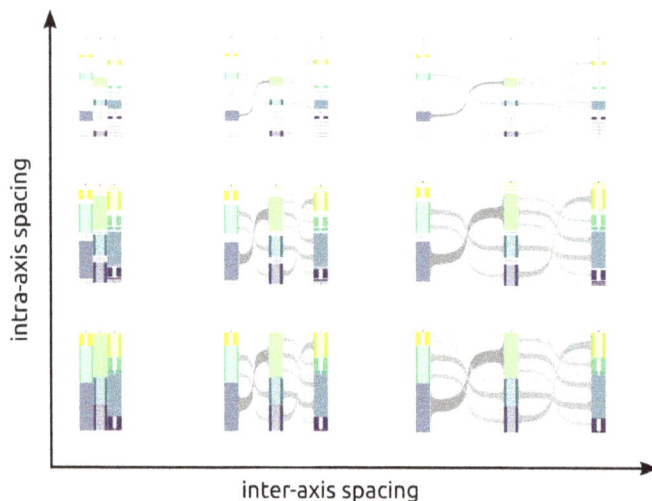

Figure 6. The modification of the two spacings, inter- and intra-axis, allows to tune the representation to get the best ratio depending on the user interest. Thus, we can go from no spaces at all, providing a stacked-histogram (bottom left), to a representation that rather focuses on relative distance between clusters and elements distribution between axes (top right).

5.3. Server-Supported Interaction

As the visualization client only stores the necessary data to display the abstracted parallel coordinate view, interactions that need to update the rendered data require for data to be transferred from the server to the client.

- Axis reordering: the use of this interaction tool is to compensate the main drawback of parallel coordinates: as axes are aligned, comparisons can only be made between pairs of attributes. Furthermore, datasets with a lot of attributes are difficult to read because of the horizontal resolution limit of screens. Moving an axis within the representation implies to update the meta-links between the moved axis and its neighbors (before and after the displacement).
- Removing or adding axis: Removing an axis is used to reduce the width of the representation by hiding unnecessary axis. As the need for an attribute can change over time and with user needs, each hidden axis can be shown again.
- Aggregate selection: This interaction allows to bring the focus on aggregates and emphasizes the distribution of the selected subset on the displayed attributes. The total number of meta-links for a given abstracted dataset is always less than $k^2 \cdot d^2$. Hence, the maximum number of different *single-aggregate selections* is $k \cdot d + k^2 \cdot d^2$, considering that subset selection can be applied to any cluster or meta-link in any axis ordering. The total number of aggregates to compute

for the operation is bounded by $k^4 \cdot d^4$. This boundary remains reasonable for moderate k (resolution parameter) and d (number of dimensions) values.

- Compound selection: This interaction has similar effect as the *Aggregate selection* (see Figure 5b) but is triggered by axis sliders that define an interval of interest on each dimension and allows the selection of several groups of consecutive clusters on different dimensions at once, corresponding to set operations between aggregates' subsets. Unlike *aggregate selection*, these selections cannot be reasonably pre-computed: multiple dimension criteria create a combinatorial explosion of different sub-selections. This is why we handle their computation in real-time.

Tracking individual items could easily be implemented using the on-demand computing scheme. However, this would require an additional medium for choosing and picking the desired item since the visualization technique does not discriminate individual items.

6. Perceptual Scalability

This section presents the effectiveness of the technique described in this article.

Our technique is based on abstract visualization and as such, functions identically regardless of the dataset size, large or not. Indeed, only the chosen resolution parameter influences the representation (see Section 5). On the contrary, traditional line-based technique do not scale visually to large data. Thus, we perform two case-studies: a comparison with traditional parallel coordinate plot using a state-of-the-art dataset (the *cars* dataset provided in [44] with 400 records) and an exploratory analysis of large data (with 1.6 billions of records).

6.1. Comparison to Traditional Parallel Coordinates

The *cars* dataset represents the characteristics of 400 cars using 20 attributes: the nine first attributes are either categorical (*constructor*) or boolean value (false is set to 0 and true to 1) while the remaining are integers and reals. The numerical attributes are clustered using a resolution parameter $k = 15$ while the categorical and boolean attributes are just aggregated by exact match (alphabetical or true/false values).

6.1.1. Gain Overview

At first sight, one can notice on Figure 7 that it is rather easy to figure out the distribution of clusters per attributes with our technique since clusters and meta-links are depicted as thick as the number of elements they represent. Thus, the distribution of elements over clusters for an attribute is one of the first information we obtain when visualizing the clusters. For example if we consider the boolean attributes in Figure 7b (the 8 attributes starting from the second one on the left), elements are clustered in two aggregates (except for the *Pickup* category), and it is rather straight forward to identify the distribution within each categories. Similarly, it is quite simple to find out the meta-links between one cluster and its neighborhood and in which proportion since meta-link thickness represents the number of elements. It is difficult to obtain the same information on Figure 7a as many lines overlap. This is a well-known limit of classical parallel coordinates and modern state-of-the-art techniques do present solutions to this issue (using density [12], curves[14] for example). If we consider the two attributes of *retail price* and *dealer cost* (surrounded in red on Figure 7a,b), we observe with both techniques, line-based and abstracted, that meta-links between the two axes are parallel. This indicates that both attributes of *retail price* and *dealer cost* are, to some degree, correlated. On the contrary, the two pairs of axes *HorsePower (HP)-City MPG* (surrounded in blue) and *Hwy MPG-weight* (surrounded in yellow) present meta-links that cross in a dense area. That tends to indicate an anti-correlation: high (resp. low) value for one attribute induces low (resp. high) value for the second attribute.

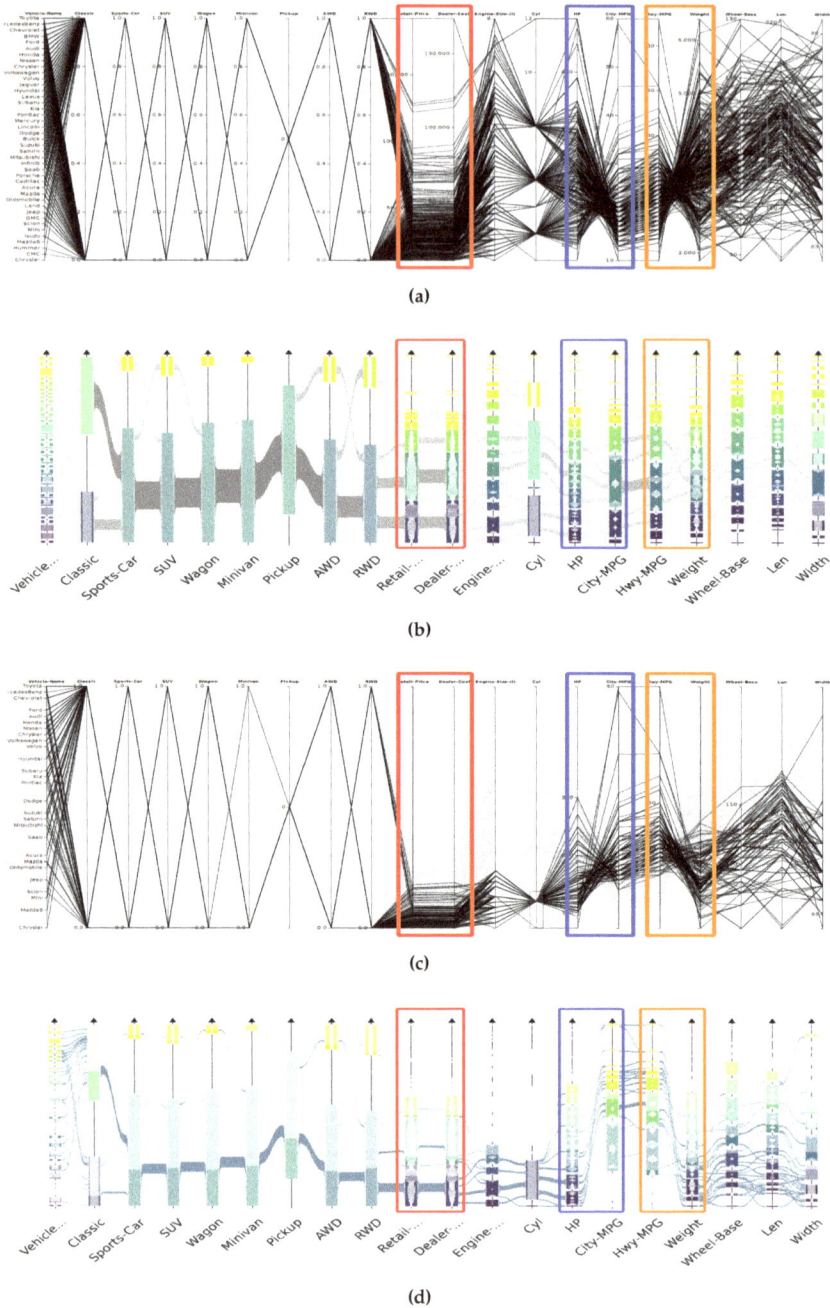

(a)

(b)

(c)

(d)

Figure 7. *Cars* dataset: (**a**,**b**) show an overview of the dataset; (**c**,**d**) are selections of the cars with less than 4 cylinders; (**a**–**c**) use the traditional parallel coordinates implementation of the Tulip software [45] and (**b**–**d**) use our technique.

6.1.2. Subset Highlighting

Both techniques make possible to highlight a subset corresponding to continuous values on a dimension and trace the distribution of its elements over all others. While Figure 7c suffers from clutter due to overlaps, our technique makes possible to highlight the selected subset for every attribute and the proportion it represents for each cluster of the plot. For example, selecting cars with less than 4 cylinders (see Figure 7c,d) emphasizes cars with a low *retail price, dealer cost, engine size, horsepower, weight, wheel base, length* and *width*. The analysis also indicates that the subset matching the selection also tends to have a lower mileage consumption (city and highway) and matches with city cars which are small and light-weight cars.

6.2. Large Dataset Visual Analysis

The C2C dataset contains 1.6 billions elements of web traffic data on a C2C (Consumer to Consumer) website (see Figure 8). This dataset contains 9 dimensions where each item is a sequence of pages visited by the same user in a given time. The three first axes represent the user's browser, OS and device type; the following three are its screen properties (resolution height, width and total resolution). The last ones correspond to the number of visited pages during the session, the session duration and the average time spent per page.

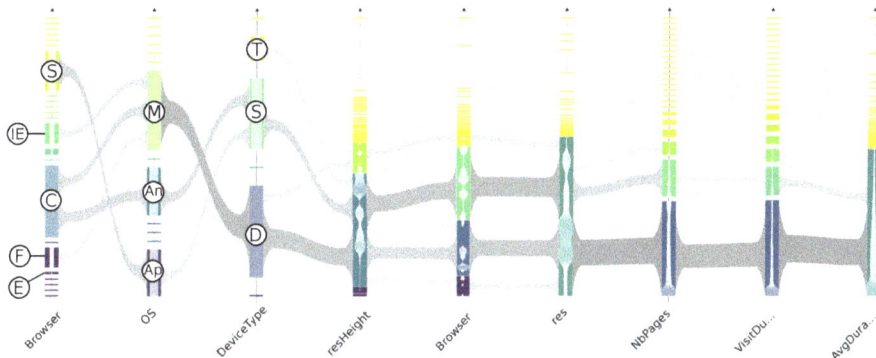

Figure 8. *C2C website* dataset overview with $k = 30$ (max. number of clusters per attributes). For this dataset, each item is a sequence of pages visited by the same user in a given time. The dataset represents various information collected during visitors navigation : system and device information (OS, browser, device, screen resolution) and navigation information (number of page visited, visit duration and average time spend per pages). The labeled clusters correspond to clusters described in the use case for the browser axis (Edge—E, Firefox—F, Chrome—C, Internet Explorer—E and Safari—S), for the OS axis (Apple—Ap, Android—An and Microsoft—M), and for the Device type (Desktop—D, Smartphone—S and Tablet—T).

One can easily identify on Figure 9 two major devices corresponding to *Desktops* and *Smartphones* and a smaller one, the *Tablets* (respectively labeled *D*, *S* and *T*). We can notice that users can be distinguished in two equal size categories according to the device they use: either a *mobile* device (Smartphones or tablets) or desktop device. If we consider the *Desktop* devices (see Figure 10a), they relate to various resolution sizes and are mainly used with a *Microsoft* Operating System (labeled *M*). We can also identify the four major browsers, in growing order of size *Chrome, Safari, Firefox* and *Internet Explorer* (respectively labeled *C*, *S* , *F* and *IE*). If we consider the *Smartphones* (see Figure 10b), the resolutions are smaller and two main OS are highlighted, the one with the smaller gauge relates to *Apple* operating systems (labeled *Ap*) while the bigger one is *Android* (labeled *An*). The used browser is largely *Safari* (labeled *S*) for Apple and *Chrome* for Android.

If we analyze this information, we can draw a few conclusions. First, it seems that Microsoft operating systems are mostly used on desktops while their browsers (Internet Explorer and Edge) are not in the mostly used. Second, Apple's products are mainly used with mobile devices (smart-phones and tablets) and almost solely with the dedicated browser (as highlighted on Figure 9a). Third, Chrome Browser is as much used with desktops than with smart-phones and mainly with Microsoft and Android operating system as emphasized on Figure 9b.

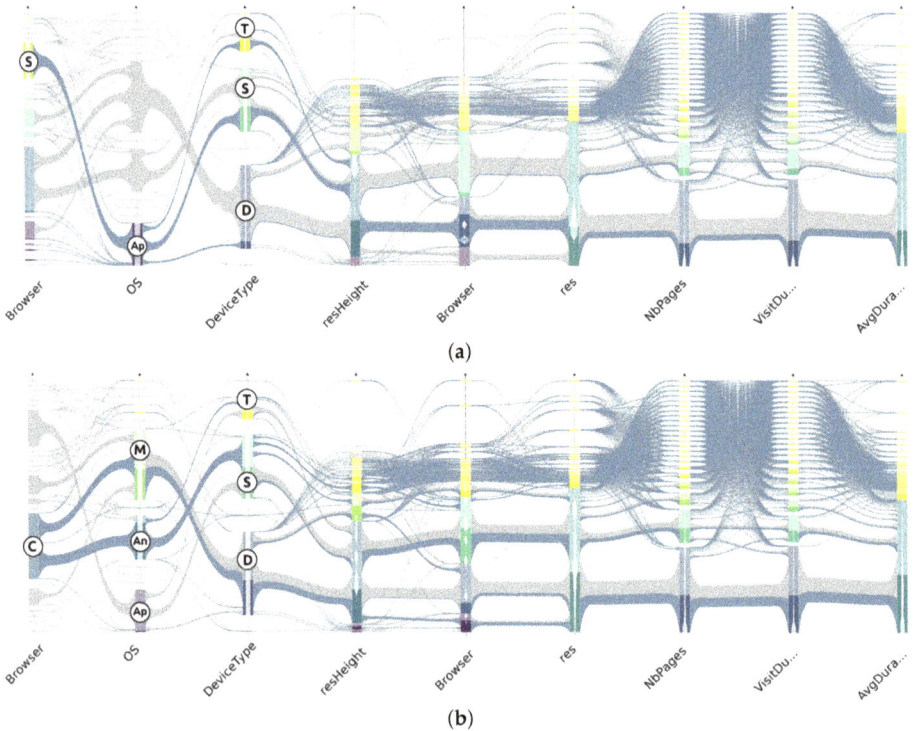

(a)

(b)

Figure 9. Subset highlighting on the *C2C website* dataset with $k = 30$: Selecting an aggregate highlights the subset over all the plot. (**a**) Selection of Apple Operating System (iOS or macOS) shows that users only use Safari browser (S) and mainly for mobile devices : tablets (T) and smartphones (S); (**b**) Selection of Chrome browser (C) highlights users using either desktop devices (D) and smartphone devices (S) and using the corresponding operating systems: Microsoft (M) and android (An).

We can also identify a low correlation between resolution width, height and total. The low strength of this correlation is understandable as only higher (resp. lower) total resolution result from high (resp. low) width and height. Medium total resolution can result from high width and low height (or the opposite) which decreases the correlation strength.

These two case studies demonstrate that analyses usually done with traditional parallel coordinates plot can also be performed using our novel abstract-parallel-coordinates technique. While analyses with classical parallel coordinates are element-oriented, using abstracted parallel coordinates, they are aggregate-oriented. This makes sense as abstracted parallel coordinates are dedicated to big data analysis and rather focus on major trends rather than single element analysis. As for any abstract method, the aggregation used, in our case the clustering algorithm and resolution parameters chosen, strongly affect the representation and thus, the possible analyses. Pre-computing the same dataset at various resolution parameter or/and using various algorithms makes possible to refine the analysis but at the expense of storage. Furthermore, using a higher resolution parameter has direct impact on the data transfer size and thus, can affect responsiveness negatively.

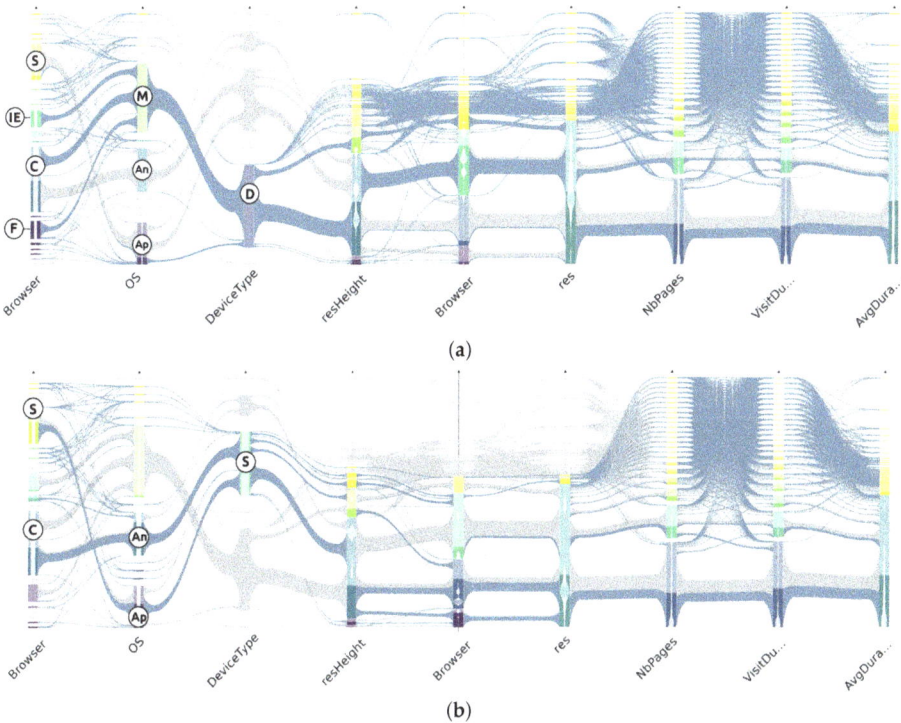

(a)

(b)

Figure 10. Subset selection: selecting an aggregate highlights the subset over all the plot. (**a**) Selection of desktop devices (D) highlights users using mainly Microsoft (M) operating system and Chrome (C), Internet Explorer (IE) and Firefox (F) browsers; (**b**) Selection of smartphone devices (S) highlights users using almost equally Android and Apple operating systems and the dedicated browsers (resp. Chrome (C) and Safari (S).

7. System Scalability

In our solution, two types of computation occur: pre-computing and on-demand query computation. Pre-computing and on-demand computation are what makes the system interactive, therefore we evaluated response time for pre-computed queries as well as on-demand queries. We also examined the scalability of the pre-processing step and on-demand queries execution times relative to the allocated resources.

Our benchmarks were made using a self-hosted platform composed of 16 computing units, each having 64 GB RAM and 2 × 6 hyper-threaded cores (2.1 GHz). Network capacity within the platform is 1 Gbit/s. Data are transferred on a local network between the server and the platform, as well as between the client and the server. Consequently, all response times were measured from the client perspective, i.e., they include the local network transfer cost from the platform to the server and from the server to the client.

7.1. Implementation Details

On the client side we use WebGL, a web variation of OpenGL, to perform GPU-based computation (using GLSL vertex and fragment shaders) and render visualizations in browsers. More precisely, the client is compiled from C++ to Javascript using emscripten which also binds OpenGL calls to WebGL instructions.

We implemented our back-end system in an Hadoop environment, a data-intensive infrastructure providing distributed storage (HDFS), computing (Spark-MapReduce) and database system (HBase) with horizontal scalability. For on-demand computing, we use Elasticsearch [46], a search system which runs along with the Hadoop components (see Figure 2).

Pre-computation operations (abstraction and prepared selections) are implemented with Spark. The basic of the method is to produce RDDs (for Resilient Distributed Dataset, the main data structure in Spark), where rows represent records with appropriate (key, values) and to retrieve the desired metrics (minimum, count, etc) by reducing values by keys.

In the following, d designates the number of dimensions and n the number of records. Computing the abstraction consists in computing the clusters and meta-links properties from the per-dimension aggregations. These aggregations provide, for each individual record an associated cluster identifier. From there, cluster extrema are obtained by reducing the $n \cdot d$ raw values indexed by a pair $(dimension, cluster)$.

Using these extrema, we built an RDD (called `clusterRDD`) by mapping each of the n rows of raw values to a d-tuple of $(cluster, bin)$, where the tuple order indicates the dimension. We then transform each row of the `clusterRDD` into d rows with key $(dimension, cluster, bin)$. The distribution and weight of each cluster are then computed by reducing the $n \cdot d$ rows by key. The number of values to process is therefore $O(n \cdot d)$. To count meta-link weights, we map each `clusterRDD` row to $\frac{d \cdot (d-1)}{2}$ rows, one for each unique pair of dimensions (i, j). These rows have keys (i, j, c_i, c_j), where c_i and c_j are cluster identifiers of dimensions i and j, and with value 1. Reducing by key the resulting RDD gives us the weights of all meta-links. The number of values to process is therefore $O(n \cdot d^2)$.

Clusters properties and meta-link weights for all the cluster selections are obtained with the same principle as for the abstraction. Since each record contributes to d different cluster selections, each `clusterRDD` will be transformed into d times more rows than for the abstraction computation. Therefore, cluster properties and meta-link weights computation processes $O(n \cdot d^2)$ and $O(n \cdot d^3)$ values respectively. Similarly, one record will contribute to $\frac{d \cdot (d-1)}{2}$ meta-links selections. Therefore, the cost to compute cluster properties and meta-link weights for those selections is $O(d^2)$ times larger, i.e., respectively $O(n \cdot d^3)$ and $O(n \cdot d^4)$.

These different steps use the `reduceByKey` Spark operation which consists in applying a reducing function onto values grouped by keys. This operation requires a shuffle step to redistribute values based on their keys between partitions before applying the reduce function. This step being memory-consuming, we segmented the cluster and meta-link selections computation in sequential steps so that each step has a computational cost comparable to the abstraction computation one.

7.2. Performance Evaluation Scope

First, we measured pre-computing time which include the clustering, the initial view computation with the preparation of axis rearrangements and single-aggregate selections. Second, we measured the performance of single and compound selection queries (respectively corresponding to prepared queries and on-demand queries). We also examined the scalability of the system by measuring the speedup obtained by allocating more resources for two operations: pre-computing step and on-demand brushing queries. The speedup for p corresponds to the ratio of the execution time using a reference number of computing units (usually one) over those using p computing units. Axis reordering queries were not evaluated as they work similarly to single-aggregate selection queries and have equivalent or better performance.

The outcomes of these experiments depend on different factors. In addition to the number of computing units and the communication overheads, they depend on the chosen dataset and abstraction properties (itself dependent on the clustering and the resolution parameter k).

We considered three types of generated datasets presenting different inter-dimension correlation factors [47], and with size ranging from 10^6 to 10^9 records for 15 dimensions and 15 clusters per dimension. The first type is *independent*: every pair of dimensions has a close to null correlation factor and close to the maximum number of meta-links between each couple of dimensions (about k^2). This type represents the worst case for our system. The second type is *correlated*, generated to obtain a correlation factor of at least 0.6 between each pair of dimensions. This dataset aims at mimicking the correlation that may be observed in *real* datasets. The last type has the minimum number of meta-links: k per couple of dimensions, and as such, is the best case. For each test, we average the measured time over three runs.

7.3. Pre-Computing Performance

Preparing a dataset consists of computing an abstraction and all *single-aggregate* (clusters and meta-links) selections. The result of this one-time operation populates HBase with abstract data and prepared queries. When using on-demand computation, an Elasticsearch index is also populated with cluster identifiers for each dimensional components of each record. Pre-computing time measurements (excluding Elasticsearch indexing) are shown in Figure 11a for different data sizes (in number of records) of the three types of dataset. The pre-computation processing is dominated by the computation of all single-aggregate selections. Although we consider the clustering method as parameter that should be chosen in regard of the studied data, we included this step in the measurement. However, it only accounts for about 0.10% of the total processing time on average.

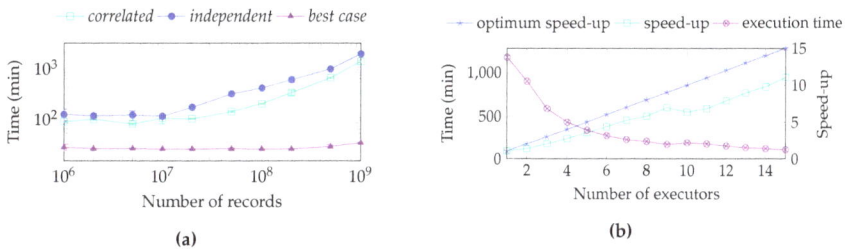

Figure 11. Performance evaluations for the pre-computation step (clustering, clusters and meta-links, single-aggregate selections and insertion into HBase). (**a**) Execution time for the three datasets and varying number of records; (**b**) Scalability evaluation for the *correlated* dataset with 10^7 records. The speedup is relative to a sequential execution.

This processing effectively occurs over pre-aggregated data composed of *identical* records which are records falling into the exact same cluster for each dimensions (as well as the same cluster histogram bin). These records are considered *identical* because they are not distinguishable given the interaction tools provided by our system. Due to its properties, the *best case* dataset is reduced to about $10 \cdot k$ such *identical* records (where 10 is the number of inner-cluster histogram bins). Additionally, the *best case* dataset presents only about k meta-links between each dimensions while *correlated* and *independent* have about k^2, that is k times more. Therefore, about k times more single-aggregate selections are pre-computed for those two later datasets. The difference in number of pre-computation and the size of their input data explains the diverging trend observed on Figure 11a between, on the one side the correlated and independent datasets and on the other side the best case dataset. Indeed, this optimization is not efficient for the correlated and independent datasets (almost no records are *identical*). On our platform, the pre-computation step takes up to 32 h for datasets of a billion of records. For the largest *independent* and *correlated* datasets, the execution time raises up to 24 h.

This computation step runs distributively, hence can be accelerated by increasing the number of computing units participating the task. Figure 11b presents the mean execution time and the corresponding speedup of this step run on the *correlated* dataset with 10^7 records. The speedup here measures the gain of allocating more executors relative to using a single one. At worst, it appears to be just over 60% of the optimum, which indicates that the communication overhead is reasonable and that the computation demonstrates good scalability. Thus, while the pre-computation appears to be costly, it is possible to allocate more resources to the platform to efficiently accelerate the processing time.

7.4. Prepared Selections Query Performance

To evaluate the performance of *single-aggregate selection* retrieval, we perform a comparative benchmark between fetching prepared results from HBase and on-demand computation with Elasticsearch (both using 16 computing units). For different dataset types and sizes, we measure response time for selection queries corresponding to each visible elements on an initial display, that is each of its clusters and meta-links. We average the results of the meta-links and clusters independently and examine results for the same operations using Elasticsearch. This way, we can characterize the gain of pre-computing compared to (distributed) on-demand computation. Figure 12 compiles these results for *correlated* datasets (*independent* datasets shows similar results). Overall, prepared queries can be retrieved in less than 0.1 s from HBase and the response time does not seem to increase with the dataset size while for Elasticsearch, the response time increases until exceeding the second for the largest datasets. This shows that past a certain size of dataset, there is a significant gain in using pre-computation.

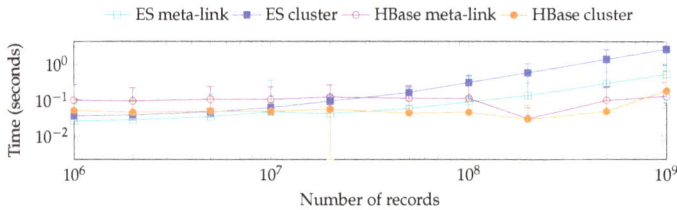

Figure 12. Prepared data fetching and on-demand computing execution times for identical queries. Here response time was measured for all possible single-aggregate selection o(clusters and meta-links) on an initial view of *correlated* datasets with varying number of records.

7.5. On-Demand Query Performance

Compound selection queries rely on on-demand processing and operate as a conditional filter followed by an aggregation. As such, they are dependent on the size of the targeted subset of records. Therefore, we are particularly interested in evaluating a cost bound for this operation to ensure good performance. To limit the cost of the aggregation, we query Elasticsearch for at most a subset of half the size of all the dataset by choosing to filter the dataset complement of the original query when it is preferable. In these cases, the server uses the pre-computed full abstraction to compose the original query's result. We choose to evaluate a higher upper bound for this type of query: the selection of all aggregates at once which means aggregating all clusters and meta-links. This case is presented on Figure 13a. The response times increase with the dataset size, which is expected. For datasets of up to 10^8 items, the response time is lower than one second, however, for upper sizes, the responses times are getting very important. It indicates that the number of computing units is not sufficient to keep up with the aggregating costs for *correlated* and *independent* datasets. To ensure sub-second compound selections for this type of datasets and configuration, more resources have to be allocated. The surprisingly low latencies observed for the *best case* dataset result from the same pre-aggregation treatment of *identical* records mention above: when populating Elasticsearch indices, only *distinct* records are inserted. Consequently, all aggregation are applied on smaller data.

As shown, the testing platform shows its limits for the largest datasets used. We measure the speedup gained by using different number of instances in Elasticsearch cluster relative to using only two. Figure 13b, presents the result of the execution time of all possible cluster selection queries on the 2×10^8 version of the *independent* dataset. Here, the speedup appears better than the optimum as it is relative to non-sequential execution. The results indicates that the computing capabilities of the platform are linked to the number of executor with no major loss of performances.

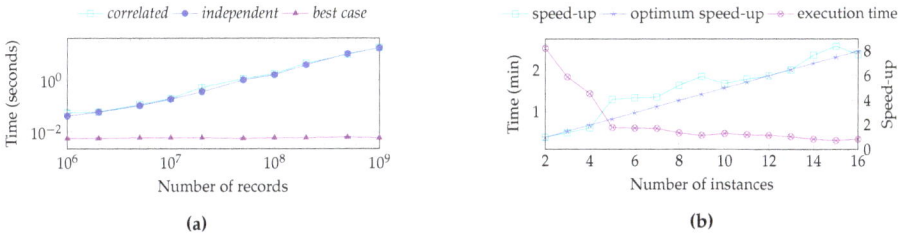

Figure 13. Performance evaluations for the on-demand computation model using Elasticsearch. (a) Upper bound for all on-demand queries, corresponding to computing the total initial view for varying dataset types and number of records; (b) Scalability of on-demand queries, tested on cluster-selection queries for an *independent* dataset with 2×10^8 records. The speedup is relative to using two instances.

7.6. Discussion

For these experiments, all components (client, platform and server) use local area networks and thus the system is not subjected to world network contingency that could add rather high latencies to transfers. While it is reasonable to consider the server-to-platform connection to be of controlled quality, the main limitation of our experiments is that we cannot not realistically assume a similar connection quality between clients and the server. Nevertheless, bounding data transfer addresses, at least partially, this concern.

However, the results obtained provide some good hints about the capabilities of the system and the pros and cons of beforehand and on-demand computing. Indeed, there is a trade-off between the time consumed by pre-computation and the gain in execution time of using prepared or partially prepared results. In our case, as the data abstraction requires pre-processing, we leveraged that first necessary step by also preparing query results for several interactions. As demonstrated, using pre-processing, we ensure lower response time compared to on-demand computation for *single-aggregate selection* queries on the largest datasets of our experimentation set. Choosing one strategy or the other for a given query type generally depends on three factors: the available storage, the time allowed for pre-processing and the needs of interactivity i.e., how fast the response-time should be. For the particular case of *compound aggregate selections*, there is a combinatorial explosion in the number of possible queries which motivates the choice of on-demand computation.

Overall, the experiments shows that on our testing platform, we achieve the targeted sub-second performance for all pre-computed queries: single-aggregate selections and axis reordering queries. Despite pre-computation taking up to 32 h for the largest dataset, the performances depends on several factors: they are closely related to the platform resources, the properties and number of dimensions of the tested data, and the resolution parameter. Since our experimentation demonstrated the horizontal scalability of the system, providing more resources would induce better performances seamlessly.

8. Conclusions & Future Work

In this paper, we present a system for interactive visual exploration of multidimensional data with a parallel coordinates representation. In order to support the visual exploration of large data sets, our system addresses three aspects of scalable visualization systems: perceptual scalability, interactive scalability and remoteness. We employ pre-processing and on-demand computation to bring interactivity to an abstract parallel coordinates representation, integrated in a client-server visualization system. A distant infrastructure handle distributed computing and data pre-aggregation and on-demand aggregation while a web-based remote visualization client for parallel rendering provides an interactive visualization that respect the perceptual scalability. The client is provided with various interaction tools to handle equivalent data exploration and data analyzes capabilities as state-of-the-art techniques. Among theses interactions, two of them (axis reordering and subset highlighting) imply data transfers between the infrastructure and the rendering client. Our system guarantees upper bounds for these transfers, ensuring interactivity and responsiveness of the rendering client. We present and evaluate an implementation of the back-end using using the Hadoop ecosystem (HDFS and HBase), Spark and Elasticsearch. The results indicate a good scalability and reasonable responsiveness of our system on up to a billion of items.

A first interesting work to pursue is thoroughly studying real-time distributed computing scalability of different technologies to ameliorate the pre-computing step in our solution. Lower preparation times would let users change dataset and tune clustering parameters in a more responsive way. Secondly, to bypass interaction latencies, prediction and pre-fetching [48] could be investigated. Other selection interactions could be implemented to further refine the highlighted subset: for instance selecting multiple ranges along one dimension or make logic operations between those selections. We also intend to examine how abstract parallel coordinates could be extended to a multi-scale visualization. Namely, how to create different levels of details and enable their exploration while guaranteeing a bounded number of displayed and transferred items.

Acknowledgments: This work has been carried out as part of SpeedData project supported by the French Investissement d'Avenir Program (Big Data—Cloud Computing topic) managed by Direction Générale des Entreprises (DGE) and BPIFrance.

Author Contributions: G.R., T.J., J.S. and R.B. performed the pipeline design, implementation and client-side data processing. F.L. performed the server-side data processing, system interface and management, and benchmark experiments. The overall work have been done under the supervision of R.B.. G.R. and J.S. wrote the paper and R.B. and D.A. performed the internal reviewing process.

Conflicts of Interest: The authors declare no conflict of interest.

References

1. Elmqvist, N.; Dragicevic, P.; Fekete, J. Rolling the Dice: Multidimensional Visual Exploration using Scatterplot Matrix Navigation. *IEEE Trans. Vis. Comput. Gr.* **2008**, *14*, 1148–1539.
2. Alpern, B.; Carter, L. The Hyperbox. In Proceedings of the 2nd IEEE Computer Society Press: Los Alamitos Conference on Visualization '91; San Diego, CA, USA, 22–25 October 1991; pp. 133–139.
3. Kandogan, E. Star coordinates: A multi-dimensional visualization technique with uniform treatment of dimensions. In Proceedings of the IEEE Information Visualization Symposium, Salt Lake City, UT, USA, 8–13 October 2000; Volume 650, p. 22.
4. Andrews, D.F. Plots of high-dimensional data. *Biometrics* **1972**, *28*, 125–136.
5. Inselberg, A. The plane with parallel coordinates. *Vis. Comput.* **1985**, *1*, 69–91.
6. Wegman, E.J. Hyperdimensional Data Analysis Using Parallel Coordinates. *J. Am. Stat. Assoc.* **1990**, *85*, 664–675, doi:10.1080/01621459.1990.10474926.
7. Heinrich, J.; Weiskopf, D. State of the art of parallel coordinates. *STAR Proc. Eurogrph.* **2013**, *2013*, 95–116.
8. Raidou, R.G.; Eisemann, M.; Breeuwer, M.; Eisemann, E.; Vilanova, A. Orientation-Enhanced Parallel Coordinate Plots. *IEEE Trans. Vis. Comput. Graph.* **2016**, *22*, 589–598.
9. Ellis, G.; Dix, A. Enabling Automatic Clutter Reduction in Parallel Coordinate Plots. *IEEE Trans. Vis. Comput. Graph.* **2006**, *12*, 717–724.
10. McDonnell, K.T.; Mueller, K. Illustrative Parallel Coordinates. *Comput. Graph. Forum* **2008**, *27*, 1031–1038.
11. Baldi, P.; Sadowski, P.; Whiteson, D. Searching for exotic particles in high-energy physics with deep learning. *Nat. Commun.* **2014**, *5*, 4308, doi:10.1038/ncomms5308.

12. Zhou, H.; Cui, W.; Qu, H.; Wu, Y.; Yuan, X.; Zhuo, W. *Splatting the Lines in Parallel Coordinates*; Blackwell Publishing Ltd.: Oxford, UK, 2009; Volume 28, pp. 759–766.
13. Nhon, D.T.; Wilkinson, L.; Anand, A. Stacking Graphic Elements to Avoid Over-Plotting. *IEEE Trans. Vis. Comput. Graph.* **2010**, *16*, 1044–1052.
14. Zhou, H.; Yuan, X.; Qu, H.; Cui, W.; Chen, B. *Visual Clustering in Parallel Coordinates*. Blackwell Publishing Ltd.: Oxford, UK, 2008; Volume 27, pp. 1047–1054.
15. Theisel, H. Higher Order Parallel Coordinates. In Proceedings of the 5th International Fall Workshop Vision, Modeling and Visualization, Saarbrücken, Germany, 22–24 November 2000; pp. 415–420.
16. Graham, M.; Kennedy, J. Using curves to enhance parallel coordinate visualisations. In Proceedings of the 7th International Conference on Information Visualization, London, UK, 18 July 2003; pp. 10–16.
17. Ellis, G.P.; Dix, A.J. A Taxonomy of Clutter Reduction for Information Visualisation. *IEEE Trans. Vis. Comput. Graph.* **2007**, *13*, 1216–1223.
18. Fua, Y.; Ward, M.O.; Rundensteiner, E.A. Hierarchical Parallel Coordinates for Exploration of Large Datasets. In Proceedings of the IEEE Visualization '99, San Francisco, CA, USA, 24–29 October 1999; pp. 43–50.
19. Andrienko, G.; Andrienko, N. Parallel Coordinates for Exploring Properties of Subsets. In Proceedings of the Second IEEE Computer Society International Conference on Coordinated & Multiple Views in Exploratory Visualization, Washington, DC, USA, 13 July 2004; pp. 93–104.
20. Artero, A.O.; de Oliveira, M.C.F.; Levkowitz, H. Uncovering Clusters in Crowded Parallel Coordinates Visualizations. In Proceedings of the 10th IEEE Symposium on Information Visualization (InfoVis 2004), Austin, TX, USA, 10–12 October 2004; pp. 81–88.
21. Johansson, J.; Cooper, M.D. A Screen Space Quality Method for Data Abstraction. *Comput. Graph. Forum* **2008**, *27*, 1039–1046.
22. Johansson, J.; Ljung, P.; Jern, M.; Cooper, M.D. Revealing Structure within Clustered Parallel Coordinates Displays. In Proceedings of the IEEE Symposium on Information Visualization (InfoVis 2005), Minneapolis, MN, USA, 23–25 October 2005; Stasko, J.T., Ward, M.O., Eds.; IEEE Computer Society: Washington, DC, USA, 2005; p. 17.
23. Luo, Y.; Weiskopf, D.; Zhang, H.; Kirkpatrick, A.E. Cluster Visualization in Parallel Coordinates Using Curve Bundles. *IEEE Trans. Vis. Comput. Graph.* **2008**, *18*, 1–12.
24. Siirtola, H. Direct manipulation of parallel coordinates. In Proceedings of the IEEE International Conference on Visualization, London, UK, 19–21 July 2000; pp. 373–378.
25. Beham, M.; Herzner, W.; Gröller, M.E.; Kehrer, J. Cupid: Cluster-Based Exploration of Geometry Generators with Parallel Coordinates and Radial Trees. *IEEE Trans. Vis. Comput. Graph.* **2014**, *20*, 1693–1702.
26. Van Long, T.; Linsen, L. *MultiClusterTree: Interactive Visual Exploration of Hierarchical Clusters in Multidimensional Multivariate Data*; Blackwell Publishing Ltd.: Oxford, UK, 2009; Volume 28, pp. 823–830.
27. Palmas, G.; Bachynskyi, M.; Oulasvirta, A.; Seidel, H.P.; Weinkauf, T. An edge-bundling layout for interactive parallel coordinates. In Proceedings of the IEEE Pacific Visualization Symposium, Yokohama, Japan, 4–7 March 2014; pp. 57–64.
28. Novotny, M.; Hauser, H. Outlier-Preserving Focus + Context Visualization in Parallel Coordinates. *IEEE Trans. Vis. Comput. Graph.* **2006**, *12*, 893–900.
29. Kosara, R.; Bendix, F.; Hauser, H. Parallel sets: Interactive exploration and visual analysis of categorical data. *IEEE Trans. Vis. Comput. Graph.* **2006**, *12*, 558–568.
30. Lex, A.; Streit, M.; Partl, C.; Kashofer, K.; Schmalstieg, D. Comparative analysis of multidimensional, quantitative data. *IEEE Trans. Vis. Comput. Graph.* **2010**, *16*, 1027–1035.
31. Liu, Z.; Jiang, B.; Heer, J. imMens: Real-time Visual Querying of Big Data. *Comput. Graph. Forum* **2013**, *32*, 421–430.
32. Rübel, O.; Prabhat.; Wu, K.; Childs, H.; Meredith, J.S.; Geddes, C.G.R.; Cormier-Michel, E.; Ahern, S.; Weber, G.H.; Messmer, P.; et al. High performance multivariate visual data exploration for extremely large data. In Proceedings of the ACM/IEEE Conference on High Performance Computing, Austin, TX, USA, 15–21 November 2008; p. 51.
33. Perrot, A.; Bourqui, R.; Hanusse, N.; Lalanne, F.; Auber, D. Large interactive visualization of density functions on big data infrastructure. In Proceedings of the 5th IEEE Symposium on Large Data Analysis and Visualization (LDAV), Chicago, IL, USA, 25–26 October 2015; pp. 99–106.

34. Chan, S.M.; Xiao, L.; Gerth, J.; Hanrahan, P. Maintaining interactivity while exploring massive time series. In Proceedings of the IEEE Symposium on Visual Analytics Science and Technology, Columbus, OH, USA, 19–24 October 2008; pp. 59–66.

35. Piringer, H.; Tominski, C.; Muigg, P.; Berger, W. A Multi-Threading Architecture to Support Interactive Visual Exploration. *IEEE Trans. Vis. Comput. Graph.* **2009**, *15*, 1113–1120.

36. Elmqvist, N.; Fekete, J.D. Hierarchical Aggregation for Information Visualization: Overview, Techniques, and Design Guidelines. *IEEE Trans. Vis. Comput. Graph.* **2010**, *16*, 439–454.

37. Wu, K.; Ahern, S.; Bethel, E.W.; Chen, J.; Childs, H.; Cormier-Michel, E.; Geddes, C.; Gu, J.; Hagen, H.; Hamann, B.; et al. FastBit: Interactively searching massive data. *J. Phys.* **2009**, *180*, 012053.

38. Card, S.K.; Robertson, G.G.; Mackinlay, J.D. The information visualizer, an information workspace. In Proceeding of the CHI Conference on Human Factors in Computing Systems, New Orleans, LA, USA, 27 April–2 May 1991; Robertson, S.P., Olson, G.M., Olson, J.S., Eds.; ACM: New York, NY, USA, 1991; pp. 181–186.

39. Godfrey, P.; Gryz, J.; Lasek, P. Interactive Visualization of Large Data Sets. *IEEE Trans. Knowl. Data Eng.* **2016**, *28*, 2142–2157.

40. Steinley, D. K-means clustering: A half-century synthesis. *Br. J. Math. Stat. Psychol.* **2006**, *59*, 1–34.

41. Ester, M.; Kriegel, H.; Sander, J.; Xu, X. A Density-Based Algorithm for Discovering Clusters in Large Spatial Databases with Noise. In Proceedings of the Second International Conference on Knowledge Discovery and Data Mining (KDD-96), Portland, OR, USA, 1996; pp. 226–231.

42. Riehmann, P.; Hanfler, M.; Froehlich, B. Interactive Sankey Diagrams. In Proceedings of the IEEE Symposium on Information Visualization (InfoVis 2005), Minneapolis, MN, USA, 23–25 October 2005; Stasko, J.T.; Ward, M.O., Eds.; IEEE Computer Society: Washington, DC, USA, 2005; p. 31.

43. Wegman, E.J.; Luo, Q.; High Dimensional Clustering Using Parallel Coordinates and the Grand Tour. In *Classification and Knowledge Organization: Proceedings of the 20th Annual Conference of the Gesellschaft für Klassifikation e.V., University of Freiburg, Baden-Württemberg, Germany, 6–8 March 1996*; Klar, R., Opitz, O., Eds.; Springer: Berlin/Heidelberg, Germany, 1997; pp. 93–101.

44. Ward, M.O.; Grinstein, G.G.; Keim, D.A. *Interactive Data Visualization—Foundations, Techniques, and Applications*; A K Peters: Natick, MA, USA, 2010.

45. Auber, D.; Chiricota, Y.; Delest, M.; Domenger, J.; Mary, P.; Melançon, G. Visualisation de graphes avec Tulip: Exploration interactive de grandes masses de données en appui à la fouille de données et à l'extraction de connaissances. In Proceedings of the Extraction et Gestion des Connaissances (EGC'2007), Actes des Cinquièmes Journées Extraction et Gestion des Connaissances, Namur, Belgique, 23–26 January 2007; pp. 147–156.

46. Elasticsearch, 1999.

47. Börzsönyi, S.; Kossmann, D.; Stocker, K. The Skyline Operator. In *Proceedings of the 17th International Conference on Data Engineering*; IEEE Computer Society: Washington, DC, USA, 2001; pp. 421–430.

48. Doshi, P.R.; Rundensteiner, E.A.; Ward, M.O. Prefetching for Visual Data Exploratio. In Proceedings of the Eighth International Conference on Database Systems for Advanced Applications (DASFAA '03), Kyoto, Japan, 26–28 March 2003; pp. 195–202.

MDPI

St. Alban-Anlage 66

4052 Basel, Switzerland

Tel. +41 61 683 77 34

Fax +41 61 302 89 18

http://www.mdpi.com

Informatics Editorial Office

E-mail: informatics@mdpi.com

http://www.mdpi.com/journal/informatics

www.ingramcontent.com/pod-product-compliance
Lightning Source LLC
Chambersburg PA
CBHW051729210326
41597CB00032B/5659